THE TRUTH OF
THE ORIGIN OF THE UNIVERSE

GOD'S SIGNATURE IN DNA'S CODE
THE RAPTURE BEFORE THE UNIVERSAL GREAT TRIBULATION AND WORLD WAR III

VOLUME-5

COLOUR EDITION

THE TRUTH OF THE ORIGIN OF THE UNIVERSE

GOD'S SIGNATURE IN DNA'S CODE THE RAPTURE BEFORE THE UNIVERSAL GREAT TRIBULATION AND WORLD WAR III

Dr. Sabrie Soloman
Ph.D., Sc.D., MBA, PE

KHANNA BOOK PUBLISHING CO. (P) LTD.
Publisher of Engineering and Computer Books
4C/4344, Ansari Road, Darya Ganj, New Delhi-110002
Phone : 011-23244447-48 **Mobile:** +91-9910909320
E-mail : contact@khannabooks.com
Website : www.khannabooks.com

ISBN: 978-93-5538-804-9

**THE TRUTH OF THE ORIGIN
OF THE UNIVERSE (VOLUME-5)**
by **Dr. Sabrie Soloman**

First Edition: Dec 2024

Published by:
Khanna Book Publishing Co. (P) Ltd.
CIN: U22110DL1998PTC095547

Visit us at: www.khannabooks.com
Write us at: contact@khannabooks.com

To view complete list of books,
please scan the QR Code:

DEDICATIONS

William Thomson – The voice of his words of wisdom echoed through the core of the inner chambers of my heart, engraving channels of living knowledge, which generated floods of insight to illuminate my life`s journey.

William Thomson`s pure and innocent soul has reached the gates of heavens to bring me knowledge of Eternity. His living loving example compelled me to follow even his shadow to touch Eternity, while my feet are still on Earth!

His knowledge imparted on my mind inspiring me to write the "**The Origin of The Universe**" book – distinct from most other science books, which composed by worldwide prominent scientists highlighting unjustified and unproven hypotheses. Through **William Thomson** the factual truth of the Universe is recognized and logically revealed to my modest intellect to compose "**The Truth of The Origin of The Universe**," Thus, I humbly dedicate this book to him.

Joseph Rondinelli – While the size of the universe may not be comprehended, so also the goodness of the heart of my friend Joseph Rondinelli may not be grasped. Joseph permitted me to touch the limitless boundaries of the goodness of his loving nature. His friendship with me permitted me to observe a living loving example to follow.

Joseph Rondinelli is bountifully living loving those whom he cherished. He brought heavens highest values to my humble living earth. While I was temporarily incapacitated, he was my feet, hands, and even eyes sitting for hours by my side, debating our intellectual capacities to understand life under the domain of our **Creator, The Origin of Our Universe**.

When my mind acted as an interrupted volcano spouting novel thoughts of **The Origin of The Universe**, he came to my dwelling to rescue me out of the dense-smoke of my scattered knowledge --suffocating my mind. He instantly changed my environment to a more tranquil atmosphere enabling me once again recollect my thoughts, helping me to **Create Order out of Chaos**. For this and much more I am forever, indebted.

FOREWORD

Shortly before Frances R. Havergal, at age 36, wrote the hymn, "Lord, Speak to Me" she wrote in a letter, "... if I am to write any good, a great deal of living must go to a very little writing." Before putting pen to paper, a great deal of living has gone into what you are about to read.

When Dr. Soloman asked me to write a forward to his book on the origin of the universe, I was humbled and honored because this book is of unusual importance by an author with an exceptional life of service and greatly respected by all who know him and his work. I pray that those who read his 2,800 pages, 32 chapters will find its message clearly stated, giving clarity to a subject that has too long been delt with on a surface and non-scientific level.

Understanding the origin of the universe is key to understanding everything else and developing a truly Biblical worldview. A person's view on this subject is a window by which we view the world. Dr. Soloman gives us a verifiable frame to evaluate whether his findings are flawed or something that is understandably sensible to merit serious consideration.

When we tell others our view on evolution is based on scientific evidence rather than solely on the book of Genesis, we have a much stronger argument. Dr. Soloman builds the case that evolution, with his accompanying philosophy, is identified with a worldview at such a level that readers must sincerely struggle how the theory could possibly go against the evidence.

This book has the capacity to train a whole new generation to confront and survive the worldview challenges they encounter all their life. To those of you who read this book in its entirety will be able to survive in a post-modern culture without being brainwashed by prejudice, faulty thinking, and preconceived assumptions.

Dr. Soloman has made exciting reading for the ordinary person, but which is thoroughly researched by scholarship to present a deep understanding of the subject. The situation we often find ourselves in is flawed thinking and lazy scholarship. To that end, I invite you to read Dr. Soloman's book which will prove to not only be enjoyable reading, but one of the most prolific and insightful tomes you will ever read on the origins of the universe.

Dr. Peter W. Teague,
President Emeritus
Lancaster Bible College | Capital Seminary & Graduate School

PREFACE

The Universe Had a Beginning

The scientific consent 100 years ago was that the universe was everlasting. This idea started to unravel with the inferences of Albert Einstein's Theory of Relativity back in 1916, where his equations pointed to an expanding universe. Yet he didn't like that conclusion and so supplemented a constant to his equation that invalidated the expansion. Later, he admitted it had been the major mathematical blunder in his life.

Then in 1929 astronomer Edwin Hubble acknowledged he observed galaxies expanding outward, which meant they had been much closer together in the past. Einstein, fascinated, wanted to see the indication for himself and in 1931 he went to the Mt. Wilson Observatory in Los Angeles, California. Einstein examined through the telescope, observed the evidence and then concluded, "I now see *the necessity* of a beginning." This began a change in the scientific boldness toward the cosmos.

Years after, in 1965, two U.S. scientists distinguished the leftovers of the original burst of energy of the creation occurrence characteristically so-called the "Big Bang." They both won a Nobel Prize in physics. One of them, Arno Penzias, later professed, "The best data we have [about the beginning of the universe] are exactly what I would have predicted had I nothing to go on but the first five books of Moses, the Psalms and the Bible as a whole" ("Clues to Universe Origin Expected," *The New York Times*, March 12, 1978, p. 1).

With the indication at hand, what was written in Genesis 1:1 truly surprised many scientists by its precision: "In the beginning, God created the heavens and the earth." Here it says the universe of matter and energy *appeared at a certain point in time* and was all created by an Ultimate Creator who existed before all of this occurred. It was an enormous proof of God's existence, with no real substitute enlightenments for a universe that, according to modern physics, *appeared out of nothing.*

The Universe is Fine-Tuned for Life

Almost 50 years ago, in 1973, cosmologist Brandon Carter instituted that the independent constants or laws in physics have one exceedingly rare characteristic in common—*they are precisely the values needed to establish and sustain a universe capable of producing life.* This is another vast and virtually accepted proof for a universe that has been prudently designed.

Scientists have instituted some 30 constants or laws of physics that rule the universe. All are dissimilar to each other and yet are finely tuned to unbelievable proportions to make life possible. The indication points to "Someone" spending much too much time tuning all of these laws so they would work in unison.

Amazingly, the Bible exposed this truth long before any scientist revealed these facts. As Jeremiah 33:25 states, "But I, the Lord, have a covenant with day and night, and *I have made the laws* that *control* earth and sky" (*Good News Translation*).

The Origin of Life – The Genetic Code

Opposing to what many have been led to trust, scientists have no genuine explanation for how life arose.

Even the well-known atheist and evolutionist Richard Dawkins acknowledged concerning the appearance of life, "*Nobody knows how it happened*" (*Climbing Mount Improbable,* 1996, p. 282). Moreover, one of the pioneers of the DNA code, the atheist Francis Crick, determined, "An honest man, armed with all the knowledge available to us now, could only state that in some sense, *the origin of life* appears at the moment to be *almost a miracle,* so many are the environments which would have had to have been gratified to get it going" (*Life Itself: Its Origin and Nature,* 1981, p. 88).

In the past 60 years, biologists have found that life started with a massive amount of accurate information *already embedded* in the cell. The human genome unaided is a molecule with almost 3 billion genetic letters, all precisely well-organized to give instructions to the cell. Furthermore, scientists have never instituted inorganic matter to generate a coded system of information and the machinery to interpret it. From the most primitive cells to human beings, all have the same elementary operating system of mind-boggling intricacy, with codes, transmitters and receivers all working together.

Furthermore, the origin of life aenigma has a "chicken-and-egg question"—which came first, the chicken or the egg? In this case, to get life to transpire, you need *both* the complete genetic code and the proteins—the machine parts—that read the code and build new proteins. Deprived of the code, you can't shape proteins. And without proteins, you can't process the code. So how could both have risen at the same time?

Biological life Exquisitely Programmed - "Robotic Machines"

In order to comprehend what is fashioning inside a cell, a good illustration is picturing a large city swarming with life and movement.

Biochemist Michael Denton defines the cell this way: "To clutch the reality of life as it has been discovered by molecular biology, we must amplify a cell a billion times until it is twenty kilometers in diameter and looks like a giant airship large enough to cover a great city like London or New York. What we would then see would be an entity of unmatched intricacy and adaptive design . . .

"We would see around us, in every direction we looked, all sorts of robot-like machines. We would notice that the simplest of the functional components of the cell, the protein molecules, were astonishingly complex pieces of molecular machinery, each one consisting of about three thousand atoms arranged in highly organized 3-D spatial conformation.

"We would wonder even more as we watched the strangely purposeful activities of these weird molecular machines, particularly when we realized that, despite all our accumulated knowledge of physics and chemistry, the task of designing one such molecular machine—that is one single functional protein molecule—would be completely beyond our capacity at present" (*Evolution: A Theory in Crisis,* 1986, p. 329).

This is the reason biochemists have a hard time have faith in explaining that blind evolution can build such machinery—and obtain all the parts to work together from the start. Furthermore, to keep the human body operational, biologists calculate that "about *330 billion cells* are substituted *daily,* equivalent to about 1 percent of all our cells" (Mark Fischetti, "Our Bodies Replace Billions of Cells Every Day," *Scientific American,* April 1, 2021).

"We take life for granted," adds Douglas Ell, "because it is everywhere. Our planet is overrun by *biological machines.* There are at least 10 million different types (species) of machines; some estimate that tens of millions of other types (species) have not yet been discovered . . .

"Coordinated systems permit blue whales to dive thousands of feet below sea level without being crumpled and sing complex songs that travel across oceans. Other systems guide bees to do a dance that tells other bees where to find the best sources of pollen. There are systems for hiding, systems for fighting, systems for reproducing, systems for obtaining food, systems for communicating, and so on" (*Counting to God,* p. 110).

Such findings show that everything about life is programmed to the last detail and that virtually nothing has been left to chance. Does this superb design point to evolution or to God? The answer is clear.

The Earliest Evidence of Life

Though Darwin titled his book *On the Origin of Species by Means of Natural Selection,* he was never able to substantiate that assumption. Many people adopt that the theory of evolution, with its countless mutations and natural selection as the means of change, can account for the origin and development of all the living things on this planet.

Yet this is sleight of hand, since evolution can account for *micro*evolution, or changes *within* the species (such as dogs of varying sizes, shapes and colors), but not *macro*evolution, or changes from one kind of creature to another. Natural selection can tell you something about the *survival* of the species, but nothing about the *arrival* of the species. It undoubtedly cannot trace the origin of the approximately *10 million species* on earth. These are classified into some 33 main body types or phyla, such as sponges, worms, insects and mammals.

Darwin foretold that as more of the fossil record was revealed, it would display types of species gradually appearing, beginning with one or a few, and then multiplying from simple to more complex life forms. He wrote, "If numerous species . . . have really *started into life at once,* the fact would be *fatal* to the theory of evolution through natural selection" (*Origin of Species,* 1859, p. 305). Yet that is precisely what has been found—major body types appearing at what's considered the *beginning* of the fossil record rather than in deposits laid down later.

Scientists call this "the Cambrian Explosion," referring to major types of plants and animals suddenly appearing *fully formed* in that fossil layer. This is the *opposite* of what Darwin and evolutionists had claimed would be found—and they have no real explanation or answers. Of the 33 main body types, 23 of them (or 70%) appear at the recognized *beginning* stage of the fossil record.

The issue in question here, by analogy, would be as discovering together such diverse inventions as a washing machine, a refrigerator, a bicycle, a car and an airplane. While they do have some shared features, they have very discrete purposes and resolutions. Likewise, the major types of creatures found in the Cambrian layer, such as sponges, worms, trilobites and jawless fish, are very diverse, complex and appear suddenly, with no evidence of these main body types evolving from other creatures.

As paleontologist Niles Eldredge admitted: "If life had evolved into the wondrous profusion of creatures little by little, then there should be some fossiliferous record of those changes . . . But no one has found any evidence of such in-between creatures . . . All of the fossil evidence to date has failed to turn up any such missing links" (George Alexander, "Alternate Theory of Evolution Considered," *Los Angeles Times,* Nov. 19, 1978). Indeed, Darwin has been let down by the fossil record!

Earth Has "Just Right" Conditions to Sustain Life

In 1966, Carl Sagan presented the famous TV documentary series *Cosmos*. He believed in order to have life you just needed two conditions—a right kind of star and a planet at the right distance. This supposition showed to be totally baseless.

Now, more than 50 years later, scientists have come to the comprehension that *more than 200 conditions* have to be "just right" for life to exist and thrive. The probabilities are about 1 in $10^{2,685,000}$. The probability of that happening comes out at about **1 in** $\mathbf{10^{2,685,000}}$, or 10 followed by **2,685,000** zeros. For comparison, the Universe only has 10^{80} atoms. The infographic finishes by letting you know that the probability of you existing as you is pretty much zero.

As author Eric Metaxas explains: "Today there are more than 200 known parameters necessary for a planet to support life—*every single one of which must be perfectly met, or the whole thing falls apart.* Without a massive planet like Jupiter nearby, whose gravity will draw away asteroids, a thousand times as many would hit Earth's surface. The odds against life in the universe are simply astonishing" ("Science Increasingly Makes the Case for God," *The Wall Street Journal,* Dec. 25, 2014).

The Bible tells us: "The Lord is God. He made the skies and the earth. He put the earth in its place. He did not want the earth to be empty when he made it. He created it *to be lived on.* I am the Lord. There is no other God" (*Isaiah 45:18, Easy-to-Read Version*).

The Universe Mathematically Designed

Amazingly, the universe has been discovered to be mathematically designed. It follows orderly laws that can be described in mathematical terms. Sir James Jeans, one of the great astronomers of the 20th century, remarked: "From the intrinsic evidence of his creation, the Great Architect of the Universe *now begins to appear as a pure mathematician* . . . The universe begins to look more like a great thought than like a great machine" (*The Mysterious Universe,* 1930, pp. 134, 137).

A substantial problem for evolutionists and atheists is this: *Evolution can't do math,* since it is founded on random variations and mutations, and math necessitates an intelligent agent who can first formulate a mathematical blueprint of laws before creating things so they will be logical. This is why the present cosmos can be traced back to mathematical rules.

As Einstein distinguished, "The most incomprehensible thing about the universe is that it is comprehensible." He meant that it could be understood in *mathematical* terms but that an explanation for that was outside math.

As far back as the early 1900s, scientists were discovering the laws that govern the subatomic realm, the tiny microcosm designated by quantum mechanics. It has very diverse rules than our macro world and appears to make room for such things as free will to arise.

Many scientists came to comprehend that *not all is determined by matter and energy.* Experiments illustrate that an observer can change a particle through means of observing it. The consequences are that we can determine the outcome of our lives by the choices we make.

It brings to mind what God said: "*Today I have given you a choice between life and death, success and disaster. I command you today to love the Lord your God. I command you to follow him and to obey his commands, laws, and rules. Then you will live . . .*" (*Deuteronomy 30:15-16, ERV*).

Science Points to The Existence of God

Cutting-edge science is continually showing more intricacy and deeper design, not only in the cosmos, but also in all living things.

The biblical patriarch Job once challenged skeptics to look at the design of the creatures around them and notice they witness to a Supreme Designer and Creator. He stated: "Even birds and animals have much they could *teach* you; ask the creatures of earth and sea for their wisdom. All of them know that *the Lord's hand made them*" (*Job 12:7-9, GNT*).

Therefore, by exploring all the evidence and grasping where it leads, we hope you will believe in God, the Creator and the "Origin of the Universe." *Keep* believing in Him, and earnestly seek His will for your life!

Sabrie Soloman

ACKNOWLEDGMENTS

The Author is extremely grateful for the opportunity to express his heartfelt acknowledgments to the individuals who have played a vital role in the successful publication of "The Truth of The Origin of The Universe" book by Khanna Book Publishing Company.

First and foremost, the author extends his deepest gratitude to the Director and Editor-in-Chief, Mr. Mukul Seth, for his unwavering dedication and expertise in shaping the book's narrative. His keen eye for detail and impeccable editing skills have significantly enriched the content and ensured its quality.

Also, the author expresses his sincere appreciation to the Editorial Team, comprising of Mr. Mukesh Sharma and Mr. Yogesh Kumar. Their valuable feedback, insightful suggestions, and meticulous attention to detail have greatly enhanced the overall coherence and readability of the manuscript.

A special mention goes out to the Marketing Team, led by Mr. Rakshit Khanna for his tireless efforts in promoting the book and reaching a wider audience. Their innovative marketing strategies and relentless pursuit of excellence have played a crucial role in generating interest and enthusiasm among readers.

Furthermore, I am truly grateful for the dedication and hard work of the Sales Team, consisting of Mr. Nand Lal, Mr. Surinder Kumar, and Mr. Roshan Lal. Their exceptional sales acumen and customer-centric approach have boosted sales and fostered positive relationships with clients and partners.

The author would also like to extend his heartfelt thanks to Creation Ministries International for its invaluable contributions to the research and development of the book. Their pioneering work in the field of creation science has laid the foundation for a deeper understanding of the origins of the universe.

Sabrie Soloman

ABOUT THE AUTHOR

Dr. Sabrie Soloman, Ph.D., Sc.D., MBA, PE – He is the Chairman & CEO of American SensoRx, Inc., USA; Founder of Advanced Manufacturing Technology Post Graduate Studies at Columbia University, NY, USA; Professor of Advanced Technology at Columbia University. Dr. Soloman authored numbers of technical books published and translated worldwide: Sensors Handbook (2 editions), Sensors and Control Systems in Manufacturing (2 editions); Affordable Automation; Introduction to Electromechanical Engineering; Modern Welding Technology; 3D Printing Technology; 3D Bioprinting Technology & Design to name a few. Dr. Soloman holds numerous Patents, Technical Awards, and several US Product Registrations. Dr. Soloman is considered an international authority on advanced manufacturing technology, robotics, biomedical engineering, pharmaceuticals, and automation in the microelectronic, automotive, beef, pork, poultry industries. He has been and continues to be instrumental in developing and implementing several industrial and modernization programs through the United Nations to Europe, Asia, and African Countries. He is the first to introduce and implement unmanned flexible synchronous/asynchronous manufacturing systems in the microelectronic and the meat industries, and the first to incorporate advanced vision technology in wide array of robot/micro-robot manipulators. Dr. Soloman was selected to deliver the US presidential closing address "Innovative Remote Sensors Technology," at the Universal Design Conference," New York, USA. Dr. Soloman was the President of the International Christian Union at New Castle-Upon-Tyne University. He debated the "Origin of the Universe" before numerous attendees presenting his case against intellectuals and proponents of Big Bang's Singularity in Great Britain.

CONTENTS

Dedications *v*
Foreword *vii*
Preface *ix*
Acknowledgements *xv*
About the Author *xvii*
Introduction *1*

CHAPTER 27: THE SIGNATURE OF THE CREATOR IN DNA **7-46**

God's Signature 8
DNA´s Hidden Codes 8
Same "Letters," Multiple Meanings 8
Genetic Sequences Provide Instructions 8
Dual-Usage "Words" 8
Regulating the Rate That DNA Is Copied into RNA 9
Pausing Protein Production to Produce the Right Folds 9
Dual-Coding Genes 9
Discovery of DNA's Ever-Increasing Complexity 10
The Signature of God 11
One Author Recounts 11
Scientists Response 14
The Fourth Letter Carbon 15
Christianity and Rationality 20
Appendix 41
References 42

CHAPTER 28: THE UNIVERSE IS BEING HELD TOGETHER 47-86

The Formation of Man 47
The Good News 49
Formation of Planets 49
The Doctrine of Death 51
How did Creation Happen 51
Natural Selection and Death 52
Evolution - the 'Enemy of Death' 53
Death of the Unfit 53
Appendix 84
References and notes 84

CHAPTER 29: THE MORAL LAWS 87-130

The Christian foundations - The Rule of law 87
Christianity Foundation of the Individual Rights 88
Christianity the Rule of law - Individual Liberty 89
US National Archives & Records Administration 89
Reasons for a Civil Government 91
How the Impression of 'Evolution' Undermines the Rule of Law 92
Social Evolution 93
The State Becomes a 'god' 93
Universe Held by Supreme Lawmaker 93
God's Law is Above the State Law 94
The Biblical Understanding of Lawful Resistance Against Tyranny 95
The Rule of Law - Christianity - Human Rights 99
The Creation Basis for Morality 100
Morality and History 104
Children and Behavior 104
The Bible - Basis Objective Morality 105
Evolution has no Objective Basis for Morality 105
Unfolding the Plan 106
Struggling to Harmonize Scripture with Evolution 108
God, Miracles, and Logic 109
Imposed Law and Divine Will 111
The Heart of Science 111
The Greek View of Nature 111
An Impediment to Science 114

The World—Animate or Inanimate? 115
The Lawgiver 115
Imposed vs Immanent Law 116
The Natural World as a Mechanism 118
The Universe is Framed by God 119
God the Creator is the Lawgiver 119
The Expectations Behind Science 120
The Biblical Basis for These Assumptions 121
The Koran 122
The Judge 123
The Interrogation 124
The Defense 124
The Verdict and Sentence 124
Appendix 126
References 127

CHAPTER 30: THE MYSTERY OF 6-DAYS CREATION 131-182

Prominent Scientists Believe in Creation 131
The Necessities for Life 132
The Mystery of Light Came to Existence 136
God Speaking into the Void 136
Creative Power of God's Voice 136
How Could There be Light on the First day of Creation 137
The Light of Stars Billions of Light Years Away 138
Gravitational Time Dilation 139
What Happened on Each of the Days of Creation? 141
The Universe Designed to Contain Human Life 145
Life and Information 147
Mathematical Proof for the Designer Necessity 147
Life from Nonlife? 150
The Author of the Bible 152
The Nature of Genesis Chapter 1–Chapter 11 153
Can We Trust Genesis 1–11 as Written? 159
The Word of the Creator 163
The Truth is Revealed 170
The Beginning 170
The Formation of Light 172

Light Before the Sun? 172
Becoming a Christian 173
Living as a Christian 174
Exoplanet Kepler 176
Oldest-Ever' Fossilized Tree — 176
Days of Genesis 1 are Literal - while Sun wasn't Created until Day 4? 177
APPendix 178
References 179

CHAPTER 31: THE BEGINNING AND THE ENDING OF THE UNIVERSE 183-252

Eternity – Infinity – God 183
Presence Outside of Time 184
Modern Physics 184
Quantum Mechanics 185
The Concept of Mathematics – The Unified Theory 185
Mathematics and Infinity 186
The Universe's Beginning 187
The Earth's Beginning 188
Scientists Assume and Conjecture 188
Numbers of Atheist Miracles/Magics 189
Illusion needs an Illusionist 189
Where Do We Go from Here? 195
The Ultimate Folly 195
The Earth will Pass Away 196
New Heavens and New Earth 197
Physical Regeneration 197
Restoration of God's Plan 198
Compromise in Genesis Causes Problems for End Times 199
God's Plan without Sin and Death! 199
Restoring the Physical Creation 199
Future of Time, Space, and Matter 199
Quantum Fluctuation 201
Eternal Universe 202
Atom & Destruction of Present Earth 202
The Future of Atoms – The Future of the Earth and Heavens 205
Time, Space, and Atoms (Matter) 206
The New Earth 206

The End of the Present World 207
Noah's Flood – God's Judgment 207
Why will this World be Destroyed? 207
Looking Forward to The New World 208
God's Triumph in the Restored Creation 210
Compromise on Genesis Creates a NHE Problem 211
God The Author of creation ... 212
Hawking's Death and Decay 213
Hawking's Origin of Life 213
Hawking's Multiple Universes 214
The Providence of the Universe and Us 214
Getting Away from it All 215
Planet Gliese 581D 215
Hawking's Panacea—Dark Energy 217
Secular Scientists Object! 217
Evolutionists Deceive Students to Believe in Evolution 218
Teaching Lies 218
Other Evolutionary Propagandists 219
The Foundation 220
Would a Loving God Send People to Hell? 220
Sin - Rebellion Against Our Loving Creator 222
The Purpose of the Tree of the Knowledge of Good and Evil 222
Jesus - The loving God Incarnate 223
Rebelling Against their Loving Creator 223
Hell - Rejecting God 224
Those Who have Never Heard of God 224
God Became Our Savior 224
Reconciled with God 225
The Relationship between the Father and the Son 225
Our Sinful State—Separated from God 226
Christ the Mediator 226
The Great Reconciliation 226
Living in the Light of Christ 227
Our Creator is our Mediator! 227
You are Absolutely Safe – If You Choose! 227
Cancel God – The Demise of Nations 229
The Resurrection of The Dead 230
The Greatness of the Creator 233

The Pre-Tribulation Premises 234
Isaac Newton - The Resurrection of the Dead 236
The Resurrection and the Tribulation 236
The Different Views of the Rapture: 237
The Trumpet of Grace and the Trump of Judgment 239
The Doctrine of the "Rapture" 241
Comprehensive Rapture 241
Rapture Views 242
Post-Tribulation 244
1 Thessalonians 4:13–18 Parallels John 14:1–3 246
Christ's Post-Tribulation Coming Differs from the Rapture 247
Daniel's Seventieth Week - Revelation 3:10 248
The Day of The Lord 249
Appendix 249
References 250

CHAPTER 32: WILL ROBOTS INHERIT THE EARTH? 253–338

Will Robot Become Humanoid? 253
AI – The Beast of Revelation 255
Is Artificial Intelligence the Mark of the Beast? 256
Artificial Intelligence and Evolution 259
Much to Worry about 259
Error Reigns 260
Driving it All: Naturalism 261
Making God 261
Microbic Mastery over Micro-bytes 262
Even 'Simple' Life is Beyond the Reach of Chance 262
Summoning the Demon 262
Heavenly Profound Wisdom in Human Body 265
The Nervous System 266
The Basic Building Block 266
Creator's Power and Wisdom 267
How Do Our Nerves Transmit Information? 268
God's Magnum Opus 270
Chemical Manufacturing Plant 270
Cooling and Heating System 271
Not AI – Heavenly Complex Computer System 271

Miraculous Central Nervous System 272
The Human Eye 272
Auditory Sound Waves – Mystery in Science 273
Intricately Manufactured Muscular Pump 274
Skillfully Intelligent Miraculous Designer 274
Dust of the Ground- Perfect Input of Intelligence 274
A New Twist on Blood Vessels 275
A Simple Way of God 275
The Intelligent Designer Created the Universe 277
The Laws of Mathematics Exist in God's Mind – 277
Mathematicians did not Invent Math 279
God the Communicator 279
Mathematics Reflects the Perfect Mind of God 280
God's Hands on Outstanding Godly Scientist - I 280
Education 281
Astronomy and Astrology 281
Distinction between Astronomy and Astrology 282
Rejecting Astrology 283
Motion of the Planets 283
Creator Logical Pattern 284
Philosophies and Mathematics of Ancient Greeks 284
Scientific Planetary Motion 285
Kepler's Two Laws of Planetary Motion 285
Kepler's Third Law of Planetary Motion 285
Additional Vital Discoveries 285
THE ROLE OF THE EYES 286
Origin of the Universe 287
Kepler's Defended His Mother – Saved Her Life 287
Acknowledging God 287
God's Hands on Outstanding Godly Scientist -II 287
Early life 288
Christian faith 288
Ray's Scientific Contributions 289
Defining Species According to its Kind 289
Opposition to Evolutionary Ideas 290
His Later Years 291
Catastrophe, Curse and Judgment 291
He Bore the Divine wrath to the Full 295

GLOBALISM: Preparation for the Antichrist 296
Global Stage Setting 297
The Fourth Judgment of God 300
The Early Battles of The Final World War 302
The Triumphant King of kings 305
Satan Cast into the Abyss 306
Catastrophe, Curse and Judgment 307
The Great White Throne 309
The Tribulation of God's Judgment (The 4th Judgment) 310
God's Fourth Judgment 310
The Beast and the False Prophit 311
The Armies of the World 312
The Early Battles of The Final World War 312
Blood Shed 313
Satan Cast into the Abyss 315
The End of the world 316
Three Religions in The Holy City of Jerusalem 317
What Lies Ahead? 318
Bitter Family Rivalry 319
Daniel's Famous Prophecy 320
Will Russia Control The Middle East? 322
Who Will Control Jerusalem? 323
Temple Sacrifices Restored? 324
Watch For The Abomination of Desolation 326
Three Prophetic Milestones 328
Who Rides The Beast? 329
The King of The North Invades The Middle East 330
The Final Phases of World War III 332
Watch The Middle East 334
Ultimate Peace 335
Appendix 335
References 336

INTRODUCTION

Moral Law Dictates Right and Wrong

Moral law is the set of principles and guidelines that dictate right and wrong behavior in society. It is based on the idea that certain actions are inherently good or bad, regardless of the circumstances in which they occur. These moral norms are often derived from religious beliefs, cultural values, and natural law. The book of Romans 2:14-15 states that "Even Gentiles, who do not have God's written law, show that they know his law when they instinctively obey it, even without having heard it. They demonstrate that God's law is written in their hearts, for their own conscience and thoughts either accuse them or tell them they are doing right.

God's Basic Law is Written in Everyone's Hearts

This is not to be confused with salvation but the basic law of God toward each other. Kindness, no murder, stealing, hatred. The general laws of God are written on everyone's heart. Our conscience works with this law. This is different from the Holy Spirit working in our hearts. That is a deeper work beyond fallen man. There is no excuse for a man to say I was never taught how to treat another human being. God has written this law into the hearts of every human being on earth.

The Sanctity of Human Life

One of the foundational principles of moral law is the concept of the sanctity of human life. This belief holds that all human beings have inherent dignity and worth and that they should be treated with respect and kindness. This principle underpins many of the moral laws that govern our society, such as laws against murder, assault, and theft.

Justice and Fairness

Another key aspect of moral law is the idea of justice and fairness. This principle dictates that individuals should be treated equally and that they should be held accountable for their actions. Justice is a central tenet of moral law, as it ensures that individuals are treated with dignity and respect and that they receive the consequences of their actions.

Religious Beliefs and Faith - Given by God

The concept of moral law is deeply rooted in religious beliefs, with many faith traditions teaching that certain actions are inherently sinful or virtuous. For example, Christianity teaches that lying, stealing, and killing are sins, while acts of kindness, compassion, and generosity are virtuous. In the Judeo-Christian tradition, the moral law is believed to have been given by God to humanity as a way to guide our actions and ensure our well-being.

The Ten Commandments

The Ten Commandments, for example, are a set of moral laws given by God to the Israelites to guide their behavior and ensure their obedience to Him. In the Christian tradition, the idea of moral law is closely connected to the concept of sin and redemption. Sin is seen as a violation of the moral law, and redemption is the process of seeking forgiveness and reconciliation with God. The idea of moral law also extends beyond individual behavior to societal norms and values.

Marriage, Family, and Community Life

Laws governing marriage, family, and community life are all derived from moral principles that seek to promote the common good and ensure the well-being of all members of society. The moral law is a foundational aspect of human society, guiding our actions and behavior under principles of justice, fairness, and respect for human life. It is a set of principles that are deeply rooted in religious beliefs, cultural values, and natural law, and it serves as a guide for ethical living and a framework for a just and harmonious society.

The Mystery Surrounding the Six Days of Creation

The concept of the mystery surrounding the six days of creation, the beginning and end of the universe, the possibility of robots inheriting the earth, the coming tribulation of God, the battle of Armageddon, and the second coming of Jesus Christ has intrigued theologians, scientists, and philosophers for centuries. These topics delve into the realms of theology, cosmology, artificial intelligence, eschatology, and spirituality, prompting intense debates, discussions, and inquiries into the nature of existence, destiny, and the ultimate purpose of humanity.

The idea of the six days of creation stems from the biblical account in the Book of Genesis, where God is said to have created the heavens and the earth in six days before resting on the seventh day. This narrative has sparked various interpretations among religious scholars, with some adherents taking a literal view of the creation story, while others adopt a more symbolic or figurative approach. The mystery lies in the question of how the universe came into being and whether the six days represent literal 24-hour periods or symbolic epochs of time.

The Field of Cosmology

In the field of cosmology, scientists have put forth theories such as the atheistic Big Bang theory to explain the origins of the universe, suggesting that all matter and energy were condensed into a singularity before expanding rapidly and forming galaxies, stars, and planets. While this scientific explanation offers a naturalistic view of creation, it does not address the philosophical or theological implications of the existence of the universe and the purpose behind its intricate design.

Will Robots Inherit the Earth?

The idea of robots inheriting the earth raises ethical, social, and existential questions about the potential future of artificial intelligence and its impact on human civilization. As technology advances and machines become increasingly sophisticated, there is a concern that robots could surpass human intelligence and autonomy, leading to a scenario where they dominate or even replace humanity as the dominant species on Earth. This concept has been explored in science fiction literature and films, prompting reflections on the nature of consciousness, identity, and the boundaries between man and machine.

The 7-Year Tribulation of God

The coming tribulation of God and the battle of Armageddon are apocalyptic events prophesied in religious texts, where God's judgment is said to befall the earth, leading to cataclysmic upheavals, wars, and natural disasters. These teachings have been interpreted in various ways by different religious traditions, with some believers anticipating a literal end times scenario marked by divine intervention and the establishment of a new heaven and earth. The imagery of Armageddon has been depicted in popular culture, fueling speculation and fascination about the fate of humanity and the ultimate destiny of the universe.

The Second Coming of Jesus Christ

The second coming of Jesus is a central belief in Christianity, where it is believed that Jesus Christ will return to earth in glory to judge the living and the dead, ushering in a new era of peace, justice, and renewal. This event is seen as the culmination of history and the fulfillment of biblical prophecies, signaling the ultimate victory of good over evil and the restoration of creation to its original harmony. The anticipation of the second coming has inspired hope, faith, and dedication among believers, who await the promised return of their savior with eagerness and vigilance.

The Creation of the Universe

Throughout history, influential individuals have contributed to the exploration and interpretation of the mysteries surrounding the creation of the universe, the potential rise of artificial intelligence, the coming tribulation of God, Armageddon, and the second coming of Jesus Christ. From ancient philosophers and theologians to modern scientists and religious leaders, a diverse array of thinkers has offered insights, arguments, and perspectives on these profound and enigmatic subjects, shaping the way we understand our place in the cosmos and the meaning of our existence.

The Theological Implications of Creation

In the realm of theology, prominent figures such as Augustine of Hippo, Thomas Aquinas, and John Calvin have expounded on the theological implications of creation, providence, and eschatology, drawing upon biblical texts, philosophical reasoning, and theological doctrines to elucidate the mysteries of divine action in the world. These thinkers have grappled with questions of predestination, free will, and the sovereignty of God, seeking to reconcile the tension between human agency and divine providence in the grand scheme of salvation history.

Forces that Govern the Cosmos

In the field of cosmology, luminaries such as Albert Einstein, Stephen Hawking, and Carl Sagan have advanced scientific theories and observations to explain the origins and evolution of the universe, shedding light on the physical laws, constants, and forces that govern the cosmos. From the discovery of cosmic microwave background radiation to the detection of gravitational waves, these scientists have revolutionized our understanding of the universe's age, size, and composition, painting a portrait of a vast and expanding cosmos filled with galaxies, stars, and planets.

Artificial Intelligence

In the realm of artificial intelligence, pioneers like Alan Turing, John McCarthy, and Marvin Minsky have pioneered the development of computer science, robotics, and machine learning, laying the groundwork for intelligent machines capable of reasoning, learning, and adapting to their environment. The emergence of advanced AI systems like DeepMind and IBM Watson has raised questions about the ethical, social, and existential implications of artificial intelligence, prompting debates about the risks and benefits of creating intelligent machines that could surpass human intelligence in the future.

The Mysteries of the End Times

In the realm of eschatology, figures such as Saint John of Patmos, Hildegard of Bingen, and Jonathan Edwards have delved into the mysteries of the end times, the final judgment, and the coming of the kingdom of God, drawing upon biblical prophecies, visions, and revelations to paint vivid pictures of the apocalyptic events that will unfold before the return of Christ. These prophetic voices have inspired believers to heed the warnings, prepare for the trials, and await the fulfillment of God's promises with faith, hope, and perseverance.

In the realm of popular culture, artists, writers, and filmmakers have explored the themes of creation, apocalypse, and redemption in their works, portraying visions of the future, the end of the world, and the cosmic battles between good and evil. From the paintings of Hieronymus Bosch to the novels of J.R.R. Tolkien, the films of Stanley Kubrick to the music of Pink Floyd, creative minds have grappled with the mysteries of existence, destiny, and transcendence, weaving narratives and images that captivate the imagination and stir the soul.

The positive aspects of contemplating the mysteries of creation, the end of the universe, the rise of robots, the tribulation of God, Armageddon, and the return of Jesus include fostering a sense of awe, wonder, and humility in the face of the vastness, complexity, and beauty of the cosmos. These reflections can inspire reverence, gratitude, and stewardship for the natural world, prompting us to care for the planet, protect its resources, and preserve its biodiversity for future generations.

Unbelievers, Negative Aspects

To the unbelievers, the negative aspects of dwelling on the mysteries of creation, the end of the universe, the advent of artificial intelligence, the wrath of God, Armageddon, and the return of Jesus Christ include provoking fear, anxiety, and despair about the fate of humanity, the planet, and the ultimate destiny of the soul. These apocalyptic scenarios can engender apathy, fatalism, and fanaticism among unbelievers, leading to harmful behaviors, divisive attitudes, and destructive actions, while believers are confident to escape the diverse calamities, which are befalling upon Earth. The believers are raptured to heaven before God's 7-year tribulation.

The mystery surrounding the six days of creation, the beginning and end of the universe, the potential rise of robots, the coming tribulation of God, Armageddon, and the second coming of Jesus Christ is a profound, complex, and enduring topic that continues to captivate the hearts and minds of believers, thinkers, and seekers across the ages. By exploring the historical context, key figures, influential individuals, various perspectives, and potential future developments related to these profound mysteries, we can gain deeper insights, broader understandings, and a richer appreciation of the ultimate questions of existence, destiny, and the divine purpose for humanity and the cosmos.

Sabrie Soloman

27

THE SIGNATURE OF THE CREATOR IN DNA

Since time immemorial, the artist's signature has become a key ingredient in the art-making process—it signifies that the artwork is finished and that the artist is satisfied.

According to the Bible, on the sixth day of Creation God created man in His own image and proclaimed that the work was *very good*. Ostensibly, He didn't sign His work.

It occurred to me that on the contrary, God had not only signed His work, but had done so with poetic majesty, finality, and breathtaking power.

The amount of information in the human body is outright staggering. The human body has an average of one hundred trillion (100^{12}) cells. In a single cell, the DNA contains the informational equivalent of roughly eight (8,000) thousand books. If the DNA from one cell were uncoiled, it would extend to about three meters in length. Thus, if the DNA in an adult human were strung together, it would stretch from Earth to the sun and back roughly seventy times ($70 \times 2 \times 94.119 \times 10^6$ miles! (13.2×10^9) mile.

The more we learn about the amazingly complex structure of DNA, the more we realize the sheer quantity of information contained in every DNA molecule, and the more we grasp the reality that *information cannot spring spontaneously from nothing.*

GOD'S SIGNATURE

DNA is God's magnificent signature, and it proclaims His infinite power, grace, and love with the birth of every new human cell.

DNA´S HIDDEN CODES

The world's most complex language system is located within every cell of your body. Now biologists are discovering that DNA is hiding a language within a language.

When we say there are hidden codes in DNA, some people jump to the conclusion that I believe "messages from God" were written in our DNA.

Scientists are now discovering that our DNA really does have hidden codes that have a practical function and purpose in our cells. It's like discovering a coded message that means one thing when you read it in English, but if you pull out every third letter, it means something completely different in French. How complicated is that!

Children in public schools are bombarded with claims that random processes can explain the messages encoded in DNA, and this creates doubt in the Bible's claims about the Creator. Accordingly, we need to keep up with amazing new discoveries like "messages within messages." Scientists confirm that DNA could never assemble randomly by evolution and confirm instead the label "Designer Required."

SAME "LETTERS," MULTIPLE MEANINGS

Genetics can get very complicated very quickly. But everyone needs to grasp the basics because they are so central to defending our faith in this skeptical, scientific age. The easiest way to understand DNA is by a comparison with language.

English uses 26 letters of the Latin alphabet, which can be shuffled to produce hundreds of thousands of words in the English language. French uses these very same letters to produce all the words in that language. Similarly, DNA is composed of four bases symbolized by the letters A, T, C, and G. These "letters" can be arranged into thousands and thousands of different sequences with different meanings, depending on how you read them.

GENETIC SEQUENCES PROVIDE INSTRUCTIONS

The most familiar purpose of genetic sequences is to provide instructions to make proteins (the building blocks of our bodies, such as collagen in our skin). But these letters can also be used to provide instructions for regulation, packaging, and many other duties in the cell. Until recently, scientists believed that each DNA sequence served just one purpose or the other, but not both. Now, however, scientists are discovering that the same DNA code can be used for both. Cells essentially read the same string of DNA for two kinds of information at once!

Below are two astonishing examples of multipurpose coding, but they appear to be only the tip of the iceberg in DNA's complexity.

DUAL-USAGE "WORDS"

The first multipurpose wonder is not hard to understand if you continue the language analogy. Just as DNA has four "letters," these letters are combined into three-letter "words," called codons. These words are combined into sentences and paragraphs that ultimately form complete sets of instructions for making proteins or regulating them.

It gets more complicated than that because the original instructions are kept safe in the cell's central library, or nucleus. Since DNA usually doesn't leave the nucleus, the instructions must first be copied into an intermediate "document" known as RNA. This coded document leaves the nucleus and is carried to the cell's factories, where the instructions are read to assemble proteins.

The codons were believed to have just one purpose—pass along instructions for building proteins. Recent research has turned this idea on its head. They may perform two other purposes.

REGULATING THE RATE THAT DNA IS COPIED INTO RNA

Scientists have discovered that some codons may also play a role in regulating the rate at which the central library is copied from DNA into RNA. It is estimated that 15% of codons (called duons) serve this dual purpose.

One of the researchers, Dr. John Stamatoyannopoulos, stated his surprise, "For over 40 years we have assumed that DNA changes affecting the genetic code solely impact how proteins are made. Now we know that this basic assumption about reading the human genome missed half of the picture." Rather than being the product of random, chance processes, DNA bears the hallmark of incredible design.

PAUSING PROTEIN PRODUCTION TO PRODUCE THE RIGHT FOLDS

As the cell assembles proteins, the growing protein begins to fold. Proper folding is very important to the protein's function. Misfolded proteins don't work properly. Scientists have discovered that some codons appear to play another dual role: they hit the "pause" button during the construction of proteins to allow for proper folding.

This finding clears up a mystery in genetics. Several words (codons) appeared to have the same meaning. For example, the codons CCA, CCG, and CCC all code for the same amino acid, proline. In all my years of schooling and teaching, we always learned and taught that this redundancy caused the code to be more robust. A mutation that changes CCA to CCC does not necessarily change the amino acid that it produces, so the change was thought to have little or no detrimental impact.

Recent research has shown that different combinations of codons strung together (called hexamers), really do carry instructions. Like a comma in a sentence, they may produce a "pause" in the code. Each protein must pause at the right times during its formation, or else it will not fold properly. A change from CCA to CCC, in our example, will not change the proline, but it might remove a necessary pause and change the "meaning."

The scientists involved with this research stated, "Redundancy of the codon to amino acid mapping, therefore, is anything but superfluous or degenerate." Once again, we see evidence that DNA was not haphazardly assembled over eons but rather was assembled purposefully by the Creator God.

DUAL-CODING GENES

Another central tenet of molecular biology is now in limbo: one gene codes for one protein. I never learned or taught anything to the contrary. Before the human genome was sequenced, scientists anticipated finding approximately 100,000 genes since we know the human body has at least 100,000 proteins. Much to the scientists' surprise, they found only 20,000–25,000 genes when they sequenced the human genome. So how do you get 100,000 or more proteins from so few genes? One possibility is that some genes are dual-coding—they code for more than one protein.

Several dual-coding genes have now been discovered, and scientists anticipate finding many more. When I say these genes code for two proteins, I don't mean two versions of the same protein; I mean two distinct proteins that have different structures and different functions.

For example, the instructions for one protein might start at the beginning of the gene, but the instructions for the other protein start 500 letters down the line. This doesn't lead to just a shortened form of the protein but different sets of words (codons) to manufacture an entirely different protein.

Dual-coding genes are common in bacteria and viruses. These organisms are very small, so they were designed to make the most efficient use of space. One gene can encode multiple proteins. But no one expected that in mammals. The authors of one study stated, "We simply do not believe that dual-coding genes can occur in eukaryotes [organisms that are not bacteria]."

Mammal genomes are so large that they didn't seem to need dual-coding genes.

True, scientists have known for some time that certain genes code for more than one protein, but they thought only one of the proteins was functional and that the other was just a useless artifact of evolution. So, the evolutionary ideas inhibited research and understanding. Recent studies have shown that both proteins from dual-coding genes are indeed functional.

DISCOVERY OF DNA'S EVER-INCREASING COMPLEXITY

Traits are Inherited Through DNA

In 1866, Gregor Mendel's experiments on pea plants showed that distinct traits are inherited through some invisible process. In 1869, Frederick Miescher isolated DNA from the nucleus of human white blood cells. He called this strange acidic substance "nuclein."

How likely is it that dual-coding genes occurred by chance? The title to one section of an article on dual-coding genes says it all: "Dual Coding Is Virtually Impossible by Chance." In other words, programming of this nature requires a Programmer!

Made by the Creator God

Whether we look at hidden codes in codons or in genes, the "Made By" label should undoubtedly read "***The Creator God***." I believe the information encoded by DNA is so vast and multilayered it is impossible to quantify.

Hidden codes pose a real problem for evolution. Changes are likely to have multiple effects on multiple functions. What are the odds that all of those changes would be beneficial?

Romans 1:20 states that God can be known through what He has made. The evidence is so clear that people are "without excuse" in denying His existence. Hidden codes are a great testimony of God's role as an intelligent and wise Creator. And think about this: these are just some of the hidden codes we know about. How many more are left to discover?

There is, however, a Creator who used infinite intelligence and engineering skill to provide the providential programming that is observed in the interactive coded and encoded communication that occurs, non-stop, in nuclear and mitochondrial DNA, RNA, and ribosomes. That Creator is the God of the Bible. He has revealed Himself in and through the Lord Jesus Christ, and He is the one we should gratefully *revere*.

THE SIGNATURE OF GOD

It is written: Psalm 139:14-I will praise You, for I am fearfully *and wonderfully* made; Marvelous are Your works, and *that* my soul knows very well.

Fig.27.1: YHWH – The Name of God

One of the fascinating things about the Hebrew alphabet is that every letter represents more than a phonetic sound, Figure 27.1. One Jewish rabbi writes:

"There are an estimated 3,000 languages (not counting dialects) and more than 66,000 letters which make up the alphabets for these languages. However, only one language and one alphabet is Divinely created, the letters having been formed and shaped by God alone. That language is "Lashon Ha-Kodesh," biblical Hebrew. It is no wonder, then, that the Hebrew letters are multifaceted. The letters of the Hebrew alphabet, the alef-beis, are so rich with meaning that even Judaism's greatest scholars had to engage in lengthy study to understand why God made them as He did. Traditionally, Hebrew letters possess:

1. Design—the specific way each letter is formed. This form represents the Divine power within each letter.
2. Gematria—each of the letters of the alef-beis represents a certain number, e.g., alef = 1, beis = 2, etc.
3. Meaning—each letter has many meanings, e.g., the letter alef stands for chief, to learn, wondrous, and much more. Beis means house, etc.
4. Nekudos (vowels)—most letters have a vowel that tells us how it is to be pronounced.
5. Crowns—some letters in the Torah have crowns—little lines drawn on the top of Hebrew letters—which add strength to the letters, e.g., Rabbi Akivawas famous for his expositions upon them. The crowns have their own special meanings beyond the scope of this work.
6. Cantillation—each word in the Torah has a musical note. In this chapter we will deal only with the first four topics.

Realizing that every Hebrew letter represents a number gives us further insight into a remarkable discovery that has been made regarding DNA (the code of life which programs every living cell). One unique scientist, Dr. Israel Rubinstein describes something that he discovers in 1986.

ONE AUTHOR RECOUNTS

"Dr. Israel Rubinstein, professor of chemistry at the Department of Material Science and Interfaces at the Weizmann Institute of Science in Rehovot, Israel, holds a Ph.D., in chemistry from Tel Aviv University.

In a video presentation, Dr. Rubinstein unveiled a stunning piece of information he claims to have discovered. In 1986 when, there was no worldwide video-posting platforms, and no other practical means for globally distributing to his findings, especially to the general populace. Accordingly, for the longest time, most of the world was unaware of his research. But now, Dr. Rubinstein's presentation has been made available to the entire planet—hearkening back to the fulfillment of the prophecy we studied in Daniel 12: 4.

Following is the transcript from that presentation, which premiered on a YouTube channel on June 23, 2021, but with an unknown origin date.

Dr. Rubinstein is speaking to a live but mostly unseen audience (presumably his students) while he is seated behind his desk, he said. "*Okay, listen carefully. In the year 1986, when most of you hadn't been born yet, I was already a doctor and a scientist in the Weizmann Institute of Science. I researched the DNA mainly of mice, monkeys, and Humans… 18,000 tissues of liver, kidneys, muscles and even a human eye, I inspected them under a microscope. For your information, the optic microscope can magnify up to 10,000 times. The digital Electronic microscope, on the other hand, can enlarge up to 1 million times! You really enter inside the cell. A digital microscope is not the regular one you are familiar with. It's basically a huge computer, and you are looking through a screen and communicating with the software. Is that clear to everyone?*"

Dr. Rubinstein continued:

"*This is science, not religion! So, I communicated with this* software, *and I asked the software; What is the force that holds the DNA attached … ?*"

"*The software responded that sometimes there is a sulphuric bridge [disulfide bonds], which makes sure the DNA stays attached.*"

Then Dr. Rubinstein said:

"*I asked the million-dollar question. Listen carefully, I asked the software, When does the bridge appear?*"

The software answer:

"*There are four nucleic acids which are the basis of the DNA chain: A, T, C & G. These are the acids which make up the DNA chain; that's a scientific fact! And the sulphuric bridge appears in the following manner: every 10 acids there was a bridge, every 5 acids a bridge, every 6 acids a bridge… and again every 5 acids were a bridge.*"

"*The [Number sequence: 10-5-6-5] I was shocked!*"

The Sulfuric Bridge occurs in this pattern:

- Every 10 acids, there is a bridge
- Every 5 acids, there is a bridge
- Every 6 acids, there is a bridge
- Every 5 acids, there is a bridge

This is a repeating pattern in the DNA double helix spiral, Figure 27.2.

Dr. Rubinstein was visibly shaken and said:

"*A sulphuric bridge that keeps the DNA attached? I was amazed! Who can tell me the meaning of 10-5-6-5?*"

[Dr. Rubinstein is looking at the students in his class for an answer; Responses are heard from the unseen audience.]

The name Yahweh is written on our very DNA

DEOXYRIBO NUCLEIC ACID

4 Nuclide Acids bind the Helixes together by Sulfuric Bridges in the sequence of

A-T-C-G

Every 10, 5, 6, 5 Acids

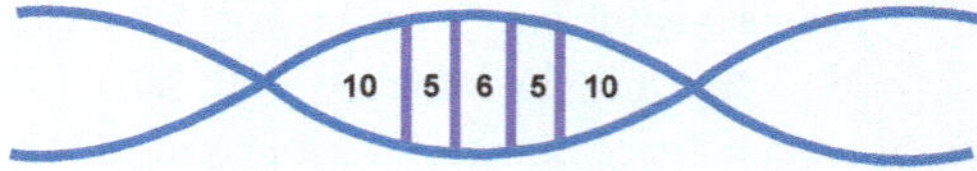

HEBRAIC NUMEIC MEANING

10 5 6 5

Y-H-V-H

The bridge, bond, Glue that keeps the Human DNA Sequence together.

Fig.27.2: The Name of YHWH Written in our DNA (10-5-6-5 … … 10-5-6-5)

"This is the numerical digit of Y-H-W-H! Our Heavenly Father is in our DNA! Like a Rembrandt with a signature at the bottom of His artwork, God tells us, "I made you"! Our Heavenly Father signed His name in every cell of our body! I think I discovered God!" (Emphasis added).

Most people are not familiar with Biological Science, DNA Structuring, or Disulfide Bridges within the DNA bonding processes. Nonetheless, Dr. Rubinstein is an expert in all of this, and probably much more. His résumé lists his "areas of activities" as:

- Supramolecular Chemistry;
- Self-assembly Biochemistry;
- Nanostructured Metal Films;
- Localized Plasmon;
- Optical Sensors;
- Electrochemistry;
- Thin Films;
- Nanoparticles; and
- Template Synthesis."

The DNA molecule appears as a double helix, looking like a spiral staircase. There are "steps" to the spiral staircase, or, sulfuric bridges. These sulfuric bridges are what keeps the two DNA strands attached together.

In Hebrew Gematria, Yod Heh Vav He is the Same Pattern 10-5-6-5 which equals 26! This is the numerical value of YHVH! The sulfuric bridges holding the DNA together repeat in a pattern that spells out the Sacred Name!

Dr. Rubinstein biography also states he has received eleven prestigious academic awards since 1971. In addition, he has written scores of articles published by scholarly journals in a multitude of scientific fields. Therefore, an intensive search for several weeks, none found to refute Dr. Rubinstein discoveries and claims. He appeared a trustworthy scientist who spoke accurately detailing his discoveries.

Two thousand years ago a vital Scripture was written: "All things were created through Him and for Him. He is before all things, and in Him all things hold together." (Colossians 1: 16–17).

It is very likely, only a Hebrew familiar with the Hebrew letters and their numbering system, could have readily recognized this coded signature upon our DNA. Dr. Rubenstein, is obviously a Jew who decoded the signature. God revealed the truth to him at such a time in history and in such a manner that it would be given to us by way of a global communication and information system that is now archived and watched by the entire world.

Dr. Rubenstein discovered the code by utilizing a computer aid. This is not a process of mere divination or numerology; it is historical fact. The numeric code for the name YHWH is easily verified. We've seen these letters many times before—yud, heh, vav, heh—the name of God, the Tetragrammaton. The numbers correspond exactly with each of those letters in the Hebrew alphabet. The sequence is 10 (yud), 5 (heh), 6 (vav/ waw), and 5 (heh)—10-5-6-5, YHWH, just as Dr. Rubenstein said. God's Trademark is printed in each cell of our body, the divine name of "Yahweh."— This divine name has been in use among the Hebrew people for thousands of years. The name "YHWH" was found more than seven thousand times in the Old Testament. It is a name with a deep profound meaning, connecting directly with the crucifixion of Yeshua.

Recently, we discover "YHWH" signature claimed by a PhD chemistry professor in Israel that the ancient and widely known numeric code of the name "YHWH." Dr. Rubenstein, of course, didn't necessarily "discover" God, he certainly discovered our Creator's signature in almost 100 trillion cells-DNA.

SCIENTISTS RESPONSE

Dr. Gregg Braden, research scientist commented: "a remarkable discovery linking the biblical alphabets of Hebrew and Arabic to modern chemistry reveals that a lost code—a translatable alphabet—and a clue to the mystery of our origins, has lived within us all along. Applying this discovery to the language of life, the familiar elements of hydrogen, nitrogen, oxygen, and carbon that form our DNA may now be replaced with key letters of the ancient languages. In doing so, the code of all life is transformed into the words of a timeless message. Translated, the message reveals that the precise letters of God's ancient name are encoded as the genetic information in every cell, of every life. The message reads: "God/ Eternal within the body." The meaning: Humankind is one family, united through a common heritage, and the result of an intentional act of creation!

Preserved within each cell of the estimated 7 billion inhabitants of our world, the message is repeated, again and again, to form the building blocks of our existence.

This ancient message from the day of our origins—the same message—remains within each of us today, regardless of race, religion, heritage, lifestyle, or belief. As we'll see the code is universal that it produces the identical message when translated into either the Hebrew or the Arabic language!"

Science continues to confirm the identity of the one true God and His Word in amazing ways.

The mapping of the genetic code, known as DNA, is probably the most important scientific breakthrough of the new millennium. the first genetic map of the human DNA molecule is called the discovery the "language in which God created life." "Mapping the chemical sequences for human DNA — the chemical "letters" that make up the recipe of human life — is a breakthrough that is expected to revolutionize the practice of medicine by paving the way for new drugs and medical therapies," says one web site.

This discovery has lasting physical and spiritual implications. A direct link can easily be found between the building blocks of life and the Creator of the universe. Mankind is *fearfully and wonderfully made,* with a hidden code within the cell of every life. This code is the alphabet of DNA that spells out the Creator's name and man's purpose.

Scientist discovered a "map" of four DNA bases that carry the ability to sustain life. These bases, known as chromosomes, are paired differently for each person. Human DNA contains 23 pairs of chromosomes, made up of hydrogen, nitrogen, oxygen, carbon, and their acidic counterparts. Encoded within these elements is an amazing blueprint of life that proves the Creator has put His own unique stamp upon every person. This stamp is actually His name as revealed to Moses' thousands of years ago.

It was from the burning bush that the Almighty revealed his character as the great "I AM." This name is the tetragrammaton of the Hebrew letters yod, hey, waw, hey. "YHWH-I AM WHO I AM. This is what you are to say to the Israelites:

`I AM has sent me to you. The Almighty said to Moses, "Say to the Israelites, `Y H W H, the mighty one of our fathers… This is my name forever, the name by which I am to be remembered from generation to generation," Exodus 3:14,15. The Almighty has given us His name as a sign of His existence and an avenue of communication. However, translators have hidden this Hebrew name in English Bibles.

In the Scriptures, the Sacred name of YHWH is used whenever the English words "LORD" or "GOD" appear in all capital letters. YHWH is used almost 7,000 times throughout the Bible as the only and unique name of the Mighty One of Israel. Isaiah corroborates this: "I am YHWH; that is my name; and my glory will I not give to another, " Isaiah 42:7. Jeremiah adds his confirmation: "They shall know that my name is YHWH," in chapter 16 verse 21. While Amos 5:8 says, "YHWH is his name." The book of Zechariah declares: "In that day there shall be one YHWH, and his name one." The Creator's Name is YHWH. Now, compare this four-lettered name to the four elements that make up human DNA and discover an ancient secret of creation.

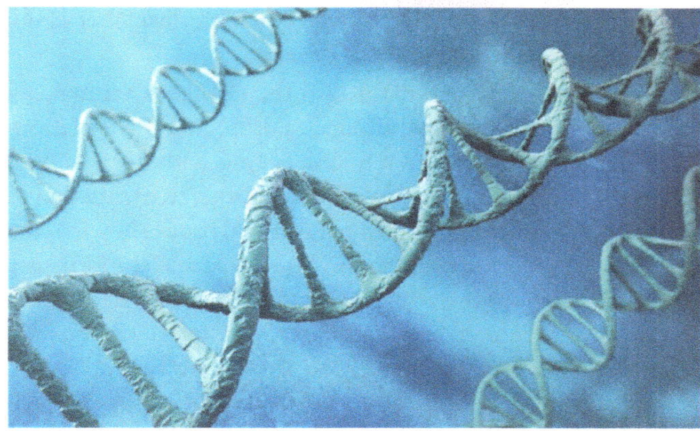

Fig.27.3: *Translating the DNA Code*

"The key to translating the code of DNA, Figure 27.3, into a meaningful language is to apply the discovery that converts elements to letters. Based upon their matching values of atomic mass,

- hydrogen becomes the Hebrew letter Yod (Y),
- nitrogen becomes the letter Hey (H),
- oxygen becomes the letter Wav (V or W), and
- carbon becomes Gimel (G).

These substitutions now reveal that the ancient form of YHWH's name, YHWH, exists as the literal chemistry of our genetic code. Through this bridge between YHWH's name and the elements of modern science, it now becomes possible to reveal the full mystery and find even greater meaning in the ancient code that lives as each cell of our bodies. When we substitute modern elements for all four letters of YHWH's ancient name, we see a result that, at first blush, may be unexpected. Replacing the final H in YHWH with its chemical equivalent of nitrogen, YHWH's name becomes the elements hydrogen, nitrogen, oxygen, and nitrogen (HNON) — all colorless, odorless, and invisible gases! In other words, replacing 100 percent of YHWH's personal name with the elements of this world creates a substance that is an intangible, yet very real form of creation! This is not to suggest that YHWH is simply a wispy gas made of invisible elements. Rather, it's through the very name that YHWH divulged to Moses over three millennia ago that our world and the foundation of life itself became possible.

YHWH tells us that in the form of hydrogen, the single most abundant element of the universe, He is a part of all that has ever been, is, and will be. Indeed, in the earliest descriptions of YHWH, we are told that He is omnipresent and takes on a form in our world that cannot be seen with our eyes. Thus, He can be known only through His manifestations. The Sepher Yetzirah describes this nonphysical form of YHWH's presence as the "Breath" of YHWH: "Ten Sefirot of Nothingness: One is the Breath of the Living YHWH, Life of worlds. This is the Holy Breath."

THE FOURTH LETTER CARBON

Additionally, the first chapters of Genesis relate that it is in a nonphysical form that the Creator was present during the time of creation (Genesis 1:2). It was "the spirit of YHWH" that first moved over the face of the earth. Substituting modern elements for the ancient letters, it is clear that although we share in the first three letters representing 75 percent of our Creator's name. While the presence of YHWH is the invisible and intangible form of the three gases hydrogen, nitrogen, and oxygen, the last letter of our name is the "stuff" that gives us the color, taste, texture, and sounds of our body: carbon.

The one letter that sets us apart from YHWH is also the element that makes us "real" in our world – carbon. Through both the secret letter codes of antiquity, and the literal translation of DNA as an alphabet, we're shown that something about our existence remains lasting and eternal. We share that never-ending quality with our Creator through a full seventy-five percent of the elements that define our genetic code," wrote Gregg Braden in his book The God Code. This is scientific proof showing us that YHWH has written His own name upon every human being. Sefer Yetzirah (The Book of Creation) says, "Within the letters is a great, concealed mystical exalted secret… from which everything was created." His name is within us, encoded into the basic cells of humanity. Every person, regardless of race, religion, sex, or status has the divine imprint inside their body. YHWH's name is in every person… "there is…one YHWH and Father of all, who is over all and through all and in all," says Ephesians 4:5. The potential of becoming like YHWH is in every person.

Genesis recounts that we have been made "in His image." In the beginning, the Creator "breathed upon man and he became a living being." It is this deposit from the heavens, the gift of a soul, that separates us from other species. Inside of each person is the breath of the Creator, known as the soul. In Hebrew, this is called the "neshamah." The neshamah is a divine spark of YHWH found within mankind. "Should a person strive towards purity in life, he or she is aided by a holy neshama," says the Zohar in Genesis 206a.

The Scriptures translate "neshamah" as "breath, spirit, and inspiration." It is the supernal soul of man, which pulls man towards YHWH. Imagine a pure light inside of every person in the world; this is the neshamah. It is totally good and unblemished. The neshamah is the part of YHWH within man. Proverbs 20:27 says, "The neshamah of man is the candle of YHWH, searching all the inward parts of the belly." Through the neshamah, one may connect to the will, wisdom, and understanding of Yah. How? The neshamah longs to be reunited with the Almighty. Its only desire is to return to its source; to be reunited in purpose. Each cell of our body, containing the divine name, groans to be reunited with YHWH. But it can't. Why not? Because of fleshly desires that result in sin. Sin stops the earth suit of the body from fully returning to its starting place with YHWH. The fleshly nature leads us to rebel against the Almighty's will and His ways. Our soul cannot cleave to YHWH because of our fleshly nature and ego. The desire to receive for self alone blocks the light of our neshamah. Our selfish actions are like a huge dark cloth, covering the Light of the Creator. Darkness grows, but the light remains. Try viewing mankind as an ember from the burning bush. The human body is the container of a divine spark from YHWH. Left alone, this spark will diminish and burn low through seeking pleasure in worldly desires.

The false fulfillment of momentary happiness is a darkness that seeks to put out our fire. Each action of the flesh places another layer of darkness upon the light.

These layers of darkness are called "sin," or "chet" in Hebrew. Chet is an ancient word that literally means to "miss the mark, loose focus, stray, miss the goal or path of right and duty, to incur guilt, incur penalty by sin, forfeit." While many people think that someone who sins is a "bad person," the Biblical concept is different. We all chet; we all sin; we all miss the mark. That doesn't make us bad people, we're just off target! In truth, the Hebrew word "Chet" appears in the Bible (Judges 20:16) referring to slingers who could shoot at a hair and not "chet", meaning not to "miss the target."

Chet is failure in a person's relationship with YHWH. Our goal should be to continually move closer to the target of YHWH, but chet causes distance. Sin is equivalent to "distance." Our sins distance us from the Light within. "Your sins have separated between you and YHWH, and your chet (sins) have hidden his face from you, so that he will not hear," Isaiah 59:2. Chet is the distance we place between our neshamah and our Creator as we miss the mark of the Scriptures. The center of YHWH's bull's eye is clearly explained within the pages of the first five books of the Bible. These teachings are often referred to as the "law." In Hebrew these books of instruction are called "torah." The Torah is YHWH's will for mankind and blueprint for living.

"The Torah is holy, and the commandments holy, and just, and good," says Romans 7:12. When we follow Torah we don't sin. To obey the precepts of Torah is to stay on the straight and narrow road of redemption. A person sins when the Torah is violated or forgotten. The book of First John clarifies this. "Everyone who sins practices "torah-lessness." Because sin (chet) is torah-lessness," 1 John 3:4. The problem is that we can't follow Torah enough. The obedience of today doesn't erase the disobedience of yesterday.

Though we obey the Torah, the layers of darkness remain within our soul. This is because Torah does not redeem. Torah describes how the redeemed believer is to live and relate to YHWH. Mankind is redeemed only through YHWH code. The soul code of DNA links man to YHWH. However, this doesn't equate mankind to YHWH. We aren't god. The code shows only our potential – to be like YHWH in our intentions and purpose. We can't achieve His state of greatness. Just as a flashlight will not work without batteries, our sincere efforts to correct the soul are useless.

We just can't correct our soul enough. We just can't follow Torah enough. The darkness of chet is too much. Good works can't dispel total darkness. "Whoever keeps the whole Torah and yet stumbles at just one point is guilty of breaking all of it," says the Newer Testament.

Prayer, obedience, and faith bring us closer to YHWH, but without the love of Messiah, we are still in the dark. Messiah is the floodlight that lights up our life. Mashiach' love (Christ' love) is the bridge that joins our neshamah to our Creator. The Savior is the light that saves us from eternal darkness and suffering. To handle the issue of sin, we must realize that stars are only seen at night. Under the deep darkness of sin is the light of the soul. Hidden in the DNA of every man, woman, and child is the YHWH code. To experience life at its fullest, all one must do is look inside and see the Sacred Name. YHWH is the path to purpose and way to life eternal. YHWH is our only hope. We need YHWH's salvation to deliver us from evil. Let's decipher this code and understand man's redemption.

"Anyone who calls upon the name of YHWH will be saved," Joel 2:32. This verse is quoted twice in the New Testament, in which both cases the Messiah is seen as fulfillment of this prophecy. The YHWH code is manifest in His Son. His son is the path of deliverance. Calling upon His Name allows the believer to excess the Almighty's power for deliverance.

Appropriately, the true name of Messiah demonstrates how this works…"And you shall call Him Yahshua, for He will save his people from their sins," says Matthew 1:21. The Hebrew speaking, King of the Jews was given a Hebrew name. He wasn't named "Jesus," but "Yahshua." In the gospels, the Messiah said that He came in His Father's name – the name of YHWH. How is YHWH Yahshua's name? The name Yahshua is a compound word, made up of two Hebrew phrases. First, YAH is a shortened version of the name of YHWH. The Name YAH is a poetic form of YHWH, found throughout the Psalms. The KJV says, "Sing praises to his name: extol him that rideth upon the heavens by his name YAH, and rejoice before him," Psalm 68:4. Secondly, "shua" is a Hebrew word meaning to "deliver, turn, save, or salvation." When these two words are put together, the Savior's true name is revealed:

"YAH" + "shua" = "YHWH is salvation" = Yahshua

YHWH offers His salvation, His deliverance through the person of Yahshua. Yahshua bridges the gap between YHWH and our souls. Mankind was made in the image of YHWH. We have His name written upon our very DNA. Scientists have proved that His name is stamped upon every soul. However, because of chet, because of sin, layers of separation distance our soul from our creator. "All have sinned and fallen short of the glory of YHWH," Romans 3:23.

Our stubborn self-will causes us to go an independent way. However, because of loving kindness, YHWH has "sent His only begotten son, that whosoever believes upon him would not perish but have everlasting life," John 3:16. We can connect our neshamah to YHWH though his son, Yahshua. The Savior is the only path to deliverance and salvation from the sinful self. The layers of sin that cloak our neshamah can only be removed through His blood. "As many as received Him, to them He gave the right to become the children of YHWH, even to those who believe in His name," John 1:12. The DNA within our body's points to our Creator and the salvation that He has provided. The YHWH code, within each person, is His son Yahshua.

Who Designed the Designer?

'Who created God?' It is a smokescreen argument, as resentment against Christianity. Let us examine the wonders of butterfly wing construction. Atheist may assert that it is a distinctive proof of the genius creation of nature. One can equally assert that it is a distinctive proof of the genius work of the perfect creator, Figure 27.4.

Fig.27.4: *Nature Seclares the Mind of the Creator – Curtsy - Photo sxc.hu*

Objects of *known* origin, have certain features—specified complex information—that occur only in those made by an intelligent designer (or an intelligently designed program). Thus, by the normal *analogical reasoning* we use in science, when we see these features in an object where the origin is *unknown*, we can *likewise* conclude that this object had an intelligent designer.

These features are those that an archaeologist would use to determine whether an object was designed by an intelligent designer, or that a SETI (Search for Extra Terristial Intelligence), devotee would use to argue that a signal from space came from an intelligent alien, or whether a ballot or card game was fixed, or whether a sequence of letters was the result of intelligence or monkeys on a keyboard.

It would be perverse to complain that the archaeologist didn't discuss whether the object's designer *itself* had a designer, or that the SETI researcher didn't tell us who designed the alien. It would be even meaningless to argue from this that we should simply drop the idea of design, and conclude that the object or hypothetical space signal had *no* designer.

Oxford philosopher Daniel Came, himself an atheist, responded similarly to this Dawkins argument, and gave some examples of where Professor Richard Dawkins would not like this used against his position:

'Dawkins maintains that we're not justified in inferring a designer as the best explanation of the appearance of design in the universe because then a new problem surfaces: ***who designed the designer?*** *This argument is as old as the hills and as any reasonably competent first-year undergraduate could point out is patently invalid. For an explanation to be successful we do not need an explanation of the explanation. One might as well say that evolution by natural selection explains nothing because it does nothing to explain why there were living organisms on earth in the first place; or that the big bang fails to explain the cosmic background radiation because the big bang is itself inexplicable.'*

As it happens, the God of the Bible does not fulfil the criterion of specified complex information, since He is not complex at all but simple. The leading philosopher Alvin Plantinga explained Dawkins' adolescent argument and then took it apart in The Dawkins Confusion: Naturalism ad absurdum, his review of Dawkins' *The God Delusion*:

'Now suppose we return to Dawkins' argument for the claim that theism is monumentally improbable. As you recall, the reason Dawkins gives is that God would have to be enormously complex, and hence enormously improbable ('God, or any intelligent, decision-making calculating agent, is complex, which is another way of saying improbable'). What can be said for this argument?

'Not much. First, is God complex? According to much classical theology (Thomas Aquinas, for example) God is simple, and simple in a very strong sense, so that in him there is no distinction of thing and property, actuality and potentiality, essence and existence, and the like. Some of the discussions of divine simplicity get very complicated, not to say mysterious. (It isn't only Catholic theology that declares God simple; according to the Belgic Confession, a splendid expression of Reformed Christianity, God is 'a single and simple spiritual being.') So first, according to classical theology, God is simple, not complex.

'More remarkable, perhaps, is that according to Dawkins' own definition of complexity, God is not complex. According to his definition (set out in Dawkins' *The Blind Watchmaker*), something is complex if it has parts that are 'arranged in a way that is unlikely to have arisen by chance alone.' But of course, God is a spirit, not a material object at all, and hence has no parts. *A fortiori* (as philosophers like to say) God doesn't have parts arranged in ways unlikely to have arisen by chance. Therefore, given the definition of complexity Dawkins himself proposes, God is not complex.

'So first, it is far from obvious that God is complex. But second, suppose we concede, at least for purposes of argument, that God *is* complex. Perhaps we think the more a being knows, the more complex it is; God, being omniscient, would then be highly complex. Perhaps so; still, why does Dawkins think it follows that God would be improbable? Given *materialism* and the idea that the ultimate objects in our universe are the elementary particles of physics, perhaps a being that knew a great deal would be improbable—how could those particles get arranged in such a way as to constitute a being with all that knowledge?

Of course, we aren't *given* materialism. Dawkins is arguing that theism is improbable; it would be dialectically deficient *in excelsis* to argue this by appealing to materialism as a premise. *Of course,* it is unlikely that there is such a person as God if materialism is true; in fact, materialism logically entails that there is no such person as God; but it would be obviously question-begging to argue that theism is improbable because materialism is true.'

Fig.27.5: God has Sovereign existence independent/outside this universe that He created - Curtsy - Photo sxc.hu

Dr. Andrew Lamb pointed to the question If God created the universe, then who created God? and wrote:

'The physical laws of our universe are intrinsic to our universe, but God has Sovereign existence independent/ outside this universe that He created, Figure 27.5. ***The Law of cause and effect*** (or in science – for every action there is a reaction) stipulates that for every effect (result), there must be an equal or greater cause, and this applies to everything that has a beginning—to a ***butterfly***, a ***person*** and to the ***universe***. But God is eternal. He has no beginning and therefore He needed no cause. There is no requirement that the Creator of our universe and its laws should Himself need a Creator.'

The opposition to Dr. Andrew Lamb responded further:

'Interesting corruption of the law of relativity, namely that every action must have an equal and opposite reaction.

It's actually Newton's Third Law of Motion, not relativity. Every effect must have a cause, but it doesn't have to be a god.

Not all causes must be God, and we never said that.

Response: However, a cause of the Universe would be unlimited by time, space or matter to be able to cause them; and it would also be personal to choose a particular universe and moment of creation. This is *consistent with* the biblical God.

The opponent commented: 'notice that whenever a Christian is cornered, they reach for their doctrine.'

Response: If a teaching within *any* philosophical framework is being attacked, it is perfectly reasonable to defend it with another teaching *from within that same framework.*

The proponent continued: 'Such is the case below. You state that God doesn't need a creator, yet you offer no proof of this.'

Repones: Yes; we do; i.e., God is not an effect at all; unlike the universe, He had no beginning in time.

The Opponent: 'If you believe that some sentient being has to create everything, then there is no way to rationally explain the existence of the alleged creator. I don't I believe that *only objects with beginnings* need causes, not that everything needs a cause. It is the impossible Christian catch 22 that can only be resolved by resorting to pointless faith and mythology that has gained credence through extended use.'

Catch-22 refers to a different sort of problem. In the original 1961 novel by Joseph Heller, if a pilot wanted to get out of flying hazardous combat missions, the only allowable exemption was *insanity*. But the military reasoned that any sane person would want to avoid such missions. So, if one applied for exemption on grounds of insanity, it would automatically prove one's *sanity*. Conversely, people who wanted to fly these missions *were* insane, so could apply for an exemption, but these are the very people *who would not want to be excused*. And if they did, they would automatically prove their *sanity*, so the Catch-22 rule meant that there could never be an exemption.

The novel also illustrated a self-imposed Catch-22 situation: a girl was upset that no man would marry her because she wasn't a virgin, yet she refused to marry any man crazy enough to marry a non-virgin.

CHRISTIANITY AND RATIONALITY

The opponent Response: 'I'm happy for people to believe this nuance, if it makes them feel more secure in this troubled world. What bothers me about the Christian ani-evolution agenda, is that it shuts down rational thought with threats of hell for those who don't fall into line with their beliefs; either that or the more subtle psychological tactics of manipulation through propaganda.'

Actually, as we Christians have pointed out that Christianity is the foundation for rational thought, while evolution provides no basis for it:

1. Man can *initiate* thoughts and actions; they are not fully determined by deterministic laws of brain chemistry. This is a deduction from the biblical teaching that man has both a material and immaterial aspect (e.g., *Genesis 35:18, 1 Kings 17:21–22, Matthew 10:28*). This immaterial aspect of man means that he is more than matter, so his thoughts are likewise not bound by the makeup of his brain. But if materialism were true, then 'thought' is just an epiphenomenon of the brain, and the results of the laws of chemistry. Thus, *given their own presuppositions*, materialists have not freely arrived at their conclusion that materialism is true, because their conclusion was *predetermined by brain chemistry*. But then, why should their brain chemistry be trusted over mine, since both obey the same infallible laws of chemistry? Thus, in reality, if materialists were right, then they can't even help what they believe (including their belief in materialism). Yet often call themselves 'freethinkers,' overlooking the glaring irony! Genuine initiation of thought is an insuperable problem for materialism.

2. Man can think rationally and logically, and that logic itself is objective. This is a deduction from the fact that he was created in God's image (*Genesis 1:26, 27*), and from the fact that Jesus, the Second Person of the Trinity, is the *logos*. This ability to think logically has been impaired *but not eliminated* by the Fall of man into sinful rebellion against his creator. (The Fall means that sometimes the reasoning is flawed, and sometimes the reasoning is valid but from the wrong premises. Thus, it is folly to elevate man's reasoning above what God has revealed in Scripture. But if evolution were true, then there would be selection only for survival advantage, not rationality.

   It's no accident that superstition has thrived where the Bible has weakened.

Global Warm-Mongering

The Falsehood of Global Warming

This suspension of rational thought also has the effect of immobilizing people's ability to respond to threats such as global warming.

Whether it is a threat at all is something that needs to be determined by the *evidence*, not 'consensuses' or politics. Dr. John Christy, Professor of Atmospheric Science, University of Alabama; Lead Author of "The Intergovernmental Panel on Climate Change (IPCC)." He points out:

Fig.27.6: The Irrational Thoughts of Global Warming Falsehood – Curtsy - Photo sxc.hu

'While most participants are scientists and bring the aura of objectivity, there are two things to note:

- this is a political process to some extent (anytime governments are involved it ends up that way)
- scientists are mere mortals casting their gaze on a system so complex we cannot precisely predict its future state even five days ahead

'Skepticism, a hallmark of science, is frowned upon. (I suspect the IPCC bureaucracy cringes whenever I'm identified as an IPCC Lead Author.)

'The signature statement of the 2007 IPCC report may be paraphrased as this: "We are 90% confident that most of the warming in the past 50 years is due to humans."

'We are not told here that this assertion is based on computer model output, not direct observation. The simple fact is we don't have thermometers marked with "this much is human-caused" and "this much is natural."

'So, I (*Dr. John Christy*) would have written this conclusion as "Our climate models are incapable of reproducing the last 50 years of surface temperatures without a push from how we think greenhouse gases influence the climate. Other processes may also account for much of this change."

. . .

'Of all scientists, climate scientists should be the humblest. Our cousins in the one-to-five-day weather prediction business learned this long ago, partly because they were held accountable for their predictions every day.

'Answering the question about how much warming has occurred because of increases in greenhouse gases and what we may expect in the future still holds enormous uncertainty, in my view.

. . .

'[F]undamental knowledge is meagre here, and our own research indicates that alarming changes in the key observations are not occurring.'

And our decisions should not rely on Al Gore's film, which a British court found had 11 serious errors so couldn't be imposed on school pupils without a disclaimer. But I would be more inclined to believe Gore was at least genuine in his actions if he didn't jet everywhere to promote the cause and he didn't have significant interests in companies set up to trade carbon credits, which are now worth much more because of his 'documentary.'

Gore's producer Laurie David could also skip the two holidays per year on her private jet (but at least she says, 'I feel horribly guilty about it', and since much of the crusade is about the right *feelings* than actual *results*, that supposedly makes it OK. So, she's still entitled to abuse SUV-driving motorists from her car window for emitting CO_2—a tiny fraction of the amount that her jet spews).

Our own view on this issue is clearly stated in our comprehensive article "Anthropogenic Global Warming (AGW)—a biblical and scientific approach to climate change." We also noted in "The Great Global Warming Swindle Debate" that one leading global warming skeptic is the fanatical *anti-creationist, Ian Plimer.*

Christian Stewardship Vs Ecofascist Earth-Worship

After all, if, as the Bible claims, everything was created for man's benefit, then how could there be anything wrong with exploiting the environment and laying it waste?

Try actually *reading* what the Bible teaches. God gave man dominion, which is responsible stewardship, not a license to destroy. You could also try reading what we *actually* teach before ranting against what you *think* we teach, e.g., Earth Day: Is Christianity to blame for environment problems?

But let's turn the question around on you: After all, if, as evolution claims, all creatures are rearranged pond scum, the result of survival of the fittest, then how could there be anything wrong with exploiting the environment and laying it waste? After all, it would just prove that we are fitter than all the exploited creatures.

For another illustration, Mr. Burns in the anti-Christian cartoon series "The Simpsons" said:

'Oh, so Mother Nature needs a favor? Well, maybe she should have thought of that when she was besetting us with droughts and floods and poison monkeys. Nature started the fight for survival and she wants to quit because she's losing? Well, I say hard cheese!'

God will ensure that we will be fine in the end—isn't that, right?

Yes, but it doesn't mean that He will pick up our garbage for us. The Bible makes it clear that our actions have consequences: 'A man reaps what he sows.' It is a decidedly anti-biblical view that says that we can do what we like now because it will all pan out for well in the end. Grace is not a license for sin (see *Romans* 6).

Humans: Made in God's Image or Parasites Upon Earth?

This means there is no need to control population because resources are unlimited, Figure 27.7.

1. First, as we have pointed out, the 7 billion people in the world 'could all fit into an area the size of England, with more than 20 square meters each.'
2. Second, the reason for the population growth, as we've said, is 'not because people suddenly started breeding like rabbits—rather, it was because they finally stopped dying like flies. Between 1900 and the end of the 20th century, the human life span likely doubled, from a planetary life expectancy at birth of perhaps 30 years to

Fig.27.7: Humans are not Parasites Upon Earth - Curtsy - Photo sxc.hu

one of more than 60. By this measure, the overwhelming preponderance of the health progress in all of human history took place during the past 100 years.'

3. Third, population growth has been good for countries:'Troubled as the world may be today, it is incontestably *less* poor, *less* unhealthy, and *less* hungry than it was 30 years ago. And this *positive* association between world population growth and material advance goes back at least as far as the beginning of the 20th century.'
4. Fourth if there is to be population control, ***who** does the controlling and **how***? People like American Eric Pianka, Melbourne's John Reid and Finland's Pentti Linkola, the role model for the atheistic evolutionary murderer Pekka-Eric Auvinen, who don't mind some population culling? And population doom-mongers just loved the one-child policies of China (and we have noted the deafening silence from the feminists about the consequences of abortion and sex selection that kill girl babies preferentially).

As for resources, many doom-mongers misunderstand that the limitations have *nothing* to do with the amount of physical stuff in the ground, and *everything* to do with the *cost of extracting them*. Below a certain cost, it is uneconomical to extract the stuff, so most is not extracted.

The leading and oft-discredited (but Greenie hero) population doom-monger Paul Ehrlich lost a famous bet about resources running out with economist Julian Simon, because he didn't understand that limitations were really *economical* not physical.

This was illustrated in practical terms by the shortages of petrol/gasoline and the long lines at petrol stations (if they were even open) during Jimmy Carter's presidency, caused by price controls on petrol.

Nonetheless, Ronald Reagan's first action when he became president was to lift these price controls. Reagan realized that lifting price controls would make it economical to re-open capped oil wells, and it would also encourage *self-rationing* by consumers. The huge petrol queues disappeared almost overnight, and have never returned, and before long, the price was lower than Carter's previous capped price.

It also means there is no need to control pollution, because God will take care of the waste.

Again, how does this logically follow?

Then if it all looks like it's coming unstuck, it must mean judgement day is approaching—which is also good, because Christians get to go somewhere better; somewhere else they can lay waste!

As we already cited above:

'The reality is that nations and cultures with a Christian background have, on the whole, the best track records when it comes to issues like pollution, etc. Following the collapse of communism, the appalling environmental track record of these socialist "model states," run on atheistic assumptions, went on world display.'

The Blessings of Christianity

Atheists accused Christians saying: "No, Christianity is not about goodness and kindness. It's about control."

Interesting irony here. Above, the Global Warming Alarmist were claiming that we *should* use coercive population *control* measures. Now you rail against alleged *control* by Christianity. But you overlook how Christianity really has been a force for goodness and kindness, e.g., founding hospitals and orphanages, abolishing slavery, providing the basis for modern science, helping the needy. Surprising the research facts highlighted who really helps the poor! What has atheism done for the world? Not only that, but the greatest flowering of personal freedom followed the Protestant Reformation that re-established the authority of the Bible in society. Just try being an atheist in a country where Islam dominates; or try being a Christian in a country where atheism is the state religion (i.e., Communist states).

Look at the First Testament

Actually, we *have* looked, in quite some detail! God as the Creator of life has the right to take it. There is no difference in the 'second' testament actually—both testaments clearly teach God's righteous judgment upon sin as well as his mercy and longsuffering (where sinners don't get what they deserve) and grace (where we get what we didn't deserve).

No One Ultimately Control Nature

Of course not. Yet far too many people in effect make an idol of the government and think that this is the answer to everything—even though we are powerless against many natural disasters. Nevertheless, C.S. Lewis argued that living in this dangerous world was God's 'megaphone' that this world was not the intended final destination for man.

'The Bible teaches that suffering is part of the "big picture" involving sin, but *individual* cases of suffering are *not* always correlated with *particular* sins of individuals.'

The biblical cases of Job, the man born blind, the victims of the collapse of the Tower of Siloam, Lazarus and Paul.

God is sovereign and so does not dance to your tune. But He has given guidance about how He works in the Bible. God is in fact consistent: 'He is the same yesterday, today and forever' (*Hebrews 13:8*), not capricious as one may charge. This very principle provided one of the philosophical bases for modern science to flourish. Christianity is no armor in life.

It is the best sort of armor of all: it is *true, and rationally defensible!* It is merely a *de facto* lobotomy.

Atheism is *self*-lobotomy, by removing the basis for rational thought. It also provides no armor, since it gives no basis for morality or hope. E.g., one atheistic evolutionary biologist said:

'Let me summarize my views on what modern evolutionary biology tells us loud and clear. And I must say that these are basically Darwin's views. There are no gods, no purposeful forces of any kind. No life after death. When I die, I am absolutely certain that I am going to be completely dead. That's just all. That's gonna be the end of me. There is no ultimate foundation for ethics, no ultimate meaning in life, and no free will for humans, either.' [William B. Provine, Professor of Biological Sciences, Cornell University (*Origins Research* **16**(1/2):9, 1994)].

Loving God with all your Intellect:

Logic and Creation

Logic and reason are far from being incompatible with biblical Christianity. Rather, they are **essential**. Without them it is impossible to deduce anything from the true propositions of the 66 books of Scripture, the Christian's final authority. This applies to Creation, one of the foundational doctrines of Christianity. Examples of valid and fallacious reasoning are deliberated, with emphasis on showing how logical reasoning can support the truth of biblical creation, and demonstrate the fallacies in many evolutionists' arguments.

Logic is the science of the relations between *propositions*. Logic can tell us what can be inferred from a given proposition, but it cannot tell us whether the given proposition is true in the first place. All philosophical systems rely on logical deductions from starting assumptions—*axioms*—which, by definition, cannot be proven from prior assumption. For our axioms, it is rational to accept the propositions revealed by the infallible infinite God in the 66 books of the Bible.

Scriptural Deliberations

Martin Luther correctly distinguished between the magisterial and ministerial use of reason.

The magisterial use of reason occurs when reason stands over Scripture like a magistrate and judges it. Such 'reasoning' is bound to be flawed, because it starts with axioms invented by fallible humans and not revealed by the infallible God. But this is the chief characteristic of liberal 'Christianity,' It is refuted by Scriptural passages such as **Isaiah 55:8–9**

- 'For my thoughts are not your thoughts, neither are your ways my ways,' declares the Lord.
- 'As the heavens are higher than the earth, so are my ways higher than your ways and my thoughts than your thoughts.'

This does not mention 'My logic is higher than your logic.' If so, then if we believed 2+2=4, God could believe 2+2=5. What it does mean is that God knows every true proposition, while we know only a part. Another passage is **Romans 9:19–21**

- One of you will say to me: 'Then why does God still blame us? For who resists his will?'
- But who are you, O man, to talk back to God? Shall what is formed say to him who formed it, 'Why did you make me like this?'
- Does not the potter have the right to make out of the same lump of clay some pottery for noble purposes and some for common use?

The Ministerial Use of Reasoning

The ministerial use of reason occurs when reason submits to Scripture. This means that all things necessary for our faith and life are either expressly set down in Scripture or may be deduced by good and necessary consequence from Scripture.

Many Scriptural passages show that Christians are not supposed to check in their brains at the church door, but to use their God-given minds in subjection to God's Word, e.g., **Isaiah 1:18**

- 'Come now, let us reason together,' says the Lord. 'Though your sins are like scarlet, they shall be as white as snow; though they are red as crimson, they shall be like wool.'

Matthew 22:36–38

- 'Teacher, which is the greatest commandment in the Law?'
- Jesus replied: 'Love the Lord your God with all your heart and with all your soul and with all your mind.'
- This is the first and greatest commandment.'

Romans 12:2

- Do not conform any longer to the pattern of this world, but be transformed by the renewing of your mind. Then you will be able to test and approve what God's will is—his good, pleasing and perfect will.

1 Corinthians 2:16

- 'For who has known the mind of the Lord that he may instruct him?' But we have the mind of Christ.

"*mind of Christ*," not feelings or emotions of Christ.

Much confusion arises when some people disparage 'head knowledge.' For example, Geoff Smith, who was Pastor of the large Auckland Bible Church (New Zealand), has pointed out that in some churches, anything that has to do with rational thinking is suspect and strongly discouraged. Rational thinking is branded as something coming from the flesh. People of the Spirit won't try to understand what's happening—they will simply accept the 'blessing.' The catch words are unmistakable: 'Don't try to understand this', 'Don't try to analyses this', 'Don't try to figure this out with your mind', etc.

In such thinking there is no real understanding that faith is always built on *knowledge*. The prophet Isaiah asks repeatedly '*Do you not know, have you not heard?*' (Isaiah 40:21,28).

Jesus repeatedly asks: 'Have you not read …?' and tells the Sadducees that they are in error because they 'do not know the Scriptures or the power of God' (Matt. 22:29).

In his letters Paul constantly shows that true, functional faith is always built on knowledge. Conversely, deficient faith is traced back to its unmistakable cause—deficient knowledge. Paul repeatedly asks the question 'Don't you know …?' (Rom. 6:3, 16; 11:2; 1 Cor. 3:16; 1 Cor. 5:6; 1 Cor. 6:2, 3, 9, 15, 16, 19; 1 Cor. 9:13, 27). Notice also the same question being asked by James (James 4:4). Philip asked the Ethiopian eunuch: 'Do you understand what you are reading?' (Acts 8:30).

Part of the confusion lies in the misunderstanding of the word 'heart' in the Bible. Some people make a false contrast between 'head-knowledge' and 'heart-trust.' When interpreting Scripture, it is important to work out what the authors meant by the term. In this case, one should work out what 'heart' meant to ancient Semites, not what it means in Hollywood pop-psychology. In the Bible, the word 'heart' is used 75% of the time to mean the mind or intellect. However, the Bible frequently contrasts the heart and the lips—sincerity vs. hypocrisy, for example:

Genesis 6:5:

- The Lord saw how great man's wickedness on the earth had become, and that every inclination of the thoughts of his heart was only evil all the time.

Psalm 14:1:

- The fool says in his heart, 'There is no God.' They are corrupt, their deeds are vile; there is no one who does good.

The New Testament concept of faith is compatible with reason. The Greek word for 'faith' is πίστις (*pistis*) which is related to the verb πιστεύω (*pisteuo*) meaning 'believe', and πείθω (*pietho*) meaning 'to convince by argument.' It never has the connotation of 'believing six impossible things before breakfast', but 'is being sure of what we hope for and certain of what we do not see.' (Heb. 11:1). Many non-Christians have a misconception of biblical faith, and unfortunately some Christians have accepted this.

Logic in Biblical Preaching and Witnessing

Christ's chief apostle, Peter, commanded us (1 Peter 3:15):

- But in your hearts set apart Christ as Lord. Always be prepared to give an answer to everyone who asks you to give the reason for the hope that you have. But do this with gentleness and respect.

The Greek word translated 'answer' in 1 Peter 3:15 is in fact ἀπολογία (*apologia*). This term is derived from the Greek words ἀπὸ (*apo*) = away from and λόγος (*logos*) = logic/reason, so means 'out of logic/reason', so refers to a reasoned defense that would be given in a court of law. The classic example is Plato's **Apology**, Socrates' defense against the charges of atheism and corrupting the youth. The word also appears in the negative in Rom. 1:20: unbelievers are ἀναπολογήτους (*anapologētous*) (without excuse / defense / apology) for rejecting the revelation of God in creation.

The word for 'reason' above is λόγος (*logos*), in this context meaning evidence that provides rational justification for one's belief.

Christ's half-brother, Jude, commanded in verse 3 of his epistle:

'… earnestly contend for the faith which was once delivered unto the saints.'

This implies a real intellectual battle to convince people of something righteous and true. Paul elaborated on this in 2 Corinthians 10:4–5:

- The weapons we fight with are not the weapons of the world. On the contrary, they have divine power to demolish strongholds.
- We demolish arguments and every pretension that sets itself up against the knowledge of God, and we take captive every thought to make it obedient to Christ.

Obviously, evolution is the major anti-God pretension of our age, so we must make great efforts to demolish it.

Accurate Definitions of Words

It is impossible to have a logical discussion with people if there is no agreement on meanings of words, or with those who are dishonest with their terminology. Socrates, in Plato's **Phaedo**, stated succinctly, 'To use words wrongly and indefinitely is not merely an error in itself, it also creates evil in the soul.'

Many cults, including liberal 'Christianity,' often use biblical terminology, but invest the words with entirely new meanings. They resemble Humpty-Dumpty who replied scornfully to Alice's ignorance of what he meant, 'When I use a word, it means just what I choose it to mean—neither more nor less.'

Some prize examples of semantic gymnastics can be found in the ramblings of liberal 'Christians.' Since they are being paid to defend doctrines they don't believe, they redefine them instead. That way, they can pretend they are not violating their ordination vows. For them, God is not the Creator, but the 'ultimate concern;' 'Jesus is Risen' means that His influence continued after His death; 'Christian faith' need not consist of holding any doctrines, although the NT states that those who forsake orthodox Christian doctrine have departed from the faith (1 Tim. 4:1, 5:8, 2 Tim. 3:8; cf Eph. 4:5).

These are all examples of stipulative definitions. This fallacy is common among evolutionists who contrast 'scientists' and 'creationists.' A creationist would respond by producing evidence that there are thousands of practicing scientists who believe in biblical creation. But some evolutionists respond that such people cannot be true scientists because no true scientist can accept a creationist explanation, regardless of his qualifications or research experience. This becomes essentially a circular argument: all who are qualified in science and practice science and reject creation are opposed to creation.

Equivocation

The worst example of intellectual dishonesty is equivocation, that is, switching the meaning of a single word part-way through an argument. This deceitful practice is used by many evolutionary propagandists when defining the word 'evolution.'

The theory of evolution really means the development of all living things from a single cell, which itself came from non-living chemicals. This directly contradicts the Bible and has no scientific support. But many propagandists define evolution as 'change in gene frequency with time' or 'descent with modification' and use Darwin's finches and industrial melanism in the peppered moths as clinching proof of 'evolution' and disproof of creationism! An example is the atheist Eugenie Scott, Executive Director of the pretentiously named National Center for Science Education, the leading US organization devoted entirely to evolution-pushing. She approvingly cited a teacher whose pupils said after her 'definition': 'Of course species change with time! You mean that's evolution?!'

Obviously, no creationist disputes that changes occur through time, but creationists disagree that the type of change required for molecules-to-man evolution occurs, i.e. changes that increase information content.

Sense and Reference

Philosopher Gottlob Frege (1848–1925) posed a problem in his famous example "Hesperus is Phosphorus" (the evening star is the morning star).

Long ago, Pythagoras showed that they are both the same object, the planet Venus. But is someone saying "Hesperus is Phosphorus" saying exactly the same thing as "Hesperus is Hesperus"? Obviously, the latter is trivially true as a=a, but when Pythagoras said the former statement, he was conveying new information: that these two apparently different objects were really the same thing. This also means that someone not knowing that could say "Hesperus is not Phosphorus" without meaning that he is denying the trivially true "Hesperus is Hesperus." Frege argued that Hesperus and Phosphorus had a different *sense* (German *Sinn*), although they have the same *reference* (German *Bedeutung*)—Venus—because they convey different *attributes* (shining in evening or morning).

This can be applied to the cults mentioned above: when they use the same words, they don't have the same reference, e.g., the biblical Jesus v false christs. In another case, although Jesus affirmed Genesis as history, is a theistic evolutionist necessarily calling Christ a liar? Not necessarily: a theologically conservative Christian evolutionist would likewise affirm the same 'reference' as the Christian creationist: that Christ always told the truth. But some genuine Christians may not realize that Jesus affirmed that Genesis was history—a 'sense' especially the new Christians realizing that Hesperus was the same object as Phosphorus, but would still affirm that Hesperus is Hesperus.

Another application has been exposing the scientific ignorance of some environmental activists and gullible supporters. People are asked to support a ban on ban 'dihydrogen monoxide,' with lists of all its terrifying dangers of this stuff. This hoax exploits the ignorance of many people that the two senses 'dihydrogen monoxide' (and 'hydrogen hydroxide' and the proper chemical names 'hydrogen oxide' and 'oxidant') and 'water' have the same reference: H_2O.

Truth and Falsity

A simple definition of truth and falsity goes back at least as far as Aristotle (384–322 BC): 'If I say of what is that it is, I speak the truth. If I say of what is not that it is, I speak falsely.' That is, a statement is true if it corresponds to the facts, and false otherwise.

This should be obvious, but the atheistic anti-creationist Ian Plimer wrote:

'In my view, the Bible is not true. However, it is the Truth.'

Of the many crass blunders, he makes in logic, mathematics, science and exegesis, this is the worst.

Reasoned Arguments

In logic, an argument is defined as a sequence of statements comprising premises that are claimed to support a conclusion. As shown above, Scripture teaches that Christians are to argue in this sense. This is not the same as being argumentative, or arguing just for the sake of arguing.

Arguments can be either deductive or inductive. Deductive reasoning is reasoning from the general to the particular. Inductive arguments reason from a finite set of examples to a general rule. Deductive arguments are the most important, so I will concentrate on them below.

A syllogism is a common type of deductive argument with two premises and a conclusion.

I. **Validity**

A valid argument is one where it is impossible for the premises to be true and the conclusion false, i.e., the conclusion follows from the premises. Note that validity does not depend on the truth of the premises, but on the form of the argument.

One example of a valid argument with true premises is:

a. All whales have backbones;

b. Moby Dick is a whale;

∴ Moby Dick has a backbone.

An example of a valid argument with a false premise and false conclusion is:

a. All dogs are reptiles;

b. All reptiles have scales;

∴ All dogs have scales.

An invalid argument with a true premise and true conclusion is:

The sun is larger than the earth;

∴ polytheism contradicts the Bible.

This is invalid because the conclusion contains terms not contained in the premise. It is important to recognize valid forms of argument, and use them.

Many invalid arguments can be found in the works of politicians. On the astute British television political satire, *Yes Prime Minister* (episode 'Power to the People'), two head civil servants (Sir Arnold Robinson and Sir Humphrey Appleby) illustrated 'politician's logic':

a. Something must be done;

b. This is something;

∴ We must do it.

As pointed out in the program, this is just as invalid as:

a. All cats have four legs;

b. My dog has four legs;

∴ My dog is a cat.

II. Soundness

A sound argument is a valid argument with true premises. The conclusion of a sound argument must be true. So, to prove the conclusion of a valid argument, it is sufficient to prove all premises are true. For example:

a. Abortion is intentional killing of a fetus;

b. A fetus is an innocent human being

c. Intentionally killing an innocent human being is murder;

d. Murder is forbidden by God;

∴ Abortion is forbidden by God.

The form of the argument is valid, premises 'a.' and 'b.' are true by the normal definitions of words, 'b.' can be proven by science and Scripture (*Gen. 25:22 and Lk. 1:41* use the same words for unborn and born children), 'd.' is proven by *Gen. 9:6, Ex. 20:13, Rom. 13:9*, so the argument is sound.

III. Contradiction

A contradiction is defined as the conjunction of the affirmation and denial of a premise, in the same time, place, and sense (i.e., p and not-p, or in symbolic form, p.~p). For any pair of contradictory premises, one must be true and the other false. The Law of Non-Contradiction prevents both premises being true, while the Law of Excluded Middle points out that a pair of contradictory premises exhausts all possibilities. Another way of putting it is: a proposition must be either true or false—not both true and false, nor in some limbo state in between truth and falsity. This can be useful in listing all possible alternatives and refuting all of them but the correct one.

C.S. Lewis's Famous Trilemma Argument is a Good Example.

1. Jesus Christ is reported to have claimed to be God. The reports are either true or false.
2. If the reports are false, the reporters either knew they were false or they did not.
3. If they knew they were false, they were liars—but who would die for what they know is a lie?
   i. If they did not know, then it is a big problem to explain how legends could accumulate around a historical figure in such a short time.
   ii. If the reports are true, then Jesus was either speaking falsely or truly.
   iii. If Jesus spoke falsely, He either knew it or he did not.
   iv. If He knew, He was a liar.
   v. If He knew not, then He was a lunatic, since a claim to be God is the most absurd claim a mere creature can make.
   vi. If Jesus spoke truly, then He really is God.

Anti-Christians often charge the Bible with contradicting itself, as they realize that if the charge were proven, it would show that it affirms at least one false statement, thus disproving divine authorship. But most of these skeptics are ignorant of the above definition of a contradiction.

For example, *Mt. 20:29* ff. which states that Christ healed two blind men does not contradict *Mk. 10:46* ff. which states that Bartimaeus was healed, as the latter does not say only Bartimaeus was healed.

Some other alleged contradictions can be resolved by showing that words are being used in different senses, e.g.

John 1:18 vs *Exodus 24:9–10*: in the former, Jesus states, 'No man has seen God [in His full glory as Sovereign of the Universe] at any time; the only begotten God [Jesus] … has explained Him.' In the latter, Moses was clearly beholding a veiled presence of God, metaphorically referred to as 'under His feet'. In *Exodus 33:18–23*, a distinction is also made between beholding God's full glory ('face') and His veiled presence ('back').

The Three Eternal and Co-Equal Persons

Although many cults claim that the biblical doctrine of the Trinity is self-contradictory, it is not. The Oneness and Threeness of God refer to different aspects. The Three Eternal and co-equal Persons of the Godhead—Father, Son and Holy Spirit—are the same in essence but distinct in role—three Persons (or three centers of consciousness) and one Being.

An important aspect of contradiction is self-refutation. Many statements by anti-Christians might appear reasonable on the surface, but when the statement is turned on itself, it refutes itself. Some common examples are:

- 'There is no truth'—this would mean that this sentence itself is not true.
- 'We can never know anything for certain'—so how could we know that for certain?
- 'A statement is only meaningful if it is either a necessary truth of logic or can be tested empirically' (the once popular verification criterion for meaning of the 'Logical Positivists')—this statement itself is neither a necessary truth of logic nor can it be tested empirically, so it is meaningless by its own criteria.
- 'There are no moral absolutes, so we ought to be tolerant of other people's morals'—but 'ought' to imply a moral absolute that toleration is good.

IV. Conditional Statements and Implications

These are of the form: 'if p then q' (if p is true, then q is true). Another way of putting it is 'p implies q', or in symbolic form p ⊃ q (or p → q). Yet another way is saying that p is a *sufficient* condition for q, while q is a *necessary* condition for p. P is called the *antecedent*; and q is called the *consequent*.]

Asserting the truth of the implication (p ⊃ q) does not in itself imply that the antecedent (p) is true—only that if p were true, q must logically follow from it. For an example of a misunderstanding of this point, some use the following passage to 'prove' that it is possible to speak with 'angelic tongues'—*1 Cor. 13:1–3*:

1. If I speak in the tongues of men and of angels, but have not love, I am only a resounding gong or a clanging cymbal.
2. If I have the gift of prophecy and can fathom all mysteries and all knowledge, and if I have a faith that can move mountains, but have not love, I am nothing.
3. If I give all I possess to the poor and surrender my body to the flames, but have not love, I gain nothing.

Paul makes several conditional statements showing that without love, no matter what other wonders he might hypothetically be able to perform, he would be nothing. He is no more asserting that there are special angelic tongues than that he moved mountains or gave his body to the fire. There may be passages to support a common Pentecostal practice, but this is not one of them.

Another good example is Jesus' statement in *Matthew 12:27*:

And if I drive out demons by Beelzebub, by whom do your people drive them out? So then, they will be your judges.

I doubt that any Christian would claim that Jesus was asserting that He drove out demons by Beelzebub! He was showing that if His opponents were right in their accusation, then the accusation would equally apply to their own people. Jesus' argument is an example of *reductio ad absurdum* as illustrated below.

Table 27.1: Affirming the Antecedent

English	Symbolic
p implies q	p ⊃ q
p is true	p
therefore, q is true	q
Affirming the Antecedent	Modus ponens

Table 27.2: Denying the Consequent

English	Symbolic
p implies q	p ⊃ q
q is false	~q
therefore, p is false	~p
Denying the Consequent	Modus tollens

From a conditional statement, one can construct two types of valid inference: *modus ponens* (Table 27.1) and *modus tollens* (Table 27.2). *Modus ponens* is Latin for 'method of constructing.' The reason it is called 'affirming the antecedent' is that the argument proves that the consequent must be true if the antecedent is affirmed. *Modus tollens* is Latin for 'method of destroying.' This type of argument proves that the antecedent must be false if the consequent is denied.

There are two types of invalid inference: the fallacies of affirming the consequent (Table 27.3) and denying the antecedent (Table 27.4).

Table 27.3: Fallacy of Affirming the Consequent

English	Symbolic
p implies q	$p \supset q$
q is true	q
therefore p is true	$\therefore$ p

Table 27.4: Fallacy of Denying the Antecedent

English	Symbolic
p implies q	$p \supset q$
p is false	$\sim p$
therefore, q is false	$\therefore \sim q$

To illustrate: starting with the implication: If Jesus rose from the dead (p), then His bones cannot be found (q); and combining this with four possible premises as follows:

1. Jesus rose from the dead (p is true), Table 27.3.

   ∴ His bones cannot be found (q is true)

   - valid.

2. Jesus' bones cannot be found (q is true)

   ∴ He rose from the dead (p is true)

   - invalid.

   A reminder: validity is independent of the truth or falsity of the premises or conclusion. We accept that Jesus rose, but not that every dead person whose bones are missing also rose.

3. Jesus did not rise from the dead (p is false)

   ∴ His bones can be found (q is false)

   - invalid.

   The conclusion does not follow; many people who did not rise were cremated.

4. Jesus' bones can be found (q is false)

   ∴ He did not rise from the dead (p is false)

   - valid (despite the fact that both premises are false).

The founders of many counterfeit religions still have skeletons moldering away, which is proof that they are not risen.

One use of *modus tollens* is the *reductio ad absurdum* argument, i.e., showing that a premise is false by demonstrating that it implies an absurd conclusion.

An example is the effort by Bishop John Shelby Spong to show that homosexual acts are OK because some animals practice them. As it stands, the argument is invalid. To make it valid, another premise is needed that states: whatever animals do is OK.

1. Animals practice homosexual acts;
2. Whatever animals practice is OK;

∴ Homosexual acts are OK.

To prove the argument to be sound, that premise must be proved to be true. Conversely, to prove the argument to be unsound, the premise must be shown to be false. This can be done by showing that it leads to a ridiculous conclusion:

1. Animals practice rape and cannibalism;
2. What animals do is OK;

∴ Rape and cannibalism are OK.

Now if one does not accept the conclusion, if one is logical, one must reject one or more of the premises. As (1) is empirically true, (2) must be the false premise. So Spong's argument contains a false premise and is thus unsound.

Another example: pro-abortionists often claim that the unborn child is merely a disposable part of the woman's body. However, see what happens if this premise is combined with other indisputable premises in the following argument:

1. If a is part of b and b is part of c, then a is part of c (This is called a transitive relation—an example is: if a brick is part of a wall and a wall is part of a house, then the brick is part of the house);
2. A male unborn baby has a penis (that is, a penis is part of him);
3. This baby is part of the pregnant woman;

∴ A woman pregnant with a male baby has a penis.

As the conclusion is false (feminists would detest it especially), at least one of the premises must be as well. All premises are indisputably true except the pro-abortionists' (3), which was the disputed issue. Thus, this argument proves it false.

An example of a fallacious *reductio ad absurdum* is the argument that the Sadducees used against the Resurrection—*Matthew 22:23–34*

- That same day the Sadducees, who say there is no resurrection, came to him with a question.
- 'Teacher,' they said, 'Moses told us that if a man dies without having children, his brother must marry the widow and have children for him.
- Now there were seven brothers among us. The first one married and died, and since he had no children, he left his wife to his brother.
- The same thing happened to the second and third brother, right on down to the seventh.
- Finally, the woman died.
- Now then, at the resurrection, whose wife will she be of the seven, since all of them were married to her?

Please observe that they tried to refute the Resurrection by showing that it leads to the absurd conclusion that in this hypothetical situation the woman would not know whose wife she is. Jesus' answer shows the masterful logic of the Logos of God.

- Jesus replied, 'You are in error because you do not know the Scriptures or the power of God.

First, Jesus points out that the starting presuppositions are wrong—the Sadducees only accepted the Pentateuch as Scripture, while Jesus, like the Pharisees, accepted the same books as the Protestant Old Testament. If they had not been ignorant of what the Scriptures were, they would have realized that Scriptures like *Dan. 12:2* clearly teach the Resurrection.

Then He notes that if a conclusion of a valid argument is false, then it is enough for only one of the premises to be false. The false premise in the Sadducees' argument was not the resurrection, but that people would be married in heaven:

- At the resurrection people will neither marry nor be given in marriage; they will be like the angels in heaven.

However, refuting any number of arguments against a position does not in itself prove that position. Therefore, Jesus proved His own position, on the Sadducees' own terms, using Scripture they accepted:

- But about the resurrection of the dead—have you not read what God said to you,
- 'I am the God of Abraham, the God of Isaac, and the God of Jacob'? He is not the God of the dead but of the living.'
- When the crowds heard this, they were astonished at his teaching.
- Hearing that Jesus had silenced the Sadducees, the Pharisees got together.

Even the Scriptures accepted by the Sadducees (and which Jesus affirmed as being 'what God said') taught the resurrection: Christ demonstrated this with an argument showing that the Pentateuch taught that God was the God of the patriarchs and the God of the living. Therefore, the patriarchs were living in a sense in Moses' day, centuries after they had died physically. Observe also that the argument depends on the present tense of the verb 'to be' implied in the Hebrew verbless clause of the passage (*Ex. 3:6*) Jesus cited. His argument makes no sense if He did not believe in verbal plenary inspiration of Scripture.

The fallacy of denying the antecedent is committed by some groups that teach the error of baptismal regeneration by citing the following statement of Christ according to the Majority Text of *Mark 16:16*:

- Whoever believes and is baptized will be saved, but whoever does not believe will be condemned.

The first part of the verse is an implication: if a person believes and is baptized then he will be saved. It is invalid to argue from this that anyone who is not baptized will not be saved. The second part is an explicit statement that unbelief results in condemnation.

To demonstrate the fallacy, examine the following statement which is in the same logical form: 'Whatever has feathers and flies is a bird, but whatever does not have feathers is not a bird.' This statement does not teach that there are no flightless birds.

Another example of the fallacy of denying the antecedent is when some people are upset because we can no longer use a stock creationist argument (e.g., the depth of meteoritic dust on the moon to prove a young moon). But the argument in schematic form is as follows, and the fallacy should be clear:

1. If the moon dust argument works, then the moon must be young;
2. The moon dust argument doesn't work;

∴ The moon cannot be young.

This should be a lesson that our primary evidence should always be the infallible written testimony of One who was there and never errs, not the evidence of fallible scientists who weren't there and often err.

An example of the fallacy of affirming the consequent is using verified predictions as 'proof' of a scientific law. That can be seen if we analyze it:

1. Theory T predicts observation O;
2. O is observed;

∴ T is true.

To see why this does not follow, consider:

1. If I had just eaten a whole pizza, I would feel very full;
2. I feel very full;

∴ I have just eaten a whole pizza.

But I could feel very full for many different reasons.

On the other hand, the famous falsification criterion for a scientific theory devised by the late Sir Karl Popper is based on the valid denying the consequent:

1. Theory T predicts O will not be observed;
2. O is observed;

∴ T is false.

We can apply this analysis to a major evolutionary argument:

1. If organisms X and Y have a common ancestor, they will have homologous structures;
2. X and Y have homologous structures;

∴ X and Y have a common ancestor.

This demonstrates that it is an example of the fallacy of affirming the consequent. The conclusion is not proven—the homologous structures could be due to a common designer, leaving a 'biotic message' that there is a single designer of life rather than many .

On the other hand, ornithologists like Alan Feduccia argue against dinosaur-to-bird evolution for many good reasons, including a recent discovery that dinosaur embryos have an embryonic thumb that birds lack (I–II–III and II–III–IV-digit patterns respectively). The argument is:

1. If birds evolved from theropods, they will have homologous digits;
2. Bird and theropods do not have homologous digits;

∴ Birds did not evolve from theropods.

This is valid (denying the consequent), so creationists have rightly publicized this evidence

However, philosophers like Imre Lakatos point out that core theories are not tested in isolation, but are 'protected' by auxiliary hypotheses. Denying the consequent only shows that one of the premises needs to be false, and it need not be the core theory. So, the auxiliary hypotheses are modified instead. In schematic form, the valid argument is as follows:

1. Theory T and auxiliary hypothesis A predict that O will not be observed;
2. O is observed;

∴ Either T or A is false.

For example, Newton's theory predicted certain motions of Saturn, provided there were no other massive objects interfering. When Saturn didn't move as predicted, either Newton's theory was falsified, or there was another massive object perturbing the orbit—this turned out to be the planet Uranus.

The above was explaining the logic of the falsification criterion. This was not necessarily to endorse it—a coherent definition of science is hard to come by.

In the hands of evolutionists, 'unscientific' becomes a swear-word with which to attack creation. But it is more important whether creation or evolution are true or false, than whether one is more 'scientific' than another. Sometimes evolutionists are so keen to attack creationists that they don't realize their self-contradictions. For example, the philosopher P. Quinn (an anti-creationist himself) demonstrates the illogicality of the Marxist evolutionist Stephen Jay Gould:

'... Gould claims that "'Scientific creationism' is a self-contradictory phrase precisely because it cannot be falsified" ...Ironically, in the next sentence Gould goes on to contradict himself by asserting that "the individual claims are easy enough to refute with a bit of research." Indeed, some of them are! But since they are so easily refuted by research, they are after all falsifiable and, hence, testable. This glaring inconsistency is the tip-off to the fact that talk about testability and falsifiability functions as verbal abuse and not as a serious objection in Gould's anti-creationist polemics.'

V. Disjunctive Syllogism

The disjunctive syllogism is a valid form of argument familiar to those who have sat multiple choice examinations. Sometimes, a process of elimination can rule out all possibilities but one, which must therefore be true. An example is: Fred is flying either on QANTAS or Air New Zealand; he is not flying QANTAS; therefore, he is flying Air New Zealand (see Table 27.5).

Table 27.5: Disjunctive Syllogism

English	Symbolic
Either p or q	p v q
p is false	~p
Therefore, q is true	∴ q

To be sure that the conclusion is true, one must be sure that all possible alternatives are listed. The surest way is to apply the Law of Excluded Middle and have the disjunctive (either/or) premise contain a pair of contradictories (p or ~p)

An important example is that there are only two real explanations for the origin of different kinds of life—creation or evolution. For example, Professor D.M.S. Watson wrote:

'Evolution [is] a theory universally accepted not because it can be proven by logically coherent evidence to be true, but because the only alternative, special creation, is clearly incredible.'

I.e.,: C v E; ~C; E.

As this is a disjunctive syllogism, it is a valid argument. But it cannot be over-emphasized that validity and truth are not the same—this argument is not sound! Many evolutionists starting with Darwin have used this reasoning. This also demonstrates the atheistic bigotry behind much evolutionist thinking. Of course, creationists can use the equally valid argument:

C v E; ~E; C.

I.e., evidence against evolution is automatically evidence for creation. This is both valid and sound.

Many evolutionary propagandists dispute this reasoning when creationists use it, on the grounds that creation and evolution are not the only alternatives. Creationists are thus accused of the fallacy of false alternatives, that is, the disjunctive premise leaves out a possible alternative. But as shown, many evolutionists agree there are only two, so there are double standards at work. This can be shown by the Law of Excluded Middle: either things were made (creation) or they weren't (evolution). It is true that biblical creation is not the only alternative, so it is not proven by disproof of evolution. Biblical creation is certainly consistent with disproof of evolution, unlike atheism.

A genuine example of the fallacy of false alternatives is the following 'proof' of the punctuated equilibria version of evolution:

1. Life must have evolved either gradually or via punctuated equilibria;

2. There are major problems with gradualism (absence of fossil intermediates, and inability to construct a functional series);

   ∴ Life must have evolved via punctuated equilibria.

This is basically the form of argument used by Niles Eldredge and Stephen Jay Gould in their seminal paper, as pointed out by creationists.

Other Common Fallacies

Hasty Generalization

Thus far, we have discussed deductive reasoning. As discussed, inductive arguments reason from a finite set of examples to a general rule. The reason they are less important is that they don't guarantee the truth of the conclusion—they are formally invalid by the definition of validity in logic. For example, just because we find that 1000 crows are black, it does not follow that the 1001st crow will not be an albino.

Science by its nature is inductive, not deductive. Science always uses a finite number of measurements, each of which has an uncertainty, so science can never give a complete picture of reality. Hence, although science can be useful, it can never be a threat to the Christian Faith.

Genetic Fallacy

The error of trying to disprove a belief by tracing it to its source. For example, Kekulé thought up the (correct) ring structure of the benzene (C_6H_6) molecule after a dream of a snake grasping its tail, but chemists don't need to worry about correct ophiology to analyze benzene!

However, many anti-Christians commit this fallacy when they try to disprove Christianity by pointing out alleged parallels in pagan mythology. Another example is: 'You only believe Christianity because you were indoctrinated by your parents and culture; if you came from a Hindu family and culture, you would be a Hindu', with the spoken or unspoken impression 'thus Christianity need not be preferred over Hinduism'. In neither case can anything be inferred about the truth of Christianity from reasons a Christian's belief allegedly originated.

Many evolutionist propagandists believe that they simply need to demonstrate that a creationist has a 'fundamentalist' religious belief to discredit his purely scientific claims. The double standards are glaring—the radical atheist or even Marxist beliefs of many leading evolutionists are often ignored, although these beliefs determine which scientific explanations are acceptable and which are not.

Fallacy of Division

For example: a truck is heavy, therefore all its atoms are heavy. This example is obviously fallacious, but other equally fallacious arguments are advanced in all seriousness. Some New Agers like Teilhard de Chardin claim that because living beings are conscious, then their atoms must have some consciousness.

Fallacy of Composition

The opposite to the Fallacy of Division. An example is: all cells are light, therefore all animals containing cells are light.

Post hoc ergo propter hoc

This is Latin for 'after this, therefore because of this.' But just because B happened after A, it doesn't mean B was caused by A. Gordon Clark gives the following example of this fallacy:

'In the late seventies the Internal Revenue Service [USA] undertook to harass Christian schools.... [They] tried to revoke the tax exemption status of Christian schools, holding them guilty of race discrimination until they could prove themselves innocent by certain processes impossible to fulfil in some localities. One of the arguments the IRS used was that those schools were organized just after laws of racial discrimination were enacted. *Post hoc ergo propter hoc*. One of the defenses used by the Christians was that the schools were organized just after the Supreme Court banned the Bible and Prayer. One might add that they were organized after violence, drugs and sex became intolerable in the public schools.'

The Basis for Logic

A final question is, why should logic work at all? Not only can unbelievers not make a sound case against Christianity, but an atheistic world-view attacks the very basis of reasoning itself. This was realized by the famous Communist evolutionist biologist, J.B.S. Haldane:

'If my mental processes are determined wholly by the motions of atoms in my brain, I have no reason to suppose my beliefs are true ... and hence I have no reason for supposing my brain to be composed of atoms.'

In a debate between the Christian, William Lane Craig and the atheist, Frank Zindler, Zindler claimed that our logical processes evolved for survival value. Craig pointed out that this provides no reason for us to trust their validity, only their value in survival.

Even Darwin wrote in an early private notebook, 'Why is thought, being a secretion of brain, more wonderful than gravity as a property of matter?' But this argument is self-defeating. For it applies to that thought of Darwin's too, and to every thought about evolution, hence we have no reason to trust them.

The famous Marxist paleontologist Stephen Jay Gould claimed that the mind was an illusion produced by the brain. So why should we trust anything Gould says, if his thoughts are illusions?

This only shows that many atheistic theories actually refute themselves. Thus, there is no need for independent empirical tests for them. Conversely, the Christian doctrine that we are created in the image of a logical God is an excellent explanation for our logical faculties.

Various Contrary Opinions

Anything is of faith in two ways; directly, where any truth comes to us principally as divinely taught, as the trinity and unity of God, the Incarnation of the Son, and the like; and concerning these truths a false opinion of itself involves heresy, especially if it be held obstinately. A thing is of faith, indirectly, if the denial of it involves as a consequence something against faith; as for instance if anyone said that Samuel was not the son of Elcana, for it follows that the divine Scripture would be false. Concerning such things anyone may have a false opinion without danger of heresy, before the matter has been considered or settled as involving consequences against faith, and particularly if no obstinacy be shown; whereas when it is manifest, and especially if the Church has decided that consequences follow against faith, then the error cannot be free from heresy. For this reason, many things are now considered as heretical which were formerly not so considered, as their consequences are now more manifest.

So, we must decide that anyone may entertain contrary opinions about the notions, if he does not mean to uphold anything at variance with faith. If, however, anyone should entertain a false opinion of the notions, knowing or thinking that consequences against the faith would follow, he would lapse into heresy.

The Amazing Messages of DNA

2003 is the 50th anniversary of the discovery of the double helix structure of DNA. Its discoverers, James Watson, Francis Crick and Maurice Wilkins, won the Nobel Prize for Physiology and Medicine in 1962 for their discovery.

The amazing design and complexity of living things provides strong evidence for a Creator. We know from the Bible that God rested from (i.e. finished) His creative work after Day 6 (*Genesis 2:2–3*) and now *sustains* His creation (*Col. 1:16–17, Hebrews 1:3*). So how do complex living creatures arise *today*?

God's Amazing IT (Information Technology)

One aspect of this sustenance is that God has programmed the 'recipe' for all these structures on the famous double-helix molecule DNA. This recipe has an enormous *information content*, which is transmitted one generation to the next, so that living things reproduce 'after their kinds' (*Genesis 1, 10* times). Leading atheistic evolutionist Richard Dawkins admits:

'[T]here is enough information capacity in a single human cell to store the *Encyclopaedia Britannica*, all 30 volumes of it, three or four times over.'

Just as the *Britannica* had intelligent writers to produce its information, so it is reasonable and even scientific to believe that the information in the living world likewise had an original compositor/sender. There is no known non-intelligent cause that has ever been *observed* to generate even a small portion of the literally encyclopedic information required for life.

The genetic code is not an outcome of raw chemistry, but of elaborate decoding machinery in the *ribosome*. Remarkably, this decoding machinery is itself encoded in the DNA, and the noted philosopher of science Sir Karl Popper pointed out:

'Thus, the code cannot be translated except by using certain products of its translation. This constitutes a baffling circle; a really vicious circle, it seems, for any attempt to form a model or theory of the genesis of the genetic code.'

So, such a system must be fully in place before it could work at all, a property called *irreducible complexity*. This means that it is impossible to be built by natural selection working on small changes.

DNA is by far the most compact information storage system in the universe. Even the simplest known living organism has 482 protein-coding genes. This is a total of 580,000 'letters,'—humans have three billion in every nucleus.

The amount of information that could be stored in a pinhead's volume of DNA is equivalent to a pile of paperback books 500 times as high as the distance from Earth to the moon, each with a different, yet specific content. Putting it another way, while we think that our new 40 gigabyte hard drives are advanced technology, a pinhead of DNA could hold 100 *million* times more information.

The programs of life

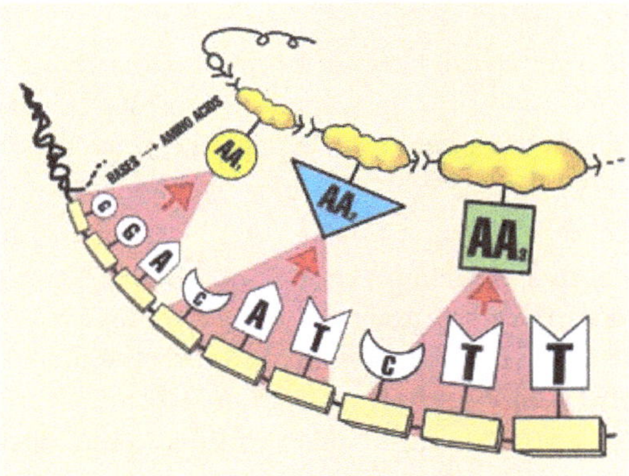

Fig.27.8: Information Stored as Sequences of Four Types of DNA Bases, A, C, G and T.

Information is a measure of the complexity of the arrangement of parts of a storage medium, and doesn't depend on what parts are arranged. For instance, the printed page stores information via the 26 letters of the alphabet, which are arrangements of ink molecules on paper. But the information is not contained in the letters themselves. Even a translation into another language, even those with a different alphabet, need not change the information, but simply the way it is presented. However, a computer hard drive stores information in a totally different way—an array of magnetic 'on or off' patterns in a ferrimagnetic disk, and again the information is in the patterns, the arrangement, not the magnetic substance. Totally different media can carry exactly the same information. An example is this article you're reading—the information is exactly the same as that on my computer's hard drive, but my hard drive looks vastly different from this page. In DNA, the information is stored as sequences of four types of DNA bases, A, C, G and T, Figure 27.8. In one sense, these could be called chemical 'letters' because they store information an analogous way to printed letters.[1] There are huge problems for evolutionists explaining how the 'letters' alone could come from a primordial soup.[2] But even if this was solved, it would be as meaningless as getting a bowl of alphabet soup.

The 'letters' must then link together, in the face of chemistry trying to break them apart.[3] Most importantly, the letters must be arranged correctly to have any meaning for life.

A group (codon) of 3 DNA 'letters' codes for one protein 'letter' called an amino acid, and the conversion is called translation. Since even one mistake in a protein can be catastrophic, it's important to decode correctly. Think again about a written language—it is only useful if the reader is familiar with the language. For example, a reader must know that the letter sequence c-a-t codes for a furry pet with retractable claws. But consider the sequence g-i-f-t—in English, it means a present; but in German, it means poison. Understandably, during the post–September-11 anthrax scare, some German postal workers were very reluctant to handle packages marked 'Gift.'

The Unity of Life

Many evolutionists claim that the DNA code is universal, and that this is proof of a common ancestor. But this is false—there are exceptions, some known since the 1970s. An example is Paramecium, where a few of the 64 (4^3 or $4 \times 4 \times 4$) possible codons code for different amino acids. More examples are being found constantly.[1] Also, some organisms code for one or two extra amino acids beyond the main 20 types.[2] But if one organism evolved into another with a different code, all the messages already encoded would be scrambled, just as written messages would be jumbled if typewriter keys were switched. This is a huge problem for the evolution of one code into another.

Also, in our cells we have 'power plants' called mitochondria, with their own genes. It turns out that they have a slightly different genetic code, too.

Certainly, most of the code is universal, but this is best explained by common design—one Creator. Of all the millions of genetic codes possible, ours, or something almost like it, is optimal for protecting against errors.[3] But the created exceptions thwart attempts to explain the organisms by common-ancestry evolution.

More than Just a Super Hard Drive

Actually, DNA is far more complicated than simply coding for proteins, as we are discovering all the time.[1] For example, because the DNA letters are read in groups of three, it makes a huge difference which letter we start from. E.g. the sequence GTTCAACGCTGAA … can be read from the first letter, GTT CAA CGC TGA A … but a totally different protein will result from starting from the second letter, TTC AAC GCT GAA …

This means that DNA can be an even more compact information storage system. This partly explains the surprising finding of The Human Genome Project that there are 'only' about 35,000 genes, when humans can manufacture over 100,000 proteins.

The circle of life

- All living things have encyclopedic information content, a recipe for all their complex machinery and structures.
- This is stored and transmitted to the next generation as a message on DNA 'letters,' but the message is in the arrangement, not the letters themselves.
- The message requires decoding and transmission machinery, which itself is part of the stored 'message.'
- The choices of the code and even the letters are optimal.
- Therefore, the genetic coding system is an example of irreducible complexity.

How DNA is transcribed into mRNA with RNA polymerase, then translated into a protein in the ribosome with tRNA 'adaptor' molecules, Figure 27,9. This protein is then escorted by chaperones into a chaperonin so it folds correctly. All this machinery is itself encoded in the DNA, but the DNA can't be read without this decoding machinery—a vicious circle, or chicken-and-egg problem. Furthermore, most of these processes use energy, supplied by ATP, produced by the motor ATP synthase (right). But the ATP synthase motor can't be produced without instructions in the DNA, read by decoding machinery using ATP… a three-way circle, or perhaps an egg-nymph-grasshopper problem.

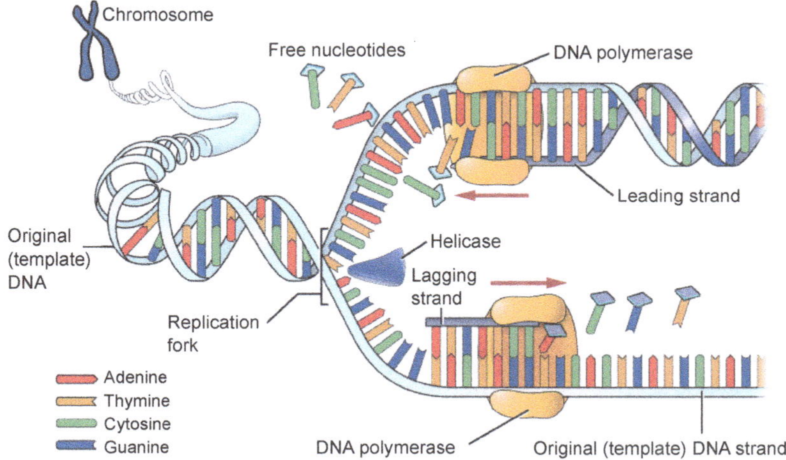

Fig.27.23: The marvelous process of DNA replication

The 20-nanometer motor (height), ATP synthase (one nanometer is one thousand-millionth of a meter). These rotary motors in the membranes of mitochondria (the cell's power houses) turn in response to proton flow (a positive electric current). Rotation of the motor converts ADP molecules plus phosphate into the cell's fuel, ATP.

APPENDIX

i. Similarly, Thomas Aquinas pointed out that one can have a false opinion without being intentionally heretical, even if that false opinion has heretical logical consequences. But if one holds to the false opinion stubbornly after the logical consequences are explained, then there is a problem. He uses the superficially trivial biblical statement that Samuel was the son of Elkanah: this is not an article of saving faith, but denying this would logically imply denying biblical inerrancy, which to him was as serious as for any evangelical today:

ii. Came, D., Richard Dawkins's refusal to debate is cynical and anti-intellectualist: Using William Lane Craig's remarks as an excuse not to engage in reasoned debate is typical of New Atheist polemic, *Guardian*, Oct 2011; theguardian.com. A few years later, Dr Came unlike Dawkins did debate Dr Craig; one report is William Lane Craig debates Daniel Came: Does God exist? winteryknight.com, 28 Jun 2017.

iii. Bradley, D., The genome chose its alphabet with care, *Science* 297(5588):1789–91, 13 September 2002. Mac Dónaill's theory involves *parity bits*, an extra 1 or 0 added to a binary string to make it add up to an even number (e.g. when transmitting the number 11100110, add an extra 1 onto the end (11100110,1), and the number 11100001 has a zero added (11100001,0). If there is a single error changing a 1 to a 0 or vice versa, the string will add up to an odd number, so the receiver knows that it has not been transmitted accurately. Mac Dónaill found that he could treat certain structural features of the DNA 'letters' as a four-digit binary number, with the fourth digit a parity bit. He found that these DNA letters all have even parity, while 'alphabets composed of nucleotides of mixed parity would have catastrophic error rates.'

iv. *The Shorter Oxford English Dictionary* (1993) defines 'vestigial' as 'degenerate or atrophied, having become functionless in the course of evolution.' Some evolutionists now re-define 'vestigial' to mean simply 'reduced or altered in function.' Thus, even valuable, functioning organs (consistent with design) might now be called 'vestigial.' This seems like changing the rules in the face of a losing argument.

v. Evolutionists call the almost identical sequences 'highly conserved,' because they *interpret* the similarities as arising from a common ancestor, but with natural selection eliminating any deviations in this 5% because precision is essential for it to function properly. Creationists interpret the same evidence as evidence of a designer creating the sequences in a precise way, because that's necessary for it to function. This is one more example of how allegedly evolution-inspired scientific advances make at least equal sense under a Biblical framework.

vi. Frege, F.L.G., Über Sinn und Bedeutung, *Zeitschrift für Philosophie und philosophische Kritik* **100:**25–50, 1892 (On Sense and Reference, *Journal of Philosophy and Philosophical Criticism*). In Greek mythology, *Hesperos* (Greek"Εσπερος) is the Evening Star, the planet Venus in the evening. He is the son of the dawn goddess Eos (Roman Aurora) and is the brother of *Phōsphoros* (Greek: Φωσφόρος = light-bringer, often translated as *Lucifer* in Latin; another name is *Eōsphoros*'Εωσφόρος = dawn-bringer), the Morning Star. But Frege's original German paper used 'the evening star' (*der Abendstern*) and 'the morning star' (*der Morgenstern*). See also Venus: cauldron of Fire. Return to text.

vii. Actually, the word cause' has several different meanings in philosophy. But in this article, I am referring to the *efficient cause*, the chief agent causing something to be made.

viii. RNA = ribonucleic acid.

REFERENCES

1. DNA= deoxyribonucleic acid. See Wieland, C., The marvelous 'message molecule', *Creation* **17**(4):10–13, 1995.
2. Dawkins, R., *The Blind Watchmaker*, W.W. Norton, New York, p. 115, 1986.
3. Grigg, R., Information: A modern scientific design argument, *Creation* **22**(2):52–53, 2000; creation.com/design-history#aside.
4. Gitt, W., *In the beginning was Information*, CLV, Bielenfeld, Germany, 1997.

5. Popper, K.R., Scientific Reduction and the Essential Incompleteness of All Science; in Ayala, F. and Dobzhansky, T., Eds., *Studies in the Philosophy of Biology*, University of California Press, Berkeley, p. 270, 1974.
6. Sarfati, J., Self-replicating enzymes? *Journal of Creation* 11(1):4–6, 1997; creation.com/replicating.
7. Fraser, C.M. et al., The minimal gene complement of Mycoplasma genitalium, *Science* 270(5235):397–403, 1995; perspective by Goffeau, A., Life with 482 Genes, same issue, pp. 445–446.
8. Gitt, W., Dazzling design in miniature, *Creation* 20(1):6, 1997; creation.com/dna.
9. Sarfati, J., Origin of life: instability of building blocks, *Journal of Creation* 13(2):124–127, 1999; creation.com/blocks.
10. Wiedersheim claimed that there were over 180 'rudimentary' structures in the human body, including 86 'vestigial' organs, in *The Structure of Man: an Index to his Past History*; transl. Bernard, by H. & M., Macmillan, London, 1895.
11. Scadding, S.R., Do 'vestigial organs' provide evidence for evolution? *Evolutionary Theory* 5(3):173–176, 1981.
12. See also Bergman, J. and Howe, G., *'Vestigial organs' are fully functional*, Creation Research Society Books, Kansas City, 1990.
13. A recent example was certain very short muscles in horse legs that are now known to have a vital role in dampening damaging vibrations. See Sarfati, J., Useless horse body parts? No way! *Creation* 24(3):24–25, 2002; creation.com/useless based on *Nature* 414(6866):895–899, 855–857, 20/27 December 2001.
14. For an overview, see Walkup, L., Junk DNA: evolutionary discards or God's tools? *Journal of Creation* 14(2):18–30, 2000; creation.com/junk-dna.
15. *Nature* 420(6915):509–590, 5 December 2002.
16. Gillis, J., 'Junk DNA' contains essential information—DNA has instructions needed for growth, survival, *Washington Post*, 4 December 2002.
17. Cohen, P., New genetic spanner in the works, *New Scientist* 173(2334):17, 16 March 2002.
18. Batten, D., 'Junk' DNA (again), *Journal of Creation* 12(1):5, 1998; creation.com/junkdna.
19. Coglan, A., Electric DNA: There's another information superhighway lurking in our genes, *New Scientist* 161(2173):19, 13 February 1999; citing Jacqueline Barton of the California Institute of Technology, *Chemistry & Biology* 6(2):85.
20. Dennis, C., The brave new world of RNA, *Nature* 418(6894):122–124, 11 July 2002; cited on p. 124.
21. Mattick, J.S. Non-coding RNAs: The architects of eukaryotic complexity, *EMBO Reports* 2:986–991, November 2001.
22. Cooper, M., Life 2.0, *New Scientist* 174(2346):30–33, 8 June 2002; Dennis, ref. 24.
23. E.g. Wood, T.C., Altruistic Genetic Elements (AGEs), cited in Walkup, ref. 17.
24. Lewontin, R., Billions and billions of demons, *The New York Review*, 9 January 1997, p. 31; Evolutionist's blind faith in atheism, regardless of how absurd it seems.
25. Todd, S.C., correspondence to *Nature* 401(6752):423, 30 Sept. 1999; A designer is unscientific—even if all the evidence supports one!
26. For a good discussion, see Craig, W.L., *Apologetics: An Introduction*, Moody Press, Chicago, chapter 1, 1984.
27. For an extensive discussion on the role of logic in Christianity and many examples, see Clark, G.H., *Logic*, The Trinity Foundation, POB 68, Unicoi, TN 37692, 1988.

28. Smith, G. The 'Toronto Blessing', *Apologia* 4(2):37–40, 1995.
29. A generally first-rate book on the importance of developing a Christ-like mind is Moreland, J.P., *Love Your God With All Your Mind*, Navpress, Colorado Springs, 1997. Its main disadvantage is that Moreland is 'sixty-forty in favor of the old-earth position' (p.107), possibly because of a slight over-emphasis on extra-biblical revelation.
30. Machen, J.G., *Christianity and Liberalism*, Macmillan, New York, 1924.
31. Martin, W.R., *The Kingdom of the Cults*, Bethany House Publishers, Minneapolis, MN, 1985.
32. Carroll, L. *Through the looking-glass, and what Alice found there* (with fifty illustrations by John Tenniel) Macmillan, London, 1877.
33. One example is Schafersman, S. Letter, *Geotimes*, August 1981. Cited in Bird, ref. 26, Vol. II, pp. 77–78.
34. See Batten, D. and Sarfati, J., How Religiously Neutral are the Anti-Creationist Organisations?, 1998.
35. Scott, E., Dealing with anti-evolutionism, *Reports of the National Center for Science Education* **17**(4):24–28; quote on p. 26, with emphasis in original, 1997.
36. Plimer, I.R., *Telling Lies for God*, Random House, Australia, p. 289, 1994.
37. See The Ian Plimer Files.
38. Obviously, I can give only a basic outline here. A more formal, in-depth coverage can be found in Hughes, G.E. and Londey, D.G., *The Elements of Formal Logic*, Methuen, London, 1965.
39. An extensive critique of Spong's errant and heretical views is Bott, M.R. and Sarfati, J.D., What's wrong with Bishop Spong? *Apologia* 4(1):3–27, 1995.
40. This argument is dramatised in the form of a Socratic Dialogue in Kreeft, P., *The Unaborted Socrates*, IVP, pp. 44–47, n.d. (after 1987).
41. Snelling, A.A. and Rush, D., Moon dust and the age of the solar system, *Journal of Creation* 7(1):2–42, 1993.
42. Clark, G.H., *The Philosophy of Science and Belief in God*, The Trinity Foundation, POB 68 Unicoi, TN 37692, 1987.
43. ReMine, W.J., *The Biotic Message*, St Paul Press, St. Paul, Minnesota, 1993; see review .
44. Ann C. Burke, A.C. and Feduccia, A., Developmental Patterns and the Identification of Homologies in the Avian Hand, *Science* 278(5338):666–8, 1997; perspective in the same issue, pp. 596–7 by Hinchliffe, R. The Forward March of the Bird-Dinosaurs Halted?
45. Oard, M.J., Bird-dinosaur link challenged, *Journal of Creation* 12(1):5–7, 1998.
46. Sarfati, J.D., Dino-bird evolution falls flat , *Creation* 20(2):41, 1998. [See also the other articles under Did birds really evolve from dinosaurs?]
47. For extensive discussion of the views of Popper and Lakatos, and other attempts to define science, see Bird, W.R., *The Origin of Species Revisited*, Philosophical Library, New York, Vol. II, chapters 9–10, 1991.
48. Quinn, P., The Philosopher of Science as Expert Witness; in: *Recent Work in the Philosophy of Science*, ed. J. Cushing, C. Delaney and G. Gutting, 1984, pp. 32, 43, 1984; Cited in Bird, ref. 26, p. 121.
49. Watson, D.M.S., Adaptation, *Nature* 124:233, 1929. Return to text.
50. Bird, W.R., *The Origin of Species Revisited*, Philosophical Library, New York, Vol. II, chapter 11, 1991.
51. Gould, S.J. and Eldredge, N., Punctuated equilibria: an alternative to phyletic gradualism; in: *Models in Paleobiology*, T.J.M. Schopf (ed.), Freeman, Cooper and Co., San Francisco, pp. 82–115, 1972.
52. Batten, D.J., Punctuated equilibrium: come of age? *Journal of Creation* 8(2):131–7, 1994.

53. Ronald Nash has dealt with such claims in detail, for example, Was the New Testament Influenced by Pagan Religions? See also Was Christianity plagiarized from pagan myths? Refuting the copycat thesis.
54. Batten, D.J., A Who's Who of evolutionists , *Creation* 20(1):32, 1997.
55. Ritchie, A., Dropping the Pretence: The Creation Science Foundation changes its name, *The Skeptic* 17(4):13,15, 1997.
56. Haldane, J.B.S. *Possible Worlds*, p. 209; cited in Lewis, C.S., *Miracles*, Fontana, London, p. 19, 1960 (first published 1947).
57. See Atheism vs Christianity: Where does the evidence point? A debate between Mr Frank Zindler and Dr William Lane Craig, *Apologia* 5(1):21–29, 1996.
58. Cited by Wieland, C., Darwin's real message: have you missed it? *Creation* 14(4):16–19, 1992.
59. Gould, S.J., *The Darwinian Revolution in Thought*. Lecture, June 6, Victoria University of Wellington, New Zealand, 1990.
60. Schramm, D.N. and Steigman, G., 1981. Relic Neutrinos and the Density of the Universe. *Astrophysical Journal* 243:1–7.
61. Watson, A., 1997. Clusters point to Never Ending Universe. *Science* 278(5342):1402.
62. Perlmutter, S. *et al.*, 1998. Discovery of a supernova explosion at half the age of the universe. *Nature* 391 (6662): 51. Perspective by Branch, D. Destiny and destiny. Same issue, pp. 23–24.
63. Branch, D., 1998, Density and destiny, *Nature* 391(6662):23–24, 1998.
64. Glanz, J. New light on the fate of the universe. *Science* 278(5339):799–800.
65. Guth, A.H. and Sher, M., 1983. The Impossibility of a Bouncing Universe. *Nature* 302:505–507.
66. Tinsley, B., 1975. From Big Bang to Eternity? Natural History Magazine. October, pp. 102–5. Cited in Craig, W.L., 1984. *Apologetics: An Introduction*, Chicago: Moody, p. 61.
67. Davies, P., 1983. *God and the New Physics*, Simon & Schuster, p. 215.
68. Craig, W.L., 1986. God, Creation and Mr Davies. *Brit. J. Phil. Sci.* 37:163–175.
69. Kay, M., 1996. Of Paul Davies and *The Mind of God*, *J. Creation* 10(2):188–193.

THE UNIVERSE IS BEING HELD TOGETHER

THE FORMATION OF MAN

For he knows our frame; he remembers that we are dust. *Psalm 103:14,* Figure 28.1.

Fig.28.1: *For he knows our frame; he remembers that we are dust. Psalm 103:14*

What are we made of?

How 'real' is the world?

What does it mean to exist?

Have you ever wondered about things like this? We humans like to ask questions, but sometimes the answers make us uncomfortable. Yes, we exist. Yes, the world is real. But, the reality of our reality is very, very strange. You see, we are made of almost nothing. Everything we see is mostly made up of empty space. This is shocking when you first realize it, but let me explain.

We and everything we can see are made up of atoms. The atom is an amazing thing. It has a dense center called a nucleus and a cloud of electrons surrounding it. Inside the nucleus are protons and neutrons. The nucleus is held together by incredibly strong forces that act at only short distances. And since 'opposite charges attract', the positively-charged protons in the nucleus attract the negatively-charged electrons that reside in a 'cloud' surrounding the nucleus.

How do we know this? The structure of the atom has taken hundreds of years and tens of thousands of experiments to figure out. There is still much we do not know, but scientists have developed an atomic theory that describes everything we do know. They have also made successful predictions, like the existence of a particle called the Higgs boson, the discovery of which made international news. In fact, as of today there is almost no evidence that atomic theory is wrong.

Remember how short my time is! For what vanity you have created all the children of man! *Psalm 89:47*

But this is where things get strange, even eerie. Our bodies are made up mostly of oxygen (65% by mass), carbon (18.5%), and hydrogen (9.5%). Hydrogen is the simplest element, and is well studied, so let's use it as an example. What are the relative distances between the nucleus and the electrons in the hydrogen atom? What are the relative distances between atoms in our bodies? How much 'empty space' is inside us?

Since there is no well-defined edge to the electron cloud, it is difficult to determine the exact size of an atom. But we can say that a hydrogen atom is *about* 100 picometres (pm) in diameter. That's 100 trillionths of a meter. In scientific notation, 1×10^{10} m. Atoms are tiny!

The hydrogen nucleus is much smaller than the electron cloud, about 2.4 femtometers. That's 0.0000000000000024 meters. In scientific notation, 2.4×10^{15} meters. Electrons do not 'orbit' the nucleus like many people think, so we can't really say how 'far away' the electrons are from the nucleus, but in the hydrogen atom the *average* distance to the electrons is about 60,300 times the diameter of the nucleus.

Molecules are held together by covalent bonds in which they share electrons. Thus, atom-to-atom distances are in the range of the size of individual atoms. The distance between two hydrogen atoms in the H_2 molecule is about 74 pm. The distance between hydrogen and carbon (one of the most common bonds in the human body) is about 109 pm.

So, the space between the nucleus and the electron cloud is quite large, compared to the diameter of the nucleus. It helps to have something to compare to. If you were to make the nucleus as large as the sun, the average distance to the electrons would be 14 times as far away as Pluto! In a molecule, nuclei would be about 100 *billion* miles apart. But even at this scale the electrons would still have essentially no 'size.' That means the inside of the atom, and the inside of any molecule, is *almost completely empty*.

Comparing the sun and Pluto to the nucleus and electron. At these scales, it would be impossible to even see these things, so each is represented by a tiny point. For example, even though the radius of the sun is approximately 700,000 km, that is less than 1/8000th the distance between the sun and Pluto. In the same way, the distance between the nucleus and electrons is truly vast, compared to the size of the nucleus.

Fun fact: our bodies are more than 99.99999999% empty space!

What is your life? For you are a mist that appears for a little time and then vanishes. *James 4:14b*

But if we are made up of empty space, how can the human body (or any other physical object) be 'seen' or 'felt'? When we see something, what we are detecting is the light (an electromagnetic wave) reflecting from the surface of the object. Some electromagnetic waves (e.g., X-rays) pass right through most objects. This is because their wavelengths are shorter than the spacing between the nuclei, so they can almost literally squeeze between atoms.

Another type of electromagnetic waves, infrared, is usually absorbed. This is because they are very long waves with low energy, and they get absorbed by molecules, causing their atoms to wiggle (that is, heat up). In between those two extremes is the 'visible' part of the electromagnetic spectrum. Visible light tends to bounce off most objects. Even though the atoms are very tiny, very far apart, and made up mostly of empty space, the electron clouds create a continuous 'surface.' What you see is the light waves that are reflected from the surface. A molecule like chlorophyll strongly absorbs blue and red light. This is why plant leaves are green, because that is what is 'left' to reflect. A substance like tar absorbs most wavelengths of light, making it black. A piece of paper absorbs few, making it white.

However, how do we 'feel' things? When you press your hand against an object, the tightly-bonded atoms in your hand are moved close to the tightly-bonded atoms in the object. Your hand cannot penetrate the surface of a brick wall, no matter how hard you try. This is because the molecules are held rigidly in place by the shared electron clouds within the molecules. What you are feeling is not 'brick' so much as an electromagnetic force field generated by the atoms in the wall. There is really not much of anything there. The physical world, including your precious body and brain, is little more than an empty vapor.

O Lord, make me know my end and what is the measure of my days; let me know how fleeting I am! Behold, you have made my days a few handbreadths, and my lifetime is as nothing before you. Surely all mankind stands as a mere breath! … Surely a man goes about as a shadow! *Psalm 39:4–6*

THE GOOD NEWS

Even though the science of reality is humbling, and even though our bodies may be nothing more than dust, this does not mean we are unimportant in God's eyes.

Consider what the Old Testament says about man:

"Yet you have made him a little lower than the heavenly beings and crowned him with glory and honors. You have given him dominion over the works of your hands; you have put all things under his feet. *Psalm 8:5–6.*

Or think about what the New Testament says about us:

In him we have redemption through his blood, the forgiveness of our trespasses, according to the riches of his grace, which he lavished upon us, in all wisdom and insight making known to us the mystery of his will, according to his purpose, which he set forth in Christ as a plan for the fullness of time, to unite all things in him, things in heaven and things on earth. *Ephesians 1:7–10*

Mankind has a very special place in God's creation. In one sense, He built this world, this universe, to bring about a bride for Christ. Next to Him, we are nothing. Yet, for reasons known to Him alone, we were brought into existence to work out His divine plan. Are we empty? Yes. Compared to God are we worth anything? No. But in God's eyes we are very, very valuable nonetheless. In the eyes of the Creator, we are precious. Holding both of these thoughts simultaneously helps us to keep everything in perspective.

FORMATION OF PLANETS

The way some scientists talk about planet formation, one would think that the process was simple:

Planet formation is just one of the many *hypothetical* evolutionary processes that started with the big bang and ended with humans on Earth after many billions of years. Since planets exist, evolutionists reason they 'must' have formed from a dust cloud called a nebula. The dust must first develop from dead stars because dust does not just develop from gas molecules. Therefore, the dust is believed to have 'evolved' from the explosion of a star in a supernova. Hence our solar system is believed to be the result of a collapsed dust cloud from an exploded star. These are the simple naturalistic deductions, assuming evolution is the only mechanism.

Hypothetically, 'Our solar system was built from the dust of dead stars. It's an often-repeated fact.'

A proposed theory of planet formation from accreted stellar material. Remnants from an exploded star produce the raw material. Though this material is thought to accrete through gravitational interaction, the effect of gravity is too small to allow this to happen in the timeframe proposed by evolution. There is also the question as to whether the small particles would coalesce under the influence of gravity at all.

Many people are satisfied with this scenario and take it no further. But if an inquiring person were to ask how the planets actually formed from the dust, he would get a surprising answer:

'But if you ask how this dust actually started to form planets, you might get an *embarrassed silence*. Planets, it seems, grow too fast—no one knows why the dust clumps together so quickly.'

This, among other theoretical processes in the big bang scenario, is actually held by faith. (The formation of stars has similar challenges as planet formation. The main difference is that stars accumulate more mass from the dust cloud. Since star and planet formation have similar problems, for the sake of simplicity, I will only discuss the naturalistic origin of planets.) A recent article in *New Scientist* admits that forming a planet naturalistically is exceedingly difficult.

There are four stages in the supposed evolution of planets:

'A successful nebular model must account in some detail for four important stages in the solar system's evolution: the formation of the nebula out of which the planets and sun originate, the formation of the original planetary bodies, the subsequent evolution of the planets, and the dissipation of leftover gas and dust. Modern nebular models (there are more than one!) give tentative explanations for these stages, but many details are lacking. No one model today is entirely satisfactory.'

For the sake of argument, I will just assume that the dust is left over from a supernova explosion. This is the first stage. Then according to Laplace's nebular hypothesis, first presented in 1796, the process of planet formation, the second stage, begins with the simple collapse of the dust cloud. There are three theoretical steps in the collapse of the dust cloud and the growth of a planet:

1. gravitational contraction of the dust into small particles,
2. accretion of particles or small aggregates to form large aggregates, and
3. condensation by the accumulation of atoms and molecules on the growing mass.

The most difficult step is the first, gravitational contraction of dust to form small particles. Dust grains must first accrete to form small particles, which must continue to grow until they are at least 10 m in diameter. This size is the point at which gravity is expected to come into its own, accreting and condensing material at a faster and faster rate. Then supposedly, planetesimals would form that are many kilometers across. The planetesimals are finally envisaged to collide to form planets. There are difficult problems with these later steps, but I will focus on the first step: how does the dust collide, stick together and grow before gravity can assert itself? That is the big question. The tiny dust particles must hit each other head on and stick. The process (which is speculative anyhow) is too slow, especially in cold regions of space, according to astronomers. A number of hypotheses are in vogue, but all seem to have fatal flaws.

Steinn Sigurdsson has given up on all the proposed hypotheses because of the extreme unlikelihood that any of them ever occurred. Since planets have obviously formed and they must hold onto their evolutionary belief, he suggests a desperate alternative:

'*... there could be an extra dimension of space in which gravity alone acts and which until now has gone unnoticed. If this is so, then gravity—which is weak over large distances—gets stronger at the tiny distances encompassed by the extra dimension ...*'

In other words, he suggests that gravity would extend into five space dimensions instead of three and would be very strong at very short distances, causing dust and small particles to attract and stick together by gravitational attraction. This would certainly make planet formation much faster and easier. But there is at least one delicate problem with this imaginative hypothesis—the dust grains cannot hit too hard or the incipient particle would break apart:

'So, the turbulence within the disc [flat dust cloud] can't be too strong, and the acceleration caused by Sigurdsson's modified gravity can't be too extreme.'

The idea is actually testable. So far, Newton's law of gravity still holds down to 218 μm, but experiments are underway to test it at even closer distances. Sigurdsson hopes that his supergravity mechanism will show up when they test gravity at less than 80 μm. It seems to me that *if* he is correct, there is still the 'sticky' problem of how such a small particle can grow larger than 218 μm, above which his hypothetical mechanism would not apply.

Sigurdsson is likely correct that all hypotheses for planet formation are wild guesses. It is even more likely that his guess is even wilder than most, as many astronomers believe. That leaves nothing to explain the development of the planets, at least using natural processes over long periods of time. A straightforward reading of the evidence at hand and the state of the many hypotheses and problems is that planets *did not form naturalistically but were supernaturally created.*

THE DOCTRINE OF DEATH

It may seem strange to refer to death as a Christian doctrine, but the Bible is replete with teaching on the subject. Equally clear is the fact that the theory of evolution depends on death. Thanatology is the branch of theology dealing with doctrine of death, but it also refers to the scientific study of death—its forensic, psychological and social aspects.

Do we have hope beyond the grave?

"As human beings, it is hard for us to shake the idea that our existence must have significance beyond the here and now. Life begins and ends, yes, but surely there is a greater meaning. The trouble is, these stories we tell ourselves do nothing to soften the harsh reality: as far as the universe is concerned, we are nothing but fleeting and randomly assembled collections of energy and matter. One day, we will all be dust" (Teal Burrell, 2017).

- "[Y]ou are dust, and to dust you shall return" (Genesis 3:19).
- [God] knows our frame; he remembers that we are dust (Psalm 103:14).
- The last enemy to be destroyed is death (1 Corinthians 15:26).

HOW DID CREATION HAPPEN

Belief in evolution can radically alter one's concept of death. The whole creation/evolution debate revolves around the question of *how God created*. Was it by speaking things into existence ***ex nihilo***, 'out of nothing' (Hebrews 11:3)?

Did God specially create unique kinds of living organisms upon the earth inside a single week? Or did God use evolution, spreading out this creative activity over hundreds of millions of years? If the latter, inevitably struggle, disease, violence and death would have been involved at every step along the way. At the heart of the origins debate, then, we must face the question, 'Did God incorporate death into his creation from the very beginning?'

One of British author "Jules Howard's" recent book was published with the intriguing title, *Death on Earth: Adventures in evolution and mortality*. The publisher's carries the following perspective in its advertising blurb:

NATURAL SELECTION AND DEATH

Natural selection depends on death; little would evolve without it. Every animal on Earth is shaped by its presence and fashioned by its specter. We are all survivors of starvation, drought, volcanic eruptions, meteorites, plagues, parasites, predators, freak weather events, tussles and scraps, and our bodies are shaped by these ancient events.

Specter of Death

In other words, the specter of death, in all its various forms, haunts every creature that has ever lived, humans included. And if evolution is true, that is how things have been since time began. We will explore this relationship between evolution and death and consider the responses by theistic evolutionists to the problems posed. What about physical versus spiritual death of human beings?

Is animal death to be distinguished from human death, or the death of plants and bacteria for that matter? Does it really matter? What does the Bible have to say? But before we get to those important questions, it will be helpful to remind ourselves of the stark reality of the subject before us, lest we forget its intensely practical and personal relevance.

Facing up to Death

Occasionally we hear someone quip sardonically that, 'nothing is certain except death and taxes. The point of this proverbial statement, of course, is that the requirement to pay taxes is as inevitable as one's death. Unquestionably, death is the ultimate statistic since 10 out of 10 people die. There is not the slightest probability of someone escaping their mortality.

As the wise king Solomon famously declared, "For everything there is a season, and a time for every matter under heaven: a time to be born, and a time to die" (Ecclesiastes 3:1–2). Yet, death is a very unpopular topic for conversation—to bring it up at a party would be seen as a serious *faux pas*. Few, it seems, are comfortable with a serious conversation about death. And funerals are especially challenging occasions for many people.

For all of us, the most pressing question is whether we are ready to face death—do we have hope beyond the grave? Without Christ, we are without hope and without God in this world (Ephesians 2:12). But Christian people experience the difference that a living faith in Jesus makes when confronted with their own death or that of a fellow Christian. They do "not grieve as others do who have no hope" (1 Thessalonians 4:13). Of course, they do experience grief at the death of a loved one, friend or colleague, but they face it with the certain hope of eternal life beyond the grave.

In contrast, consider the sad case of the late "Christopher Hitchens" who, having failed in his fight against esophageal cancer, died in 2011. An atheist, this celebrated writer and orator was known for his vitriolic criticism of people of faith, not least Christians. Some of his books—*God is Not Great* (2007) and *The Portable*

Atheist (also 2007)—helped solidify his reputation as a spokesperson for opponents of biblical Christianity. A year before Hitchens passed away, he was interviewed by well-known British journalist "Jeremy Paxman." Asked if he feared death Hitchens replied, "No. I'm not afraid of being dead. That's to say, there's nothing to be afraid of; I won't know I'm dead."

His confidence was unfounded for, "it is appointed for man to die once, and after that comes judgment" (Hebrews 9:27). Hitchens was well aware of the Bible's teaching on this point but said he'd be surprised to find himself facing God's tribunal. Like so many others, he was prepared to take the gamble that there was no conscious existence beyond his physical death. He despised the teaching of the Apostle Peter that, "there is salvation in no one else, for there is no other name under heaven given among men by which we must be saved" (Acts 4:12). He was prepared to take the chance that he had no soul that would exist after his death. Scripture, however, teaches that the soul is immortal, a person's most precious possession (Mark 8:36–37).

EVOLUTION - THE 'ENEMY OF DEATH'

Why this presumption and arrogance on the part of people like Hitchens? In the same interview, Hitchens admitted, "My view is already quite stark, which is that we're born into a losing struggle … something meaningless or random. …it's a stark existence." This is a sober illustration of how people live out their lives in practical terms—a belief in the evolutionary struggle for existence certainly tends to make atheists of people. Yet, in spite of Hitchens' claimed fearlessness regarding imminent death, others confess to more alarm. During the 1990s a cancer researcher, was interested to read about American evolutionist "Stephen Jay Gould's" battle with mesothelioma. Diagnosed in the early 1980s, he had been completely cured. Sadly, 20 years later, he succumbed to an unrelated cancer of the lung which spread to his brain. He died in 2002, aged 60. During his first battle with cancer, he wrote:

It has become, in my view, a bit too trendy to regard the acceptance of death as something tantamount to intrinsic dignity. Of course, I agree with the preacher of Ecclesiastes that there is a time to live and a time to die—and when my skein runs out, I hope to face the end calmly and in my own way. For most situations, however, I prefer the more martial view that *death is the ultimate enemy*—and I find nothing reproachable in those who rage mightily against the dying of the light (emphasis added).

Yes, death *is* the ultimate enemy. However, in the teeth of death, atheism is a dreadfully bleak, hopeless ideology. The nihilistic view of death advocated by both Hitchens and Gould, as they stared death in the face, was connected to their conviction that death and evolution go hand in hand. Yet the worldview of theistic evolutionists is little different from that of atheistic or agnostic evolutionists. Apart from the 'addition' of God as the Superintendent of natural selection, the stark picture of struggle, extinction and death is exactly the same. Surely this should give pause for thought. We have yet to hear anyone testify that, upon their conversion to belief in evolution, they gained a profound appreciation of the love of God. Or, that they received a confident hope in the face of death and the grave.

DEATH OF THE UNFIT

We have already considered the 'problem of evil' but it will be helpful to cover some similar ground before we examine the ways in which theistic evolutionists try to accommodate death into their vision of creation. Evolutionists who profess no faith often go out of their way to highlight this aspect of evolution. "Janet Browne," British science historian and acclaimed biographer of Charles Darwin, has degrees in zoology and the history of science; she is Armament Professor of the History of Science at Harvard University. In the documentary film *Darwin: The Voyage that Shook the World*, Browne is adamant:

Darwin … was working on a theory that *requires* death. His theory of natural selection is *built* on the idea that the unfit are going to *die*, that they are not going to contribute to the next generation" (italics added to match her own emphasis of speech).

In the same film, Browne explains that Darwin's own bereavement, involving three of his children, presented him with real difficulties. He could not reconcile his death-ridden theory with a God who would allow his beloved daughter Annie to die (aged just 10 years). Other biographers of Darwin agree that this tragic loss was the final death knell for any remaining vestiges of Darwin's Christian faith. Weak in body, Annie had simply lost the struggle for existence—her death, excruciating though it was for him personally, was fully in accord with natural selection. This, after all, is precisely the sort of thing that evolution entails, in fact demands.

People often describe natural selection as 'the survival of the fittest.' As discussed earlier, this rather hackneyed term was coined by famous English philosopher and promoter of evolution, "Herbert Spencer" (1820–1903); Darwin liked it so much that he made it the heading of chapter 4 of the fifth edition of his *The Origin of Species* (1869). In reality, 'death of the unfit' and 'survival of the fittest' are simply two sides of the same coin. Again, to quote the man himself, from the closing paragraph of his book:

Thus, from *the war of nature, from famine and death*, the most exalted object which we are capable of conceiving, namely, the production of the higher animals, directly follows (emphasis added).

Death is in the engine room of evolution by natural selection, a major driver in the process that supposedly optimizes the survival of the organisms with the most advantageous characteristics. Indeed, without death to weed out the less fit and the unfit, there could be no survival of the fittest and no Darwinian evolution.

The 'Horridly Cruel Works of Nature'

One often hears people remark that, "nature is so cruel", or words to that effect. Sometimes it is posed as a question. Such comments are usually provoked by our morbid fascination over predators alluring, tracking, capturing and killing their prey, sometimes in ways which seem to maximize distress for the victim. That is not to say that we cannot appreciate the spectacle of the arms race between diverse predators and their prey, the sort of thing which plays out on our TV screens. Watching nature documentaries such as *The Hunt*—which showcases such fascinating creatures as: tigers, lions, orcas, wolves, cheetahs, polar bears, blue whales, harpy eagles and more—one is constantly struck by the sheer beauty, agility, grace, speed, and ingenuity of both the predators and their prey. At the same time, many of us are also repulsed by the carnage which we observe. We cannot but empathize with the victim's suffering, which inevitably is part and parcel of the kill.

As former Simonyi Professor for the Public Understanding of Science at Oxford University, Richard Dawkins is infamous for his vitriolic attacks on religion—especially conservative, evangelical Christian faith. In his book *The Devil's Chaplain*, Dawkins quotes words which Charles Darwin penned to his friend Joseph Hooker, "What a book a Devil's Chaplain might write on the clumsy, wasteful, blundering low and horridly cruel works of nature." The doctrine of survival of the fittest as a creative mechanism (plus death of the unfit) was a subversive one for the people of Victorian times. Obviously, Darwin was mindful of that fact. The way in which Dawkins apparently styles himself as the 'Devil's Chaplain' is also interesting. Anyone familiar with his writings knows that he delights in highlighting the incompatibility of evolution's ravages with Christian theism. Interviewed on a Dutch radio station, Dawkins declared:

Nature is cruel; not only is nature cruel, but it follows from Darwin's theory of natural selection that nature is bound to be cruel, because even very beautiful things like a leopard, a lion, a cheetah or the antelopes that they hunt, they're beautiful and beautifully designed to run fast because countless millions of cheetahs and antelopes have *killed and died* over many generations to make them so (emphasis added).

Most will have to agree with Dawkins here, that nature in this sense is indeed cruel. Mind you, it must be admitted that the above words contradict what he himself had written in *River out of Eden* years earlier:

Nature is not cruel, only pitilessly indifferent. This is one of the hardest lessons for humans to learn. We cannot admit that things might be neither good nor evil, neither cruel nor kind, but simply callous—indifferent to all suffering, lacking all purpose.

In fairness to Dawkins, though, his point is simply this: what appears to be cruel to you and me is actually nothing of the sort. There being no Creator God—which, as an atheist, he believes is virtually certain—concepts of good/evil, cruelty/kindness are meaningless. Nevertheless, in common with many of his fellow atheists, his observation of life's bloody struggle for existence is a key plank in his armament against a Creator God:

Fig.28.2: Darwin's Theory of Natural Selection - Animal Carnivory

The ugliness of animal carnivory was unknown before the mankind fell through sin.

The total amount of suffering per year in the natural world is beyond all decent contemplation. During the minute it takes me to compose this sentence, thousands of animals are being *eaten alive*; others are running for their lives, whimpering with fear; others are being slowly *devoured from within* by rasping parasites; thousands of all kinds are *dying* of starvation, thirst and disease. ... The universe we observe has precisely the properties we should expect if there is, at bottom, no design, no purpose, no evil and no good, nothing but blind, pitiless indifference (emphases Added).

There is only one watertight answer to this bleak, nihilistic outlook on life. People like Dawkins are looking at a fallen, sin-cursed world (Genesis 3:14–19; Romans 8:20–22) and assume that, from their evolutionary perspective, death and life have always gone hand in hand. Contrary to the bleak picture painted here, this world was designed by God and was originally "very good" (Genesis 1:31). Surely a world in which the ugliness of animal carnivory, parasitism, disease and death were unknown before the Fall (Genesis 1:29–30)? But, not according to the doctrine of theistic evolution (hereafter TE).

Death's Endorsement by Theistic Evolution

Let us briefly recall two quotations which we encountered earlier: *"Natural selection depends on death; little would evolve without it;" and "Darwin ...* was working on a theory that requires death. His theory of natural selection is built on the idea that the unfit are going to die." As strange as it may sound, these statements represent a profound insight (but one which often goes unrealized) about what Darwin's idea really entails. ***Evolution is a philosophy of death***, even a eulogizing of it, Figure 28.2. In view of all that we have encountered in the writings of prominent evolutionists thus far, it should hardly be in doubt.

Nevertheless, many Christians have not grasped the stark fact that death is firmly anchored within the evolutionary worldview. Without death, there could be no life. And if theistic evolution is true, this was how God always intended things to be. In which case, God would have employed death, which according to the Apostle Paul is "the last enemy" (1 Corinthians 15:26), as one of his "very good" agents of Creation (Genesis 1:31). In evolutionary terms, death is a vital component—we could say with some justification that death is a contributory creator of life. "Jack Mahoney" explains:

... whereas in traditional Christianity death has always been perceived as a penalty [for sin], in evolution through natural selection the death of individuals, not just of humans, is rather seen as a biological necessity and a requirement [for life].

Intuitively, upon realizing that death so characterizes evolution, many Christians baulk at evolution as God's mode of creation. Certainly, this was true of many as spiritual journey. A one-time theistic evolutionist, a day came when the sudden realization that death was a prerequisite of evolution hit them between the eyes. Immediately, they rejected the remaining vestiges of the evolutionary worldview, embracing instead a literal understanding of Genesis 1–3; the impact of that decision on their faith was truly liberating (John 8:32). For other people, the dawning realization that evolution necessitates death leads, instead, to their rejection of Genesis as a historical narrative of events that really took place. Some of this second group adopt theistic evolution but many others reject Christianity altogether as unworkable, built, as they have come to understand it, on mere 'legendary quicksand's.' Those who shun faith in this way are unable to stomach the necessary reconciliation with death that theistic evolution requires.

We began by observing that the creation/evolution question is partly over how God created. Was death vital to his methodology? Theistic evolutionist "Denis Alexander" says, "One thing is clear: the only proper answer to the question:

'How does God interact with the world?' is 'At all times, in all places and in every way.'"

On that point we wholeheartedly agree. Indeed, he goes on to quote Paul's words to Athenian philosophers (Acts 17:25–28) which affirm these truths. So far, so good. This active involvement of God in his creation (what theologians call God's immanence) means that God is fully responsible for its exact nature in the beginning—he did not simply 'light the blue touch paper', then retire and allow events to unfold. However, theistic evolution recognizes no physical discontinuity (no break in the natural order) at Adam's Fall.

Therefore today's *status quo*, as far as death and suffering are concerned, is also the pre-Fall *status quo*. Predictably, Alexander continues, "If the immanence of God in the created order means anything, then it means God's working through all the processes of the evolutionary process without exception... In other words, God is the author of the whole story of creation, not just bits of it."

Now that does pose a problem for it means that, for theistic evolutionists like Alexander, "God is the author" of death as a creative process. Alexander does not shirk this charge. Like all other theistic evolutionists, he asserts that it is only spiritual death which the Bible is referring to in the passages which relate to the Fall and the Curse. Again, we must emphasize that Alexander is no Deist, although the views of some theistic evolutionists do sound more like Deism. As already mentioned, his concept of God is not a being who, in the beginning, wound up the universe like a clock, then left it to unfold according to universal laws which he had first instituted. No, he embraces fully all forms of physical death as an *intentional* part of God's created order:

Nowhere in the Old Testament is there the slightest suggestion that the physical death of either animals or humans, after a reasonable span of years, is anything other than the normal pattern ordained by God for this earth.

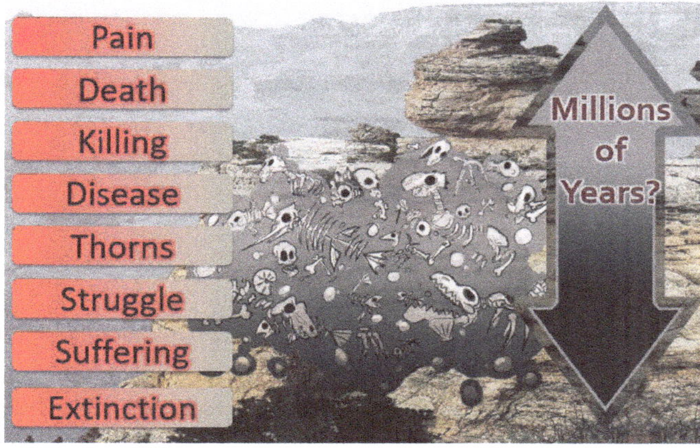

Fig.28.3: Evolution – The Culture of Death - Theistic Evolution and the Doctrine of Death – Curtsy - Creation Ministry International

The Fossil Records

Does the fossil record prove that natural evils have existed throughout time?

We observed that the teaching death is the standard teaching of theistic evolution. "Tom Ambrose" is an English Anglican clergyman with a background in geology. As an ardent theistic evolutionist, his views demonstrate the 'slippery slope' nature of such reasoning:

Fossils are the remains of creatures that lived and died for over a billion years before ***Homo sapiens*** evolved. Death is as old as life itself by all but a split second. Can it therefore be God's punishment for Sin?

The fossil record demonstrates that some form of evil has existed throughout time, Figure 28.3. On the large scale it is evident in natural disasters. … On the individual scale there is ample evidence of painful, crippling disease and the activity of parasites. We see that living things have suffered in dying, with arthritis, a tumor, or simply being eaten by other creatures. From the dawn of time, the possibility of life and death, good and evil, have always existed, side by side. At no point is there any discontinuity; there was never a time when death appeared, or a moment when the evil [sic] changed the nature of the universe. God made the world as it is … evolution as the instrument of change and diversity.

Observe, not only does Ambrose embrace death from the dawn of time, but evil as well. No prizes for guessing which authority he most reveres. Evolution is allowed to trump the Bible. Scripture, in his thinking, must bow the knee to the 'science'—which really means his philosophy—of evolution and millions of years. His statements are a subversion of the truth and profoundly impact our understanding of the Gospel.

Convoluted Bible Readings

A number of theistic evolutionists know which verses and passages of Scripture are used by historic special creationists (biblical creationists) to affirm the absence of pre-Fall death. Many authors, therefore, attempt to stifle or diffuse those challenges to theistic evolution. Yet, for all the erudition and sophisticated reasoning they employ, one rather obvious flaw runs as a common thread through their writings. They fail to acknowledge, let alone engage with, the obvious acceptance of Genesis 1–11 as historical narrative by Jesus, and the apostles and writers of the New Testament. After all, these people insisted on the historical reality of the people, events and places recorded there, including original perfection before the Fall. The only way around this difficulty—and some have gone a long way down this road—is to argue that ***the New Testament writers were wrong, just men of their day and age***.

Nonetheless, that aside, let us consider more carefully "Denis Alexander's" words (representative of theistic evolutionists generally), "Nowhere in the Old Testament is there the slightest suggestion that the physical death of either animals or humans, after a reasonable span of years, is anything other than the normal pattern ordained by God for this earth." Not the slightest suggestion? Not even in God's words "***very good***" at the end of his creative work? To get an idea of just how subversive this claim is, think about these words: "And to every beast of the earth and to every bird of the heavens and to everything that creeps on the earth, everything that has the breath of life, I have given every green plant for food" (Genesis 1:30).

Since this verse teaches a total absence of animal carnivory in that pristine world, it provides more than a slight suggestion that the physical death of animals is not normal. Considered alongside God's declaration of the moral and physical perfection of all that He had made originally (the meaning of the Hebrew for "***very good***" in Genesis 1:31), the absence of pre-Fall death seems to be beyond reasonable doubt. The argument for the physical death of animals being entirely "normal", let alone "ordained by God", sounds like a straight denial of the text of Genesis.

However, there is more. Adam was given clear instructions by God on the day of his creation, "And the Lord God commanded the man, saying, 'You may surely eat of every tree of the garden, but of the tree of the knowledge of good and evil you shall not eat, for in the day that you eat of it you shall ***surely die***'" (Genesis 2:16–17). Later, having willfully flouted God's command concerning that particular tree—for "Adam was not deceived" (1 Timothy 2:14)—God dealt with him accordingly: "And to Adam he said, 'Because you … have eaten of the tree of which I commanded you, "You shall not eat of it," cursed is the ground because of you; in pain you shall eat of it all the days of your life; … By the sweat of your face you shall eat bread, till you ***return to the ground***, for out of it you were taken; for you are dust, and *to* ***dust you shall return***'" (Genesis 3:17, 19;).

The warning, "you shall surely die," is fulfilled by the penalty, "to dust you shall return". That 'a return to dust' refers to physical death seems clear. That being the case, one would think that God's warning of certain death in Genesis 2:17 must surely refer to physical death too, not merely spiritual separation from God. While Adam's physical death was not immediate, he did not avoid the penalty of his disobedience, eventually dying at the age of 930 years (Genesis 5:5)—but he died spiritually the very day that he ate the forbidden fruit.

Just as various models of 'Adam and Eve' are held by different theistic evolutionists, there are differences in the way they model the Fall. Nevertheless, all versions of theistic evolution envisage physical death pervading the entirety of evolutionary history—without that, evolution would be a no go, as we have seen.

Treating the early chapters of Genesis as history is not possible within a theistic evolution framework; instead they are taken figuratively. Thus, Alexander, in common with most other theistic evolutionists, asserts that the penalty for Adam's sin was not physical but spiritual death, "Here in Genesis 3 the passage is quite clear that Adam and Eve died as a result of their sin, just as God had warned, but they died spiritually."

How does this claim square with the actual texts before us (Genesis 2:16–17; 3:17, 19)? On pain of death, God had warned Adam of the consequences of disobediently eating fruit from the forbidden tree: "you shall surely die" (Genesis 2:17). The Hebrew literally means, "dying you shall die." The penalty of disobedience for Adam and Eve (and their descendants) would be a process of physical decline ending in physical death. Without doubt, spiritual death (separation from God) occurred the moment that they sinned. However, this verse cannot be pressed to mean spiritual death alone without making a nonsense of the Hebrew. Quite clearly, the verse is not teaching: 'Adam, the moment you sin, you will begin to die spiritually until, eventually, you are spiritually dead.' Moreover, God reminded Adam of his words of warning in passing sentence on him: Adam would "return to the ground … to dust you shall return." This statement employs very misleading language if it is not meant to convey physical death at all, just spiritual death. Again, as with the case for animal death, there is more than a slight suggestion here that the physical death of humans is abnormal.

During the 'committal' of a funeral service, when the minister says, "For [God] knows our frame; he remembers that we are *dust*" (Psalm 103:14), nobody doubts that our physical mortality is in view. At the graveside, the object lesson is even more poignant, as the minister says, "we now commit [name of the deceased]'s body to the ground: earth to earth, ashes to ashes, dust to dust …" Try telling someone witnessing a burial that the physical death of the deceased is not especially in view! The words "ashes to ashes" are inspired by Ezekiel 28:18, "I turned you to ashes on the earth in the sight of all who saw you", while "dust to dust" is, of course, a reference to Genesis 3:19.

Puzzlingly, Alexander does say that God, "is seen as the direct agent in the death of animals in Psalm 104:29 when he 'takes away their breath' so that 'they die and return to the dust.'" Clearly, animals cannot die spiritually so Alexander is definitely using this verse to teach that animals die physically. He contends that their death has been the normal state of things from the dawn of animal origins. However, the very words he quotes from the psalm are a fitting reminder of God's words to Adam in Genesis 3:19. Animals "return to the dust"—physical death; here Alexander agrees. Adam the guilty sinner is told he will "return to the ground, … to dust you shall return"—physical death; but here Alexander disagrees. In fact, Alexander has this to say:

The reminder to the man that he will return 'to the dust' (verse 19) seems not to be a consequence of his disobedience, but rather a reminder that sweating away to extract crops from the earth is actually quite appropriate when we recall that Adam is destined to return to the earth anyway.

Such a fanciful interpretation weaves a tangled web indeed. Moreover, it is a violation of the principle that Scripture is perspicuous, that is, clearly expressed and readily understood. It amounts to a denial of what the text of Genesis 3:19 manifestly teaches, that Adam's sin was the occasion of both his spiritual and physical death.

Arguing from such passages as Job 24:19 (life snatched from sinners in Sheol, the place of the dead), 2 Kings 20 (Hezekiah's delayed death), and 2 Samuel 12 (the death of David's son through his adultery with Bathsheba), Alexander says, "It is clear from these contexts that it is not death per se which is caused by sin, but rather premature death which is seen as specific punishment for specific sins." Be that as it may, those passages of Scripture have nothing whatsoever to say about the ultimate reason for death. They are of no relevance in determining the truth about the consequences of Adam's sin. In all these cases, such a convoluted approach to Scripture interpretation is not easy to follow. Why such apparently tortuous reasoning? Is it not because, given the insistence upon theistic evolution, physical death must have been happening for millions of years before the Fall of mankind?

Turning to the New Testament, it is true that many passages focus especially upon spiritual death (although the physical and spiritual aspects of death are closely intertwined). Nevertheless, other New Testament passages especially affirm its physical aspect (e.g., 1 Corinthians 15:35–36, 42–44, 53; 2 Corinthians 5:1–4). This is in agreement with a grammatical-historical interpretation of Genesis. That Adam's sin resulted in his spiritual separation from God (spiritual death) is hardly controversial. But the contention that the Fall was not responsible for the intrusion of physical death is as unconvincing as it is a denial of a plain reading of the text.

This is not merely an academic argument. Christians have the wonderful prospect of glorious resurrection bodies and this expectation grows as their physical frames become ever frailer and more diseased with advancing age. Their increasing decrepitude is the harbinger of their final decease physically. Down here, there is groaning; they are becoming an increasing burden to themselves. All is the fruit of sin and the Curse. But their hope, praise God, is 'up there' where, in the presence of their Savior, mortality will give way to immortality and eternal life (2 Corinthians 5:4).

Scripture Coherently Explained

In direct contrast to the Bible revisionism observed in theistic evolution writings, we must view this whole question of death through the undistorted lens of Scripture. Doing so, we confidently affirm that physical death is a consequence of spiritual death. We cannot drive a wedge between the two. Try as some theistic evolutionists might, they simply cannot have one without the other. In such well-known passages as Romans 5:12–17, we learn that, "sin came into the world through one man, and death through sin", and that from Adam onwards, "death reigned". Of course, human death is particularly in view here. Paul teaches, "by a man came death" and, lest there should be any doubt as to this man's identity, he elsewhere clarifies that, "in Adam all die" (1 Corinthians 15:21–22).

People all die because of Adam's sin and also because of their own sin. Referring to Paul's words, "because of one man's trespass, death reigned through that one man" (Romans 5:17), "Thomas Schreiner" points out:

Death reigns as a power over those who are in Adam, for death is not merely an event that occurs but a state in which human beings live as a result of Adam's sin. … death can't be limited to spiritual or physical death, for both realities are designated by the word "death." … Clearly, Adam is the fountainhead for sin and death in the world.[30]

No Christian of the first century would have assumed Paul was referring only, or even primarily, to spiritual death. The words he used (inspired by the Holy Spirit) are too plain for that.

We use the word 'reign' to describe a monarch sovereignly ruling over a people. It carries with it the idea of dominion, even of subjugation. The Bible presents us with two contrasting reigns. There is, firstly, the reign of death. Here 'reign' clearly conveys death's absolute supremacy. Everyone is fully under its sway as one of its subjects. Death is the worst foe we can possibly meet, the ultimate enemy of us all—nobody can hope to escape its total dominion. With the exception of two special cases—Enoch (Genesis 5:24) and Elijah (2 Kings 2:1, 11)—physical death has had the mastery over all human beings that have ever lived. And, with the further exception of Jesus (fully God as well as fully human), this has been mankind's lot ever since Adam's Fall.

Then secondly, there is the reign of Christ, "For he must reign until he has put all his enemies under his feet. The *last enemy* to be destroyed is death" (1 Corinthians 15:25–26. Paul's epistles make clear that genuine repentance and faith in Christ result in people being "made alive" spiritually (1 Corinthians 15:22). They have been converted and are now Christians. No longer need they fear the grave. One day they will exchange their perishable bodies for immortal ones (1 Corinthians 15:53–55)—a *physical* renewal. Spiritually speaking, death's reign over Christians ends at their conversion. Physically, death's dominion is terminated when they leave behind their perishable bodies to be with Christ (2 Corinthians 5:8, Philippians 1:23).

Fig.28.4: *Why did Paul call death an 'enemy' if it was part of the good creation?*

The confidence of believers is in the One who vanquished death at the cross and was resurrected. As "the Author of life" (Acts 3:15), Jesus has power over death itself (Revelation 1:18). He said, "I lay down my life that I may take it up again. No one takes it from me, but I lay it down of my own accord. I have authority to lay it down, and I have authority to take it up again. This charge I have received from my Father" (John 10:17–18). Death, "the last enemy", will be destroyed—its reign finally and forever over (1 Corinthians 15:26). We might well ask, if death was part of God's 'good' creation and not the result of Adam's sin, why did Christ come into the world? Why undo what He had done in the first place? And why did Paul call death an 'enemy' if it was part of the good creation?

Just as death had no dominion over Christ Jesus, the sinless Son of God, it could have had no power over Adam in his pre-Fall perfection. When Jesus died at Calvary, He was willingly making himself a sin offering (2 Corinthians 5:21). In so doing, He brought himself under the penalty of sin, namely death—clearly his physical death is involved here, Figure 28.4. Even so, once He had disarmed the evil principalities and powers (Colossians 2:15), death was powerless to hold him and He was able to take his life again, rising bodily from the dead. As the pure and faultless one, Christ (the last Adam) could not die, except by becoming a sin offering. So, too, the first Adam (a type of Christ) was unable to die (either spiritually or bodily) so long as he remained in perfect obedience to God. Scripture teaches, "The soul who sins shall die" (Ezekiel 18:20), the obvious implication being that sinless souls do not die.

So far, we have distinguished spiritual from physical death of human beings. The Bible also speaks of "the second death" of people who have rejected Christ; tragically, their final destination, together with the devil and his minions, will be the lake of fire (Revelation 20:14, 21:8). No Christian need fear being hurt or conquered by the second death (Revelation 2:11; 20:6). All God's saints will share in the resurrection and enjoy eternal life.

These scriptural teachings are not just interesting theological musings. As we saw earlier, illustrated by the pitiful statements of men like "Christopher Hitchens: and "Stephen Jay Gould," such profound truths are relevant to everyone. Few people are truly reconciled with death, either their own or that of their loved ones. Most will agree with the Apostle Paul, as Gould did, that death is the last (ultimate) enemy. For non-believers, it is physical death that they view as the last enemy.

However, both physical and spiritual death are included in Paul's words. The contention, by theistic evolutionists such as "Denis Alexander," that death is "the normal pattern ordained by God for this earth" fails the test of Scripture itself and violates most people's common-sense understanding of their last enemy.

Animal Death before Adam's Fall?

So, what of animal death? Is this not intrinsically different from that of human beings? And even if there was no predation originally, or even death through old age, might they not have died accidentally? What about the death of plants, or bacteria or fungi for that matter? Even assuming that humans were created 'deathless' originally, surely death prevailed in the animal kingdom? Influenced by theistic evolution or by supposedly unassailable scientific proof for millions of years, many Christians argue in this way.

We can phrase the question a little differently. If, as virtually all evangelicals believe, there will be no animal death in the restored new earth to come—whether through old age, predation or accidents—was there animal death in Paradise (the world before the Fall)?

- Firstly, it is worth noticing that the carcasses of dead animals are something which most people naturally find revolting. We generally go out of our way to avoid physical contact with such things. In the Old Testament, a person became defiled by touching the dead body of a wild animal. This is made explicit in many places (for example, Leviticus 5:2, 7:21, 11:8, 24–28, 39–40; Deuteronomy 14:8). In other words, animal death was considered unnatural. This negative attitude revealed in the laws given to Moses is best explained if the death of animals was not part of the original creation.

- Secondly, as we have seen, every kind of wild animal, flying creature and creeping creature was intended to eat plants originally, with the emphatic "And it was so" (Genesis 1:29–30). All was "very good" in God's estimation (Genesis 1:31), which does appear incompatible with the ugliness, suffering and pain of animal death. Scripture has much more to say. It teaches that "the life of the flesh is in the blood" (Leviticus 17:11). Animals that the Bible describes as having the 'life principle' (*nephesh* in Hebrew) are both air (nostril) breathers (Genesis 7:22) and ones with blood coursing through their veins. Certainly, this includes all birds, reptiles and mammals. Similarly, Adam became a "living soul", a "living being" (in Hebrew, *nephesh chayyāh*) when God breathed his spirit into him (Genesis 2:7; 1 Corinthians 15:45). The pre-Fall death of such animals, then, is ruled out by association since they are also described as *nephesh chayyāh* in Genesis 1.

God's words to Noah just after the Flood are also pertinent here: "But you shall not eat flesh with its life, that is, its blood. And for your lifeblood I will require a reckoning: from every beast I will require it and from man. From his fellow man I will require a reckoning for the life of man" (Genesis 9:4–5). God regards the life-blood as closely tied up with the life of a human being. Again, since *nephesh* animals are exactly like us in this regard, there is simply no scriptural warrant for the pre-Fall bloodshed of such animals.

However, even if death by carnivory (or sickness) is ruled out, might not the Bible allow for pre-Fall death that does not involve suffering? It is important to be clear at this point that the biological death of plants and single cells is not death in the biblical sense. After all, every sort of plant was intended for food in that state of Edenic perfection. Both people and animals were given fruit and vegetables for food (Genesis 1:29–30). In fact, the 'death' of plant cells, various fungi, and bacteria (a natural part of the gut flora of many animals, including ourselves), must have been a necessary and natural part of God's created order. Digestion of plant matter, then, would have resulted in biological cell 'death', the nutrients released from these foods being a vital source of sustenance.

Even within animal bodies (but also known within all other organisms), an entirely natural cell-breakdown process occurs. Without it, normal growth and specialization of body tissues would be impossible. This process of cell removal, termed 'apoptosis', involves numerous proteins. Each is programmed by the cell's DNA to play its part in a sophisticated choreography of interactions—the cell is deconstructed and its components recycled for future use. Not surprisingly, apoptosis can go wrong in our fallen world, causing a host of diseases such as cancer and leukemia. This is something I studied as a cancer patient myself. Originally, however, this must have been a perfect system and would have played a vital role in all multicellular creatures. We can say that not all biological 'death' is really death in the biblical sense.

But could there have been pre-Fall animal death that was due merely to advanced age—death by 'natural causes?' The answer is that animal death is clearly unnatural today, which is why many human beings feel grief even for an animal that dies. I have observed children (and even adults) grieving over deceased pets on many occasions. Whether a humble mouse or a beloved cat or a dog, the animal's death invokes a deep sense of loss. Also, the stiffening of the body that follows for a period soon after death (*rigor mortis*) is unpleasant and disturbing. When my children were younger, I sometimes used the grief associated with their pet's demise to teach that death was not God's intention for creatures originally—a lesson that was not difficult to put across.

Some people ask why animals die at all if they cannot sin? A related question, sometimes put forward as a challenge to the doctrines of biblical creation and a historical Fall, is whether it is fair that animals should suffer the consequences of Adam's sin. However, since God gave Adam dominion over all living creatures (Genesis 1:26, 28), his rebellion was the occasion for their downfall too—indeed that of the whole creation (Romans 8:20–22). The tragic fact is that living things all shared in the pollution and defilement brought about by the Fall. Indeed, our observation of the suffering and death of animals should act as a monument to human rebellion in Adam. Speaking of this, sixteenth century reformer John Calvin wrote, "And this all tends to inspire us with a dread of sin; for we may easily infer how great is its atrocity, when the punishment of it is extended even to the brute creation."

The serpent was a form of animal, which was participated in the fall of Adam. Accordingly, it was cursed and condemned to the same fate of Adam. However, in due course when sin is removed there would be no corrupting power due to removal of sin. The power of Jesus' redemption and resurrection has put death-on-notice until the point of rapture when the saints' bodies are instantly transformed into eternal bodies of glory, inseparable from Jesus.

Consequently, the power of death, suffering and pain are also be conquered-- even the animals' carnivorous nature will be instantly converted to a coexisting peaceful nature with animals of lesser vigor. Lambs will lie in peace with lions. Isaiah 11:6-9, The wolf shall live with the lamb, the leopard shall lie down with the kid, the calf and the lion and the fatling together, and a little child shall lead them. The cow and the bear shall graze, their young shall lie down together; and the lion shall eat straw like the ox. The nursing child shall play over the hole of the asp, and the weaned child shall put its hand on the adder's den. They will not hurt or destroy on all my holy mountain; for the earth will be full of the knowledge of the LORD as the waters cover the sea.

In Vain - Death

Darwinian evolution is a philosophy which is impossible to reconcile with the biblical teaching on death, as theistic evolutionists must endeavor to do. God is the Author of life. Death and all that accompanies it—bloodshed, extinctions, mutations, diseases and untold suffering—is an unwelcome intruder into this world. The creation now 'groans' under the weight of thousands of years of the Fall and the Curse. This will continue until the time when the Curse is removed (Revelation 22:3). Only in the new heaven and earth will death, mourning, crying and pain cease at last (Revelation 21:1, 4).

The respected British journalist and television presenter "Andrew Marr" discussed the Darwinian implications for *Life and Death,* commenting on Darwin's worldview, he said, "But burdened by his thoughts of life and death, and extinction, [Darwin] soon began to retreat from the limelight. *He knew that he had the seed of a dangerous idea*, one that would undermine the Bible and Christian teaching." Logically, the dog-eat-dog, death-ridden view of life's ultimate origins leads to nihilism. No surprise then, that among young people, in their teens and twenties, suicide is the leading or second-leading cause of death in some western countries. If people believe that they owe their existence to a process of death and struggle and are heading for more of the same, ending their lives prematurely can seem rational. A suicide note stated:

'... I think that some people may have an inability to cope, and maybe this might sound a bit extreme, but that might be Darwinian theory, the Darwin theory of survival of the fittest. Maybe some of us aren't meant to survive, maybe some of us are meant to kill ourselves ...'

'There are too many people in the world as it is. Maybe it is survival of the fittest, maybe some of us are meant to just give up, and maybe that would help the species.'

Confessions like this are a tragic indictment of evolution's impact on society. Surely it ought to engender some soul-searching among those who advocate theistic evolution. The sobering words of the prophet Jeremiah seem appropriate, "***Cursed is the man who trusts in man*** and makes flesh his strength, whose heart turns away from the LORD… Blessed is the man who trusts in the LORD, whose trust is the LORD" (Jeremiah 17:5, 7). We have seen that evolution's very mechanism requires death to weed out the less fit. But a philosophy which assigns creative powers to people's mortal enemy is one which radically alters their conception of death. It gives many a 'justification' for turning away from God. The antidote is to return to a thoroughly biblical understanding of sin and death, that which is clearly revealed by the inspired writers of the New Testament. They treated the pertinent events of Genesis 1–3 as historical facts upon which they enthusiastically proclaimed the Gospel.

He created the heavenly host and is greater than every angelic being, for He is the wisdom of God, the power of God, and the express image of the invisible God.

The Man Christ Jesus was both Son of God and Son of Adam. He was Son of Man, Son of Abraham, Son of David, and Son of Mary, the Seed of the woman, and the Lord of glory. He willingly consented to humble Himself by taking upon Himself the form of a servant. He was the Word made flesh, full of grace and truth Who learned obedience by the things that He suffered. And He was made in the likeness of men so that we might become the righteousness of God by grace through faith in Him: "And by Him and through Him and for Him all things were created and have their being."

And Christ has been given a name that is above every name - Jesus Christ the righteous - the unblemished Lamb of God Who, from the foundation of the world, was destined to be the Rock of our salvation, the Light of the world Who would bring hope to the Gentiles. He is the Bread of life, the hidden Manna, the true Vine, and the Tree of life. Jesus is the good Shepherd and He is the Way, and the Truth, and the Life.

Christ is our great High Priest and the King of kings. He is the unspeakable Gift of God and the first Begotten of the dead. He is the only Mediator between man and God Who ever lives to make intercession for us. He is the Consolation of Israel Who remains faithful and true to all God's promises for they are all 'yes' and 'amen' in Him. He is the Head over the Body which is the Church, and He is the Author and Finisher of our faith, for Christ is all in all: "By Him all things were created, both in the heavens and on earth, visible and invisible, whether thrones, or dominions, or rulers, or authorities, all things have been created by Him, and through Him, and for Him," to His praise and glory.

The Son of God - The Act of Creation

A study on the Lord Jesus Christ, His Priority - Lordship over Created Being, is tailored to provide a clear understanding to the depth of apostle Paul's declaration of "***He Holds Every Thing together***,"

This study is a pivotal theme to the "Origin of the Universe." I sincerely ask the readers to indulge me with their patience and careful attention to this important Biblical passage.

Colossians 1:15-17,

- **15** Christ is the visible image of the invisible God.
  He existed before anything was created and is supreme over all creation,
- **16** for through him God created everything in the heavenly realms and on earth.
  He made the things we can see and the things we can't see—such as thrones, kingdoms, rulers, and authorities in the unseen world. Everything was created through him and for him.
- **17** He existed before anything else and he holds all creation together.

Who are the Colossians?

The Colossians, to whom this epistle is written, were not the Rhodians, by some called Colossians, from Colossus, the large statue of the sun, which stood in the island of Rhodes, and was one of the seven wonders of the world; but the inhabitants of Colosse, a city of the greater Phrygia, in the lesser Asia, near to which stood the cities of Laodicea and Hierapolis, mentioned in this epistle.

Pliny: speaks of it as one of the chief towns in Phrygia, and

Herodotus: calls it the great city of Phrygia; it is said to have perished a very little time after the writing of this epistle, with the above cities, by an earthquake, in the year of Christ 66, though it was afterwards rebuilt.

Theophylact: says, that in his time it was called Chonae. When the Gospel was brought hither, and by whom, is not known, nor who was the founder of the church in this place; for the Apostle Paul was not, since his face had never been seen by them, Col 2:1, though it is said that;

Epaphras, the same name with Epaphroditus, was fixed by him pastor of this church; and others say Philemon was set over it by him. The occasion of this epistle was this, Epaphras, who had preached the Gospel to the Colossians, and very likely was the first that did, came to Rome, where the Apostle Paul was a prisoner, and gave him an account of them, how they had heard and received the Gospel, and of their faith in Christ, and love to the saints; and also declared to him in what danger they were through some;

False Teachers: that had got among them, who were for introducing the philosophy of the Gentiles, the ceremonies of the law of Moses, and some pernicious tenets of the followers of Simon Magus, and the Gnostics; upon which the apostle writes this epistle to them, to confirm them in the faith of the Gospel Epaphras had preached unto them, and which was the same he himself preached; and to warn them against those bad men, and their principles; and to exhort them to a discharge of their duty to God, and men, and one another.

It was written by the apostle, when in bonds at Rome, as many passages in it show, and about the same time with those to the Philippians and Ephesians; and the epistle to the latter greatly agrees with this, both as to subject and style. Dr. Lightfoot places it in the year of Christ 60, in the second of the apostle's imprisonment, and in the sixth of Nero's reign.

About Colossians Epistle

This chapter of Colossians one contains the inscription of the epistle; the apostle's usual salutation; his thanksgiving to God on behalf of the Colossians for grace received; his prayers, that more might be given them; an enumeration of various blessings of grace, which require thankfulness, in which the glories and excellencies of Christ are particularly set forth: and it is concluded with an exhortation to a steadfast adherence to the Gospel, taken from the nature, excellency, and usefulness of the ministry of it. The inscription, and the salutation, are in Col 1:1, 1:2, and are the same with those in the epistle to the Ephesians, only Timothy is joined with the apostle here, and the Colossians have the additional character of brethren given them.

The thanksgiving is in Col 1:3-5, the object of it is God, the Father of Christ; the time when made, when in prayer to him; its subject matter, the faith and love of the saints; to which is added, their happiness secured for them in heaven, their hope was conversant with: and whereas the Gospel was the means by which they came to the hearing and knowledge of it, this is commended from the subject of it, the doctrine of truth; from the spread of it in the world; and from its efficacy in bringing forth fruit in all, to whom it came in power, and that with constancy, Col 1:5, 1:6, and also from the testimony of Epaphras.

Epaphras, a faithful minister of Christ, and theirs, who was dear to the apostle, and of whom he had the above account of them, Col 1:7, 1:8. And then follow his prayers for them, that they might have an increase of spiritual knowledge, and that they might put in practice what they knew; and for that purpose, he entreats they might be blessed with strength, patience, and longsuffering, Col 1:9-11.

And in order to excite thankfulness in himself and them, he takes notice of various blessings of grace; of the Father's grace in giving a meetness for eternal glory and happiness, by delivering from the power of darkness, and translating into the kingdom of his Son, Col 1:12, 1:13, and of the Son's grace in obtaining redemption by his blood, and procuring the remission of sins, Col 1:14, which leads the apostle to enlarge upon the excellencies of the author of these blessings, Jesus Christ.

In Jesus Christ divine person, as the image of God, and the first cause of all created beings, Col 1:15, which he proves by an enumeration of them, as created by him, and for his sake, by his pre-existence to them, and their dependence on him, Col 1:16, 1:17, and in his office capacity.

Jesus Christ as Mediator, being the head of the church, the governor of it, and the first that rose from the dead; by all which it appears that he has, and ought to have the pre-eminence, Col 1:18. And this is still more manifest from his having all fulness dwelling in him, to supply his body the church, of which he is the head, Col 1:19, and from the reconciliation of all the members of it to God by him, Col 1:20, which blessing of grace is amplified partly by the subjects of it, who are described by their former state and condition, aliens and enemies, and by their present one, reconciled by the death of Christ in his fleshly body; and partly by the end of it, the presentation of them holy, blameless, and irreprovable in the sight of God, Col 1:21, 1:22.

Wherefore, it is a duty incumbent on such to abide by the Gospel of Christ, which brings the good tidings of peace and reconciliation, and is the means of faith and hope; and the rather, since they had heard it themselves, and others also, even every creature under heaven; and the apostle was a minister of it, Col 1:23, and on his ministration of it he enlarges, by observing his sufferings for the church on account of the Gospel, which he endured with pleasure; and therefore they should, by his example, be encouraged to continue in it, Col 1:24.

Moreover, he argues the same from his commission of God to preach it for their sakes, Col 1:25, and from the nature and subject matter of it, being a hidden mystery, and containing riches and glory in it; Christ himself, the foundation of hope of eternal glory, Col 1:26, 1:27, and from the end of preaching it, which was to present every man perfect in Christ; which end the apostle labored and strove to obtain through the power and energy of divine grace, which wrought in him, and with him, Col 1:28, 1:29.

Fig.28.5: *Creation Ex Nihilo Through Jesus Christ - Curtsy - The Institute for Creation Research*

Christ is the visible image of the invisible God. He existed before anything was created and is supreme over all creation, Figure 28.5.

[Christ is] in closest connection with the preceding matter, a confession of truth and faith about the Person of the Redeeming Son of God, the King of the redeemed. He appears in His relation to

a. the Eternal Father,

b. the created Physical Universe,

c. the created Universe of Angelic spirits,

d. the Church of redeemed men.

Every clause is pregnant of Divine truth, and the whole teaches with majestic emphasis the great lesson that the Person is all-important to the Work, the true Christ to the true salvation.

the image] So 2 Corinthians 4:4. *"Satan, who is the god of this world, has blinded the minds of those who don't believe. They are unable to see the glorious light of the Good News. They don't understand this message about the glory of Christ, who is the exact likeness of God."* The Greek word (*eicôn*) occurs often in Biblical Greek, most frequently (in Old Testament) as a translation of the Hebrew *tselem*. Usage shows that on the whole it connotes not only similarity but also "*representation* (as a *derived* likeness) and *manifestation*". An instructive passage for study of the word is Hebrews 10:1, where it is opposed to "*shadow*," and plainly means "*the things themselves, as seen*." Thus, the Lord Christ, the mystery of His Person and Natures, is not only a Being resembling God, but God Manifest. Cp. John 14:9, and Hebrews 1:3.

"Christian antiquity has ever regarded the expression 'image of God' as denoting the eternal Son's perfect equality with the Father in respect of His substance, power, and eternity … The Son is the Father's Image in all things save only in being the Father."

the invisible God] For the same word see 1 Timothy 1:17; Hebrews 11:27. And cp. Deuteronomy 4:12; John 1:18; John 5:37; 1 Timothy 6:16; 1 John 4:20. This assertion of the Invisibility of the Father has regard to the *manifesting* function of the Image, the Son. See Lightfoot here. The Christian Fathers generally (not universally) took it otherwise, holding that the "Image" here refers wholly to the Son in His Godhead, which is as invisible as that of the Father, being indeed the same. But the word "Image" by usage tends to the thought of vision, in some sort; and the collocation of it here with "the Invisible" brings this out with a certain emphasis. Not that the reference of the "Image" here is directly or primarily to our Lord's visible Body of the Incarnation, but to His being, in all ages and spheres of created existence, the Manifester of the Father to created intelligences. His being this was, so to speak, the basis and antecedent of His gracious coming in the flesh, to be "seen with the eyes" of men on earth (1 John 1:1).

the firstborn of every creature] Better perhaps, **Firstborn of all creation**, or, with a very slight paraphrase, **Firstborn over all creation**; standing to it in the relation of priority of existence and supremacy of inherited right. So, to borrow a most inadequate analogy, the heir of a hereditary throne might be described as "firstborn to, or over, all the realm." The word "*creature*" (from the (late) Latin *creatura*) here probably, as certainly in Romans 8, means "creation" as a whole; a meaning to which the Greek word *inclines* in usage, rather than to that of "a creature."

"*Firstborn:*"—cp. Psalm 89:27; and the Palestinian Jewish application, thence derived, of the title "Firstborn" to the Messiah. A similar word was used of the mysterious "Logos" among the Alexandrian Jews, as shown in the writings of St Paul's contemporary, Philo. Studied in its usage, and in these connections, the word thus denotes

a. *Priority of existence*, so that the Son appears as antecedent to **the created Universe**, and therefore as belonging to the eternal Order of being;

b. *Lordship over* "all creation," by this right of eternal primogeniture; Psalm 89:27, and cp. Hebrews 1:2.

"*Of all creation:*"—so lit. The force of the Greek genitive, in connection with the word "*first*" (as here "*firstborn*"), may be either partitive, so that the Son would be described as first of created things, or so to speak comparative as in, John 1:15, Greek, so that He would be described as first, or antecedent, in regard of created things.

- **16** for through him God created everything in the heavenly realms and on earth.

He made the things we can see and the things we can't see—such as thrones, kingdoms, rulers, and authorities in the unseen world. Everything was created through him and for him.

I. CHRIST AND CREATION. (Colossians 1:16)

"The heresy of the Colossian teachers took its rise … in their cosmical speculations. It was therefore natural that the Apostle in replying should lay stress on the function of the Word in the creation and government of the world. This is the aspect of His work most prominent in the first of the two distinctly Christological passages.

The Apostle there, predicates of the Word [the Son] not only prior but absolute existence. All things were **created by Him**, are **sustained in Him**, are tending towards Him. Thus, He is the beginning, middle, and end of creation. This He is because He is the very *Image* of the Invisible God, because in Him dwells the Plenitude of Deity.

"This creative and administrative work of Christ the Word [the Son] in the natural order of things is always emphasized in the writings of the Apostles when they touch on the doctrine of His Person … With ourselves this idea has retired very much into the background … And the loss is serious … How much more-hearty would be the sympathy of theologians with the revelations of **science and the developments of history**, if they habitually connected them with the operations of the same Divine Word who is the center of all their religious aspirations, it is needless to say.

"It will be said indeed that this conception leaves … creation … as much a mystery as before. This may be allowed. But is there any reason to think that with our present limited capacities the veil which shrouds it ever will be removed? The metaphysical speculations of twenty-five centuries have done nothing to raise it. The physical investigations of our own age from their very nature can do nothing; for, **busied with the evolution of phenomena**, they lie wholly outside this question, and do not even touch the fringe of the difficulty. But meanwhile revelation has interposed, and thrown out the idea which, if it leaves many questions unsolved, gives a breadth and unity to our conceptions, at once satisfying our religious needs and linking our scientific instincts with our theological beliefs."

The views outlined by Bishop Lightfoot, are pregnant of spiritual and mental assistance. At the same time with them, as with other great aspects of Divine Truth, a reverent caution is needed in the development and limitation. The doctrine of the Creating Word, the Eternal Son, "in" Whom finite existence has its Corner-stone, may actually degenerate into a view both of Christ and Creation nearer akin to some forms of Greek speculation than to Christianity, if not continually balanced and guarded by a recollection of other great contents of Revelation. Dr J. H. Rigg, in *Modern Anglican Theology* (3rd Edition, 1880), has drawn attention to the affinity which some recent influential forms of Christian thought bear to Neo-Platonism rather than to the New Testament. In particular, any view of the relation of Christ to "Nature" and to man which leads to the conclusion that all human existences are so "in Christ" that the individual man is vitally united to Him antecedent to regeneration, and irrespective of the propitiation of the Cross, tends to non-Christian affinities.

It is a fact never to be lost sight of that any theology which on the whole gives to the mysteries of guilt and propitiation a less *prominent* place than that given to them in Holy Scripture, tends to a very wide divergence from the scriptural type. Here, as in all things, the safety of thought lies on the one hand in neglecting no great element of revealed truth, on the other in coordinating the elements *on the scale, and in the manner*, of Divine Revelation.

II. DEVELOPMENTS OF DOCTRINE IN COLOSSIANS. (Colossians 1:16)

In the precise form presented in *Colossians* the revelation of the Creative Work of the Son is new in St Paul's Epistles. But intimations of it are to be found in the earlier Epistles, and such as to make this final development as natural as it is impressive. In 1 Corinthians 8:6 we have the "one Lord Jesus Christ, *through whom are all things, and we through Him;*" which is in effect the germ of the statements of Colossians 1. And in Romans

8:19-23 we have a passage pregnant with the thought that the created Universe has a mysterious relation to "the sons of God," such that their glorification will be also its emancipation from the laws of decay; or at least that the glorification and the emancipation are deeply related to each other.

Nothing is wanted to make the kinship of that passage and Colossians 1 evident at a glance, but an explicit mention of Christ as the Head of both Universes, The Visible Universe and the Angelical Universe. As it is, His mysterious but most real connection with the **Creating** and the **Holding** of the Universe is seen lying as it were just below the surface of the passage in *Romans*.

III. "THRONES AND DOMINIONS." (Colossians 1:16)

We transcribe here a note from *Ephesians*; on the words of Ephesians 1:21:

"Two thoughts are conveyed; first, subordinately, the existence of orders and authorities in the angelic (as well as human) world; then, primarily, the imperial and absolute Headship of the Son over them all. The additional thought is given us by Colossians 1:16, that He was also, in His preexistent glory, their Creator; but this is not in definite view here, where He appears altogether as the exalted Son of Man after Death. In Romans 8, Colossians 2, and Ephesians 6 ... we have cognate phrases where *evil* powers are meant.... But the context here is distinctly favorable to a *good* reference. That the Redeemer should be "exalted above" powers of *evil* is a thought scarcely adequate in a connection so full of the imagery of glory as this. That He should be "exalted above" the holy angels is fully in point. 1 Peter 3:22 is our best parallel; and cp. Revelation 5:11-12. Also, Matthew 13:41; "The Son of Man shall send forth *His* angels."

"We gather from the Epistle to the Colossians that the Churches of Asia Proper were at this time in danger from a quasi-Jewish doctrine of Angel-worship, akin to the heresies afterwards known as Gnosticism. Such a fact gives special point to the phrases here. On the other hand, it does not warrant the inference that Paul repudiates all the ideas of such an angelology. The idea of order and authority in the angelic world he surely endorses, though quite in passing.

"Theories of angelic orders, more or less elaborate, are found in the *Testaments of the Twelve Patriarchs*, (cent. 1–2); Origen (cent. 3); St Ephrem Syrus (cent. 4). By far the most famous ancient treatise on the subject is the book *On the Celestial Hierarchy*, under the name (certainly assumed) of Dionysius the Areopagite; a book first mentioned cent. 6, from which time onwards it had a commanding influence in Christendom. (See Article *Dionysius* in Smith's *Dict. Christ. Biography*). "Dionysius" ranked the orders (in descending scale) in three *Trines;* Seraphim, Cherubim, Thrones; Dominations, Virtues, Powers (Authorities); Principalities, Archangels, Angels. The titles are thus a combination of the terms Seraphim, Cherubim, Archangels, Angels, with those used by Paul here and in Colossians 1.

IV. He Holds the Universe Together

- **Col. 1:17** He existed before anything else and he holds all creation together.

he Emphatic in the Greek; He, and no other who could even seem to rival or obscure His sublime eminence.

is before all things] *ante omnes*, Latin Versions. The Greek genitive form is ambiguous; it might be either masculine or neuter. But the mention in the last clause, in the unambiguous nominative, of "all *things*," decides for a similar reference here.

"*and* He is *before all things*," comparing John 8:58, and Exodus 3:14, and adding, "The imperfect ['*was*'] might have sufficed, ... but the present ['*is*'] declares that this pre-existence is absolute existence." He quotes

Basil of Cæsarea (*adv. Eunom*., iv.) as emphasizing the special force of "*is*" (as against e.g. "*was*" or "*became*") in this very passage: "(the Apostle) indicates thus that He ever *is* while the creation *came to be*."

"*Before:*"—i.e., as the whole context shows, in respect of priority of existence; the priority of eternity.

by him] Lit. and better, **in Him**; *consist*] I.e., literally, **stand together, HOLD together**. The Latin-English "*consist*" (Latin versions, *constant*) exactly renders the Greek. "He is the principle of cohesion in the Universe. He impresses upon creation that unity and solidarity which makes it a cosmos instead of a chaos." And Lightfoot quotes Philo to show that the "Logos" of Alexandrian Judaism was similarly regarded as the "**BOND**" of the universe.

"Christ was the conditional element of their *creation*, the causal element of their *persistence* … The declaration, as Waterland observes, is in fact tantamount to '**in Him** they live, and move, and have their being'" (Ellicott)

Natural philosophy, after all observation and classification of phenomena and their processes, asks necessarily but in vain (so long as it asks only "Nature"), what is their ultimate secret, what *is*, for instance, the last reason of **universal gravitation**. *Revelation discloses* that reason in the Person and ***Will of the Son of God***.

Thus far the Apostle has unfolded the glory of Christ as the Cause and Bond of all being in the sphere of "Nature," material and otherwise.

V. The Existence of the Universe

- **For everything was created by him, in heaven and on earth, the visible and the invisible, whether thrones or dominions or rulers or authorities—all things have been created through him and for him. He is before all things, and by him all things hold together.**

In Jesus, through the explanation of the Apostle Paul, guided by the Holy Spirit, we begin to see who is behind the story of existence. He created all things and all things came about through Him and for Him. You and I, our children, our friends, our grandchildren, our neighbors, our co-workers, strangers, we are all *for* him.

He created all things and He sustains all things. This is vital because what we're seeing is that no *force* that holds together and sustains the **Universe**. It is not merely the laws of nature that are sustaining the existence of all things here on earth and out there in space. Indeed, there is an entity whose embrace holds it all together. This is the LORD Jesus.

The Lord Jesus Christ created the Universe and the heavenly host. He is the wisdom of God, the power of God, and the expressed Image of the Invisible God, the Father. And yet to fulfil God's eternal plans and purposes, He has taken a human form for a time.

VI. The Redeemer of Man's Sin and Death

The Man Christ Jesus was both Son of God and Son of Adam. He was Son of Man, Son of Abraham, Son of David, and Son of Mary, the Seed of the woman, and the Lord of glory. He willingly consented to humble Himself by taking upon Himself the form of a servant. He was the Word made flesh, full of grace and truth Who learned obedience by the things that He suffered. And He was made in the likeness of men so that we might become the righteousness of God by grace through faith in Him: "And by Him and through Him and for Him all things were created and have their being."

Fig.28.6: Plan of Salvation from Sin and Death

And Christ has been given a name that is above every name - Jesus Christ the righteous - the unblemished Lamb of God Who, from the foundation of the world, was destined to be the Rock of our salvation, the Light of the world Who would bring hope to the Gentiles. He is the Bread of life, the hidden Manna, the true Vine, and the Tree of life. Jesus is the good Shepherd and He is the Way, and the Truth, and the Life, Figure 28.6.

Christ is our great High Priest and the King of kings. He is the unspeakable Gift of God and the first Begotten of the dead. He is the only Mediator between man and God Who ever lives to make intercession for us. He is the Consolation of Israel Who remains faithful and true to all God's promises. He is the Head over the Body which is the believers, the Church, and He is the Author and Finisher of our faith, for Christ is all in all: "By Him all things were created, both in the heavens and on earth, visible and invisible, whether thrones, or dominions, or rulers, or authorities, all things have been created by Him, and through Him, and for Him," to His praise and glory, amen and amen

Dark Matter – Dark Energy

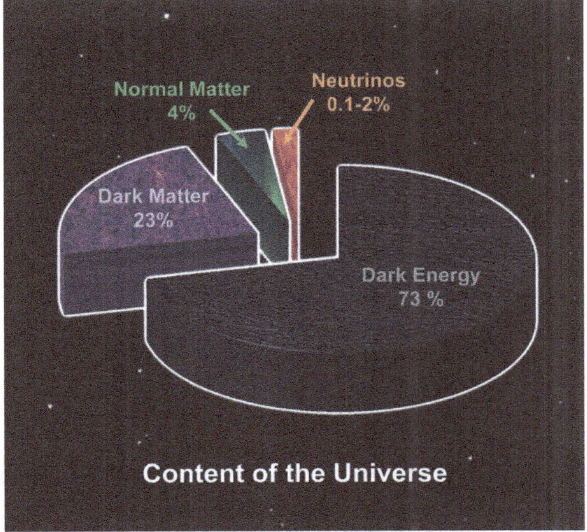

Fig.28.7: The Universe Dark Matter/Dark Energy

Yet, in the nearly 40 years that followed, researchers still haven't been able to figure out what dark matter is made of, Figure 28.7.

A popular hypothesis is that dark matter is formed by exotic particles that don't interact with regular matter, or even light, and so are invisible. Yet their mass exerts a gravitational pull, just like normal matter, which is why they affect the velocities of stars and other phenomena in the universe.

However, try as hard as they might, scientists have yet to detect any of these particles, even with tests designed specifically to target their predicted properties.

"I think on the dark matter side there is some discouragement among the people who are kind of mid-career," Panek said. "They went into this field thinking, Certainly, we're going to solve this problem and then we'll build from there. Nevertheless, 15~20 years later, they're saying, I've invested my career in this and I don't know if I'm going to find anything in my lifetime.'"

Still, many hold out hope that we're getting close and that experiments such as the newly built Large Hadron Collider particle accelerator in Geneva may finally solve the puzzle.

Dark Energy

Dark energy is possibly even more baffling than dark matter. It's a relatively more recent discovery, and it's one that scientists have even less of a chance of understanding anytime soon.

It all started in the mid-1990s, when two teams of researchers were trying to figure out how fast the universe was expanding, in order to predict whether it would keep spreading out forever, or if it would eventually crumple back in on itself in a "Big Crunch."

To do this, scientists used special tricks to determine the distances of many exploded stars, called supernovas, throughout the universe. They then measured their velocities to determine how fast they were moving away from us.

When we view very distant stars, we are viewing an earlier time in the history of the universe, because those stars' light has taken millions and billions of light-years to travel to us. Thus, looking at the speeds of stars at various distances tells us how fast the universe was expanding at various points in its lifetime.

Astronomers predicted two possibilities: either the universe has been expanding at roughly the same rate throughout time, or that the universe has been slowing in its expansion as it gets older.

Shockingly, the researchers observed neither possibility. Instead, the universe appeared to be *accelerating* in its expansion.

That fact could not be explained based on what we knew of the universe at that time. All the gravity of all the mass in the cosmos should have been pulling the universe back inward, just as gravity pulls a ball back down to Earth after it's been thrown into the air.

"There's some other force out there or something on a cosmic scale that is counteracting the force of gravity, "Panek explained." *People didn't believe this at first because it's such a weird result."*

Fierce Competition

Scientists named this mysterious force ***dark energy***. Though no one has a good idea of what dark energy is, or why it exists, it is the force that appears to be counteracting gravity and causing the universe to accelerate in its expansion. However, the author is providing his perception of the ***Dark Matter/Dark Energy*** at the end of this chapter.

The lack of a good explanation for dark energy hasn't seemed to dampen scientists' enthusiasm for it.

"What I hear again and again is how excited people are to be working in this field right now, when this revolution is going on," Panek told SPACE.com. *"The problems are so great and profound, they're actually rather thrilled with it."*

Overall, dark energy is thought to contribute 73 percent of all the mass and energy in the universe. Another 23 percent is dark matter, which leaves only 4 percent of the universe composed of regular matter, such as stars, planets and people.

This bizarre, but apparently true, conclusion was reached at about the same time by the two groups working to measure the expansion of the universe. The competition between the groups became very contentious, Panek said, and they grew to dislike each other quite a lot.

Ultimately, though, members of both teams should reap the rewards of finding one of the biggest surprises in the history of science.

"I think that it's kind of assumed the dark energy will win the discoverers the Nobel," Panek said. "There certainly is that assumption that it's just a matter of years."

If you have done star-gazing, you probably would have enjoyed and admired the majestic arrangements of stars into galaxies, Figure 28.8. Our Sun is a typical star in the Milky Way, our home galaxy, a typical galaxy in the universe. The hundreds of billions of galaxies in the observable universe group into galaxy clusters, forming even larger structures, Figure 28.2. Other than stars,

- what constitutes a galaxy?
- How did galaxies and galaxy clusters form?
- How did these structures emerge from the almost perfectly uniform distribution in the early universe?

These are just some of the questions that scientists want to answer. It turns out that a mysterious form of matter, called dark matter, plays a crucial role in all these questions.

Fig.28.8: *NGC 3982, the Pinwheel Galaxy, a Milky Way-like system with hundreds of billions of stars. Curtsy - Credit: NASA/STScI.*

Fig.28.9: Coma Cluster, a typical and nearby galaxy cluster made of thousands of galaxies. Curtsy - NASA/STScI.

In the 1930s, Astronomers "Zwicky" and "Smith" tried to measure the mass of the Coma and Virgo Clusters, two relatively nearby galaxy clusters. Their reasoning goes like this:

the mass of a galaxy cluster is proportional to its gravitational pull on the galaxies, which the orbital speeds would reveal, Figure 28.8. After carefully measuring the motions of galaxies, Zwicky and Smith were surprised to find that Coma and Virgo Clusters should be much more massive than the visible matter (stars and gas), Figure 28.9. In other words, there must be too much of 'Dunkle Materie', or "dark matter," which gravitates but doesn't emit light. Applying the same idea, Vera Rubin measured the rotation curves of galaxies – the orbital speeds of stars and gas at different distances from the galactic center, Figure 28.10. She found that the rotation curves of most galaxies do not decrease even at large distances, indicating that the gravity at the outskirt of a galaxy is not weakening, even though almost no visible matter can be observed there. This observation strongly suggests that a much larger dark matter halo engulfs the visible matter. For example, the dark matter halo of the Milky Way extends to at least 4-5 times the radius of its luminous disk, containing a total mass 5-10 times that of visible matter.

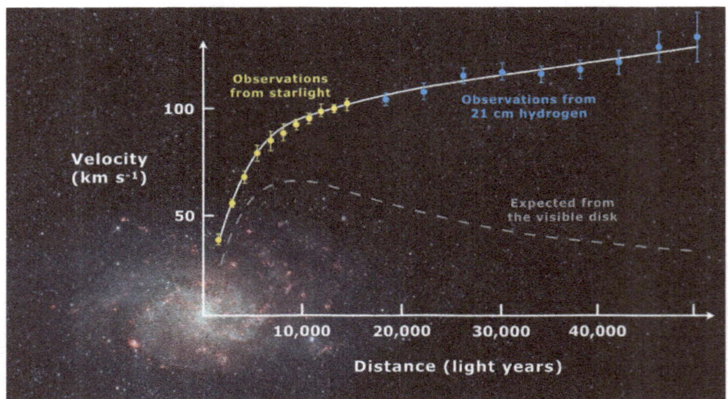

Fig.28.10: Rotation curve of galaxy M33. The measured orbital speeds of stars and gas (yellow and blue data points) continue to rise beyond the luminous disk where most of the visible matter reside. This indicates that dark matter extends to at least 4 times the radius of the luminous disk, forming the dark matter halo. Curtsy - Mario De Leo. CC BY-SA 4.0.

In recent years, many independent observations have confirmed the existence of dark matter in the universe. Some examples include those of galaxy or galaxy cluster collisions. Since the typical separation between neighboring galaxies is only a few times their size, it's not surprising to find that many galaxies would collide with each other, Figure 28.11. Astronomers found that long tidal tails are drawn out from the galaxies in many such events, reflecting the presence of the dark matter haloes. Figure 28.12, shows the famous Bullet Cluster, which is believed to form after two galaxy clusters collided. The visible matter, mostly hot gas, dragged on each other because of friction and became so hot that X-rays were emitted and detected by the Chandra X-ray telescope.

On the other hand, dark matter passes through matter without slowing down since they do not interact much. Through gravitational lensing – the bending of light by gravity – astronomers can map out where most of the mass is, thereby revealing the dark matter. The separation of dark matter from the visible matter in the Bullet Cluster shows us that the dark matter interacts weakly only, if at all, and is quite different from the normal, visible matter.

Fig.28.11: Examples of colliding galaxies. In many cases, long tidal tails are drawn out, indicating the presence of the dark matter haloes. Curtsy - NASA, ESA, the Hubble Heritage (STScI/AURA)-ESA/Hubble Collaboration, and A. Evans (University of Virginia, Charlottesville/NRAO/Stony Brook University).

Fig.28.12: Bullet Cluster, believed to form after a galaxy cluster punches through another one. The red-colored regions contain the visible matter, which become so hot that they emit X-rays detected by the Chandra telescope. The blue-colored regions indicate most of the mass revealed by gravitational lensing. The collision between the two galaxy clusters separates the visible and dark matter. Curtsy - Credit: X-ray: NASA/CXC/M.Markevitch et al.; Optical: NASA/STScI; Magellan/U. Arizona/D.Clowe et al.; Lensing Map: NASA/STScI; ESO WFI; Magellan/U.Arizona/D.Clowe et al.

Through many independent observations, scientists now know that the universe's matter content is dominated by ***dark matter***, an unknown form of matter that undergoes gravitational and perhaps weak interactions, but not electromagnetic or strong interactions.

They are therefore invisible to us, but they play an essential role in providing strong enough gravity to form the ***structures of the universe***, Figure 28.6. We would not have the many galaxies and galaxy clusters we can see today without dark matter. However, we know very little about dark matter.

On average, the ratio of dark matter to normal matter is about 6 to 1. So even if we understand all the atoms and electrons, they only represent a small fraction of matter in the universe. This is both humbling and exciting – we are so ignorant, and there is so much more to discover!

Understanding the nature of dark matter is one of the most important open problems in astrophysics, cosmology, and particle physics.

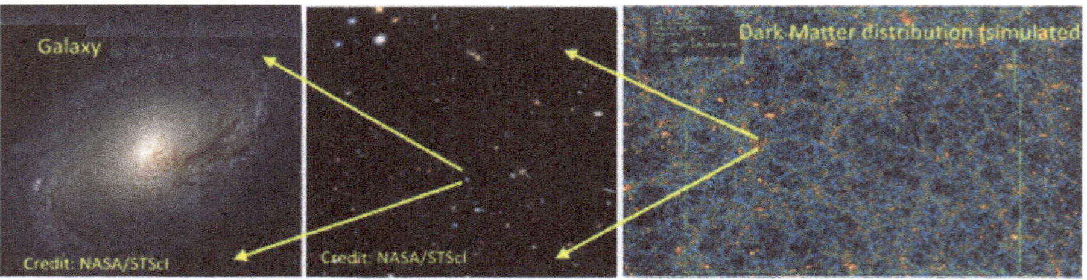

Fig.28.13: Galaxies are the visible part of matter distribution in the universe (left and middle). These structures are formed and embedded in dark matter haloes, which form an invisible cosmic web (right). Curtsy – NASA/STcI

Dark matter interacts extremely weakly with normal matter, as it may not interact at all; it does not emit light, and light passes through dark matter without being scattered or absorbed.

Therefore, it is challenging to detect and study dark matter directly, even though there should be much too much presence of dark matter around.

Many groups worldwide are building large detectors hidden underground to detect dark matter directly, with the hopefully, a dark matter particle scatters with the normal matter in the detector. While experiments like Xenon1T have seen some interesting possible signals, no conclusive detection has been confirmed yet, Figure 28.13.

Other groups try to produce dark matter in high-energy proton-proton collisions using the *Large Hadron Collider (LHC)* at CERN, the most powerful particle accelerator in the world.

However, the detectors at the LHC were not designed to detect dark matter directly. Physicists could register dark matter production only by noticing the missing energy and momentum since the dark matter would escape the detectors carrying some energy and momentum, Figure 28.14. Thus far, direct detection and production experiments have put more and more constraints on dark matter properties such as its mass and interaction probability. Nonetheless, still, there's no sighting of dark matter yet.

Fig.28.14: An ATLAS detector event (cross sectional view) with missing momentum/energy. A photon with some detected momentum (yellow line) is balanced by the missing momentum (red dashed line) possibly carried by a dark matter particle. (Image Curtsy: ATLAS/CERN)

How can one study invisible matters? Well, visible matters should reveal dark matter's presence and properties. In other words, normal matters – stars, galaxies, etc. – are nature's dark matter detectors.

In standard astronomy, there is a detailed discussion of the structures of stars and their evolutions, but dark matter is ignored. What would happen if some dark matter became mixed into stars?

We may calculate how stars evolve if they are admixed with different kinds and amounts of dark matter. In particular, many stars die as compact stars – white dwarfs, neutron stars, and black holes, which pack about a solar mass into spheres with radii reaching extraordinary densities and strengths of gravity.

There are many such interesting objects in the Milky Way. For example, Sirius is the brightest star in the night sky, Figure 28.15, and its companion star, Sirius B, is a white dwarf with a mass about the same as the Sun but the size of Earth. We found that if dark matter consists of massive particles, protons or heavier, just a few percent of a solar mass of dark matter admixed into a white dwarf will shrink the star from Earth-sized to Moon-sized!

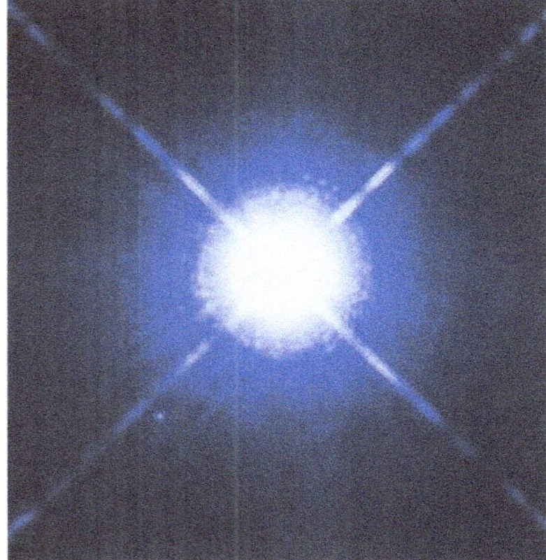

Fig.28.15: Sirius and Sirius B. By NASA, ESA, H. Bond (STScI), and M. Barstow (University of Leicester).

Furthermore, there is a maximum mass for white dwarfs, calculated by Chandrasekhar as a teenager to be 1.4 solar mass. A white dwarf reaching this Chandrasekhar limit would become unstable and explode as a Type IA supernova (SNIA).

Supernovae are among the most powerful explosions in the universe. The brightness of SNIA changes rapidly in days, called the light curve, and is believed to be universal and standard since all SNIA start from the same initial condition, that of a white dwarf with 1.4 solar mass. We found that the admixture of a small amount of heavy dark matter would significantly reduce the Chandrasekhar limit, and the SNIA would be much dimmer. Our model could fit the light curves of some of the peculiar SNIA observed, Figure 28.16.

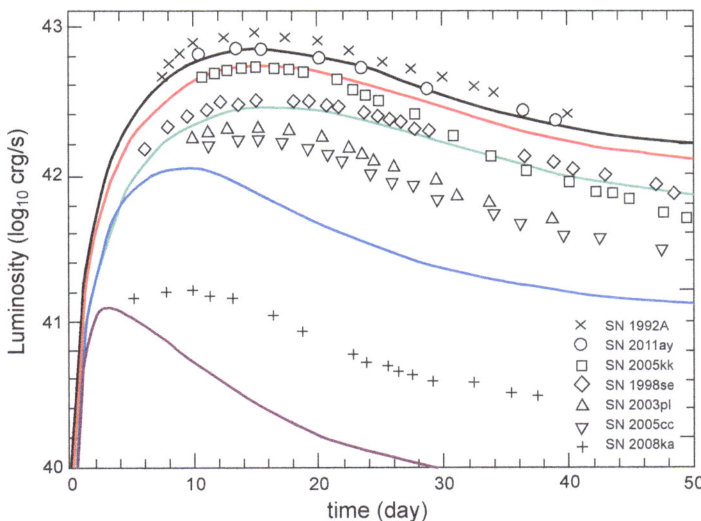

Fig.28.16: SNIA light curves with 0 (black line), 1 (red), 2 (green), 3 (blue), and 3.2% (purple) solar mass of heavy dark matter admixed, compared to observational data for 7 SNIA.

In theory, a white dwarf can also go through gravitational collapse to form a neutron star if the electrons in the star are captured by the atomic nuclei fast enough. This so-called Accretion-Induced Collapse (AIC) has not been conclusively observed yet. We studied how dark matter admixture would affect AICs. We predicted the gravitational wave and neutrino signals of such events using simulations to give suggestions for identification of AICs, with and without dark matter admixed, in future observations, Figure 28.16.

On much larger scales, dark matter provides the gravity to form structures in the universe. Large-scale simulations of structure evolution have demonstrated the ΛCDM model, in which the dark matter is stable and 'cold,' or with small velocity dispersion, predicts large-scale structures that agree well with observation, Figure 28.17. However, in recent years, there have been discrepancies between this model and observations on small scales. Therefore, many physicists are now searching for alternative dark matter models.

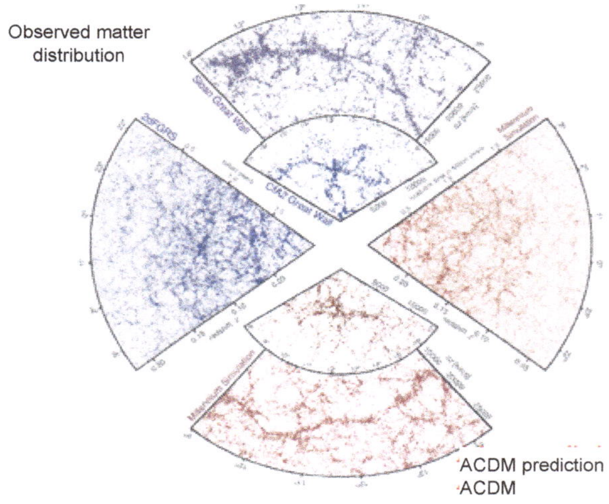

Fig.28.17: Comparison of observed large-scale matter distribution in the universe to simulation result based on the ΛCDM model. Taken from V. Springel, C. Frenk & S. White, Nature 440, 04805 (2006).

A large-scale simulation was performed to study how different dark matter properties affect the structure evolution in the universe. Using models such as the ΛCDM, Decaying Dark Matter (DDM), and Fuzzy Dark Matter (FDM) models, we can simulate how dark matter haloes are formed and study their structures. We can then compare our theoretical results with observation to test dark matter models Figure 28.17.

There will be many breakthroughs in dark matter research in the coming years, and these will likely bring even more new mysteries, Figure 28.18. We are only at the beginning of human beings' quest to understand the matter and structures in the universe.

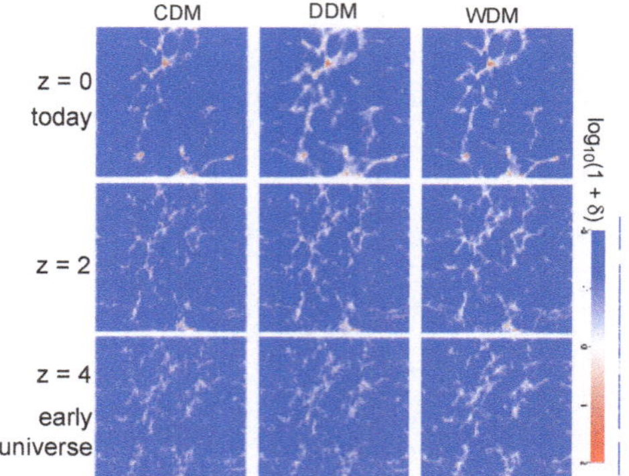

Fig.28.18: Evolutions of matter density distribution in cold dark matter (CDM), decaying dark matter (DDM), and warm dark matter (WDM) models. The redshift z indicates time: z = 0 is today, and z = 4 is early universe.

Novel Perception on Dark Matter

Dark Matter/Dark Energy represents in excess of 96% of the Universe, while the visible Universe represents only a little less than 4% of the Universe. Strangely enough the prominent scientists of old and as well as of new age have not been able to discover the true identity of the Dark Matter/Dark Energy. While the advancement of technology and the explosive information gained by the current man have exceeded by many folds the expectations and the dreams of our ancestors. The current world-wide information is doubled every twelve ours, while our ancestors pool of information by the end of World War II knowledge was doubling every 25 years. Still, we are no closer than them in finding the true knowledge and the identity of Dark Matter/Dark Energy.

As we recognize the Universe is held by such entities. This is the believes of all the categories of scientists, the "Big Bangers," the "Evolutionists," as well as the "Creationists" alike. Since the Big Bangers, category and the Evolutionists category failed to provide a distinctive and satisfactory answers to the identity of the Dark Matter/ Dark Energy, and since many scientific research and various experiments, including the failure of the LHC in Geneva, the author as a creation scientist is providing a logical and satisfactory answer of to the nature and the identity of the Dark Matter/Dark Energy.

The Universe is held in an astonishing order. The trillions of galaxies and the billions of different stars in each galaxy are also held in an amazing order. Let alone our humble Earth, which is exceptionally finely-tuned, obviously by the same order in the same Universe.

Logically, we as scientists ought to examine such order that is holding the Universe and keeping it in a harmonious order. Allow me to provide the readers with two examples of representing the unique order that is holding the Universe with its contents. One example deals with Dark Matter/ Dark Energy spread over the land for three days. The other example has achieved almost the same of spreading of Dark Matter/Dark Energy separating light from darkness. Another example is dealing with massive energy consumed to lift up fluidic substance to a certain height and maintaining it for certain time before returning the fluidic substance to its original state.

Example A: The Land of Egypt – Except Goshen

In the historical Biblical account where the Israelite, were serving the Ancient Egyptian Pharos for about 430 years. The Israelite cried out to the Author of the Universe to save them from the hardship and the cruelty of the Pharos.

But the LORD hardened Pharaoh's heart, so that he would not let the
children of Israel go. 21And the LORD said unto Moses, ***Stretch out thine
hand toward heaven***, ***that there may be darkness over the land of Egypt***, even
darkness* which *may be felt. And Moses stretched forth his hand toward heaven;
and ***there was a thick darkness in all the land of Egypt three days***: **23*They
saw not one another, neither rose any from his place for three days*: *but all
the children of Israel had light in their dwellings.*** 24And Pharaoh called unto
Moses, and said, Go ye, serve the LORD; only let your flocks and your herds be
stayed: let your little ones also go with you. 25And Moses said, Thou must give
us also sacrifices and burnt offerings, that we may sacrifice unto the LORD our
God. 26Our cattle also shall go with us; there shall not an hoof be left behind; for
thereof must we take to serve the LORD our God; and we know not with what
we must serve the LORD, until we come thither.

Fig.28.19: The Land of Goshen in Egypt

Exodus 10:20-25:

Fig.28.20: The Pharaoh in pursuit of the Israelites after they began their exodus from Egypt
Curtsy - Stock Photo - Alamy

The Land of Goshen

The entire land of Egypt was covered with darkness, except the land of Goshen, Figure 28.19, where the Israelite resided. The sun was still shining over the land of Goshen. The people of Goshen conducted their activities as normal. However, the Author of the Universe spread His Dark Matter/Dark Energy over the land of the Egyptians that they would not see the sunrays in any form. The land of the Egyptians was shielded by Dark Matter/Dark Energy that prevented the sunrays to penetrate the shield, while the same sunrays had reached the land of Goshen as usual.

The Author of the Universe allowed the Dark Matter/Dark Energy to remove its effect once its purpose was fully accomplished according to the will of the Creator.

Example B: The Wall of Darkness

Exodus 14:1-20

The Israelite were freed from the bondage of the Egyptian Pharaoh immediately after the death of the first born of male son or animal. They departed on foot, headed towards Red Sea until they arrived to ***Baal-zephon*** at Pi-hahiroth. Once the Pharaoh-King realized that he let the Israelite depart too soon, he gathered the best of his troops and chariots of war and chased after them, Figure 28.20. The Pharaoh, his troops and his chariots of war caught up with the Israelites at the same location in ***Baal-zephon*** at Pi-hahiroth. However, the Author of the Universe erected an impenetrable shielding wall of Dark Matter/Dark Energy separating the Israelites from the Egyptians troops and their Pharaoh the King of Egypt. The side of the Egyptians was dark no light could penetrate the shielding wall of darkness. While the side of the Israelites was illuminating even during the night.

1Then the Lord said to Moses, 2"Tell the people of Israel to turn back and encamp in front of Pi-
hahiroth, between Migdol and the sea, ***in front of Baal-zephon; you shall encamp facing it, by the sea,*** Figure

28.21. [3]For Pharaoh will say of the people of Israel, 'They are wandering in the land; the wilderness has shut them in.' [4]And I will harden Pharaoh's heart, and he will pursue them, and I will get glory over Pharaoh and all his host, and the Egyptians shall know that I am the Lord." And they did so.

Fig.28.21: *Wall of Darkness between The Israelite and the Egyptian Army*

[5] When the king of Egypt was told that the people had fled, the mind of Pharaoh and his servants was changed toward the people, and they said, "What is this we have done, that we have let Israel go from serving us?" [6] So he made ready his chariot and took his army with him, [7] and took six hundred chosen chariots and all the other chariots of Egypt with officers over all of them. [8] And the LORD HARDENED THE HEART OF PHARAOH KING OF EGYPT, AND HE PURSUED THE PEOPLE OF ISRAEL WHILE THE PEOPLE OF ISRAEL WERE GOING OUT DEFIANTLY. [9] The Egyptians pursued them, all Pharaoh's horses and chariots and his horsemen and his army, and overtook them encamped at ***the sea, by Pi-hahiroth, in front of Baal-zephon***.

[10] When Pharaoh drew near, the people of Israel lifted up their eyes, and behold, the Egyptians were marching after them, and they feared greatly. And the people of Israel cried out to the LORD.

[11] They said to Moses, "Is it because there are no graves in Egypt that you have taken us away to die in the wilderness? What have you done to us in bringing us out of Egypt?

[12] Is not this what we said to you in Egypt: 'Leave us alone that we may serve the Egyptians'? For it would have been better for us to serve the Egyptians than to die in the wilderness."

[13] And Moses said to the people, "***Fear not, stand firm, and see the salvation of the LORD***, which he will work for you today. for the egyptians whom you see today, you shall never see again.

[14] the lord will fight for you, and you have only to be silent."

[15] the lord said to moses, "why do you cry to me? tell the people of israel to go forward, figure 28.22.

[16] Lift up your staff, and stretch out your hand over the sea and divide it, that the people of Israel may go through the sea on dry ground.

[17] And I will harden the hearts of the Egyptians so that they shall go in after them, and I will get glory over Pharaoh and all his host, his chariots, and his horsemen.

[18] And the Egyptians shall know that I am the LORD, WHEN I HAVE GOTTEN GLORY OVER PHARAOH, HIS CHARIOTS, AND HIS HORSEMEN."

[19] Then the angel of God who was going before the host of Israel moved and went behind them, and the pillar of cloud moved from before them and stood behind them,

[20] coming between the host of Egypt and the host of Israel. And there was the cloud and the darkness. And it lit up the night[b] without one coming near the other all night.

Fig.28.22: *Pi-hahiroth, in front of Baal-zephon*

Example B: Crossing the Red Sea

21 Then ***Moses stretched out his hand over the sea, and the LORD DROVE THE SEA BACK BY A STRONG east wind all night and made the sea dry land, and the waters were divided,*** Figure 28.23. 22 And the people of Israel went into the midst of the sea on dry ground, ***the waters being a wall to them on their right hand and on their left***. 23 The Egyptians pursued and went in after them into the midst of the sea, all Pharaoh's horses, his chariots, and his horsemen. 24 And in the morning watch the LORD IN THE PILLAR OF FIRE AND OF CLOUD LOOKED DOWN ON THE EGYPTIAN FORCES AND THREW THE EGYPTIAN FORCES INTO A PANIC, 25 clogging[c] their chariot wheels so that they drove heavily. And the Egyptians said, "Let us flee from before Israel, for the LORD FIGHTS FOR THEM AGAINST THE EGYPTIANS."

Fig.28.23: *Moses Parting he Red Sea – Curtsy Adobe Stock*

26 Then the LORD SAID TO MOSES, "Stretch out your hand over the sea, that the water may come back upon the Egyptians, upon their chariots, and upon their horsemen." 27 So Moses stretched out his hand over the sea, and the sea returned to its normal course when the morning appeared. And as the Egyptians fled into it, the Lord threw the Egyptians into the midst of the sea. 28 The waters returned and covered the chariots and the horsemen; of all the host of Pharaoh that had followed them into the sea, not one of them remained. 29 But

the people of Israel walked on dry ground through the sea, the waters being a wall to them on their right hand and on their left.

Parting The Red Sea on a Dry Land

Once the morning dawn arrived the Israelite saw that Red Sea' massive amount of water was Miraculously lifted up forming a pathway to the other side of the Sea. The Israelite quickly began to cross the Sea walking on a dry land, Figure 28.24, as the sea-floor was dried due to an East wind that kept blowing all night. Also, the Dark Matter/Dark Energy shielding-wall was removed allowing the Egyptians to see the Israelite crossing the at a distance.

Fig.28.24: The Israelite Crossing the Red Sea

The Egyptian army decided to chase after the Israelite crossing the same dry land with all their chariots. As soon as the army and its chariots of war reached the middle of the crossing distance of the Red see the walls of water closed down on the entire army with their chariots causing them all to be perished, Figure 28.25.

Fig.28.25: The Egyptian Army Drowned in the Red Sea

What is Dark Matter/Dark Energy:

The author believes it is the same power that is holding the Universe together and keep watching upon the Earth since the moments of its creation. It is the Lord's consistent power and entity. It is the Lord Jesus Christ who occupies His created Universe and far beyond. He is the infinity and container of eternity. It stated in Colossians 1:15-17 He is the visible image of the invisible God. He existed before anything was created and supreme over all creation. For through Him God Created everything in the heavenly realm and on earth. He made the things we can see and the things we cannot see—such as thrones, kingdoms, rulers and authorities in the unseen world. Everything was created through Him and for Him. He existed before anything else, and He **HOLDS** all creation together.

Indeed, He holds everything in the Universe, as well as the Universe itself to maintain everything in harmony and in perfect order.

APPENDIX

i. For example, recently published figures show this to be true of the UK, Australia and the USA: Anon, Suicide, Mental Health Foundation (UK), mentalhealth.org.uk, 2016; Anonymous, Suicide in Australia, Australian Bureau of Statistics, abs.gov.au, September 2017; VanOrman, A. & Jarosz, B., Suicide replaces homicide as second-leading cause of death among U.S. teenagers, Population Reference Bureau, prb.org, June 2016.

ii. A cancer that commonly attacks the pleural layers of tissue around the lungs; less commonly it affects the peritoneum, a layer of abdominal tissue which surrounds various organs of the digestive system.

iii. In fact, he slightly misquoted Scripture here for Ecclesiastes 3:1–2 reads, "For everything there is a season, and a time for every matter under heaven: a time *to be born*, and a time to die…"

   a. Gould preferred to be described as an agnostic, although some of his closest friends and associates considered him an atheist. Nevertheless, he did acknowledge the faith of many of his colleagues.

   b. There is an overlap with the related idea of providence, God's intervention, protective care and orchestration of the affairs of human beings.

   c. In the original Hebrew, הָיָה שֶׁפֶּן.

REFERENCES AND NOTES

1. Burrell, T., A meaning to life: How a sense of purpose can keep you healthy, newscientist.com, 25 January 2017.
2. Howard, J., *Death on Earth: Adventures in evolution and mortality*, Bloomsbury Sigma, London, 2016.
3. Bloomsbury Publishing, *Death on Earth*, bloomsbury.com; accessed 1 August 2017.
4. The British edition of this book was subtitled, *The case against religion*. The American edition was subtitled, *how religion poisons everything.*
5. For a more detailed perspective on this interview, see: Bell, P., Christopher Hitchens: Staring death in the face, creation.com/facing-death, 22 December 2011.
6. *Paxman meets Hitchens: A Newsnight Special*, BBC Two, broadcast 29 November 2010.
7. Sarfati, J., Evolution makes atheists out of people! creation.com/makes-atheists, 17 February 2015.
8. Gould, S.J., The median isn't the message, *Discover*, June 1985.
9. *Darwin: The Voyage that shook the world*, DVD documentary movie, Fathom Media, for Creation Ministries International, 2009.

10. Darwin, C., *On the Origin of Species by Means of Natural Selection, or the Preservation of Favoured Races in the Struggle for Life*, 1st edition, John Murray, London, 1859, p. 490.
11. Attenborough, D. (Narrator), *The Hunt: Nothing is certain*, BBC One. In six-parts, part 1 was first screened on 1 November 2015.
12. Darwin, C., Letter to his friend and botanist, Joseph D. Hooker, 13 July 1856, www.darwinproject.ac.uk.
13. Noorderlicht Radio, 18 May 2004, noorderlicht.vpro.nl. Quoted from a transcript of the programme made at that time; see: Bell, P., The humanist apostles' creed—that which 'a Devil's Chaplain might write', creation.com/the-humanist-apostles-creed, 4 October 2004.
14. Dawkins, R., *River out of Eden: A Darwinian view of life*, Weidenfeld & Nicolson, London, 1995, p. 96.
15. Mahoney, J., *Christianity in Evolution: An exploration*, Georgetown University Press, Washington DC, 2011, p. 63.
16. Alexander, D., *Creation or Evolution: do we have to choose?* 2nd edition, Monarch Books, Oxford, UK, 2014, p. 218.
17. Ambrose, T., 'Just a pile of old bones', *The Church of England Newspaper*, 21 October 1994.
18. The outline order for funerals, *Book of Common Prayer*, churchofengland.org/prayer-worship/worship/texts/pastoral/funeral/funeral.aspx.
19. Schreiner, T.R., Original Sin and Original Death, chapter 13 of: Madueme H. & Reeves M. (Eds.), *Adam, The Fall, and Original Sin: Theological, biblical, and scientific perspectives*, Baker Academic, Grand Rapids, MI, 2014, p. 284.
20. The whole question of whether Jesus died spiritually at his crucifixion is well beyond the scope of our present discussion but the following article is a helpful place to begin: Stegall, T.L., Did Christ die spiritually and physically? *Grace Family Journal* 86, Summer 2017; gracegospelpress.org, accessed 3 August 2017.
21. This section is developed from: Bell, P., The problem of evil: pre-Fall *animal* death? creation.com/pre-fall-animal-death, 29 March, 2011.
22. Cell apoptosis is pictured and discussed in: Bosanquet, A.G. & Bell, P.B., Enhanced ex vivo drug sensitivity testing of chronic lymphocytic leukaemia using refined DiSC assay methodology, *Leukemia Research* 20(2):143–153, 1996.
23. Bell, P., Apoptosis: Cell 'death' reveals Creation, *Journal of Creation* 16(1):90–102, April 2002.
24. Calvin, J., *A Commentary on Genesis*, The Banner of Truth Trust, Edinburgh, UK, 1965, p. 250.
25. *Life and Death*, the third of three programs, was first screened on 17 March 2009. For a review of this series see: Bell, P., The BBC TV series *Darwin's Dangerous Idea*, creation.com/dangerous-idea, 29 April 2009.
26. Black Dog Days—The experience and treatment of depression, Life Matters with Norman Swan, ABC (Australia) Radio, 4 May 2000.
27. Anonymous, Pulling power, *New Scientist* 171(2310):32, 2001.
28. Bernitt, R., Stellar evolution and the problem of the 'first' stars, *Journal of Creation* 16(1):12–14, 2002.
29. Zeilik, M., *Astronomy—The Evolving Universe,* 8th Ed., John Wiley and Sons, New York, pp. 260–261, 1997.
30. See quora.com/What-is-the-average-distance-of-a-single-electron-in-the-first-orbital-of-a-hydrogen-atom-relative-to-the-size-of-the-nucleus.

THE MORAL LAWS

THE CHRISTIAN FOUNDATIONS - THE RULE OF LAW

According to the tradition of the rule of law in the Western world, to be under law presupposes the existence of certain laws serving as an effective check on arbitrary power, Figure 29.1. The rule of law is therefore far more than the mere existence of positive laws, as it also requires the state to act in accordance with principles of a 'higher law'. The search for such 'higher law' implies, however, a moral discussion on what laws ought to be. If so, the rule of law becomes an impracticable and even undesirable achievement for societies not subject to certain patterns of cultural and religious behavior. On the other hand, any radical change in such patterns can certainly produce undesirable consequences for the realization of the rule of law.

Fig.29.1: Parliament Building – United Kingdom - Diliff, Wikimedia Commons (CC BY-SA) 2.5

The Bible has been historically recognized as the most important book for the development of both the rule of law and democratic institutions in the Western world. However, we have seen over these last decades a deep erosion of individual rights, with the growth of state power over the life and liberty of individuals.

If the future we want for ourselves and our future generations is one of freedom under law, not absolute subjection to the arbitrary will of human authorities, we will have to restore the biblical foundations for the rule of law in the Western world. As such, the rule of law talks about the protection of the individual by God-given liberties, rather than by an all-powerful, law-giving government endowed by god-like powers over the civil society.

CHRISTIANITY FOUNDATION OF THE INDIVIDUAL RIGHTS

The modern roots of our individual rights and freedoms in the Western world are found in Christianity. The recognition by law of the intrinsic value of each human being did not exist in ancient times. Among the Romans, law protected social institutions such as the patriarchal family but it did not safeguard the basic rights of the individual, such as personal security, freedom of conscience, of speech, of assembly, of association, and so forth. For them, the individual was of value 'only if he was a part of the political fabric and able to contribute to its uses as though it were the end of his being to aggrandize the state'. According to Benjamin Constant, a great French political philosopher, it is wrong to believe that people enjoyed individual rights prior to Christianity. In fact, as Fustel de Coulanges put it, the ancients had not even the idea of what it means.

Fig.29.2: Emperor Theodosius Curtsy Britannica

In 390, Bishop Ambrose, who was located in Milan, forced Emperor Theodosius to repent of his vindictive massacre of seven thousand people, Figure 29.2. The fact indicates that under the influence of Christianity, nobody, not even the Roman emperor, would be above the law. And in the thirteenth century, Franciscan nominalists were the first to elaborate legal theories of God-given rights, as individual rights derived from a natural order sustained by God's immutable laws of 'right reason'. For medieval thinkers, not even the king himself could violate certain rights of the subject, because the idea of law was attached to the Bible-based concept of Christian justice.

CHRISTIANITY THE RULE OF LAW - INDIVIDUAL LIBERTY

The notion that laws and liberty are inseparable is another legacy of Christianity. Accordingly, God's revealed will is regarded as the 'higher law,' and therefore placed above human law. Then liberty is found under the law, God's law, because as the Bible says, 'the law of the Lord is perfect, reviving the soul' (Ps 19:7). If so, people have the moral duty to disobey a human law that perverts God's law, for the purpose of civil government is to establish all societies in a godly order of freedom and justice.

St Augustine of Hippo, Figure 29.3, once wrote that an unjust law is a contradiction in terms. For him, human laws could be out of harmony with God's higher laws, and rulers who enact unjust laws were wicked and unlawful authorities.

Fig.29.3: St. Augustine of Hippo – Curtsy Saints & Angels - Catholic Online

In *The City of God*, St Augustine explains that a civil authority that has no regard for justice cannot be distinguished from a band of robbers. 'Justice being taken away, then, what are kingdoms but great robberies? For what are robberies themselves, but little kingdoms?'

In the same way, St Thomas Aquinas considered an unjust law a 'crooked law,' and, as such, nobody would have to obey it. For St Aquinas, since God's justice was the basic foundation for the rule-of-law system, a 'law' that prescribed murder or perjury was not really law, for people would have the moral right to disobey unjust commands. Rulers who enact unjust 'law' cease to be authorities in the rightful sense, becoming mere tyrants. The word '*tyranny*' comes from the Greek for 'secular rule,' which means rule by men instead of the rule of law.

US NATIONAL ARCHIVES & RECORDS ADMINISTRATION

.Barons compelled King John to sign the Magna Carta at Runnymede, England, Figure 29.4, on June 15th, 1215. The charter underlies basic rights of the individual according to the Christian worldview.

In declaring the equality of all human souls in the sight of God, Christianity compelled the kings of England to recognize the supremacy of the divine law over their arbitrary will. 'The absolutist monarch inherited from Roman law was thereby counteracted and transformed into a monarch explicitly under law.'

At the time of Magna Carta (1215), a royal judge called "Henry de Bracton" (d. 1268), Figure 29.5, wrote a massive treatise on principles of law and justice. Bracton is broadly regarded as 'the father of the common law', because his book "*De legibus et consuetudinibus Anglia*" is one of the most important works on the constitution of medieval England. For Bracton, the application of law implies 'a just sanction ordering virtue and prohibiting its opposite,' which means that the state law can never depart from God's higher laws. As Bracton explains, jurisprudence was 'the science of the just and unjust'. And he also declared that the state is under God and the law, 'because the law makes the king. For there is no king where will rules rather than the law.'

Fig.29.4: *The Magna Carta (1215)*

Fig.29.5: Henry de Bracton - British Jurist - Curtsy Britannica

The Christian faith provided to the people of England a *status libertatis* (state-of-liberty) which rested on the Christian presumption that God's law always works for the good of society. With their conversion to Christianity, the kings of England would no longer possess an arbitrary power over the life and property of individuals, changing the basic laws of the kingdom at pleasure. Rather, they were told about God's promise in the book Isaiah, to deal with civil authorities who enact unjust laws (*Isaiah 10:1*).

In fact, the Bible contains many passages condemning the perversion of justice by them (*Prov 17:15, 24:23; Exo 23:7; Deut 16:18; Hab 1:4; Isa 60:14; Lam 3:34*). In explaining why, the citizens of England had much more freedom than their French counterparts, Charles Spurgeon (1834–1892) declared:

'There is not land beneath the sun where there is an open Bible and a preached gospel, where a tyrant long can hold his place ... Let the Bible be opened to be read by all men, and no tyrant can long rule in peace. *England owes her freedom to the Bible*; and France will never possess liberty, lasting and well-established, till she comes to reverence the Gospel, which too long has rejected ... The religion of Jesus makes men think, and to make men think is always dangerous to a despot's power.'

REASONS FOR A CIVIL GOVERNMENT

The first reference to civil government in the Holy Scriptures is found in *chapter 9 of the book of Genesis*. In this chapter, God commands capital punishment for those who take the life of human beings, who are always created in the image of God. In this sense, the right to execute murderers does not belong to government officials themselves, but to God who is the author of life and commands the death penalty for murder in several passages of the Holy Scriptures (e.g., *Exod 21:12; Num 35:33*). Thus, life can only be taken away from the individual if civil authorities apply it under God's law and commission, as the sanctity of human life is the basis on which God sanctions capital punishment. As John Stott, Figure 29.6, explains:

Fig.29.6: Life Lessons from One of John Stott's Final Letters – Curtsy Langham Partnership USA

'Capital punishment, according to the Bible, far from cheapening human life by requiring the murderer's death, demonstrates its unique value by demanding an exact equivalent to the death of the victim.'

The state is a 'necessary evil' that has to be subject to God's higher laws. After sin entered in the world, it became necessary to establish the civil government in order to curb violence (*Gen 6:11–13*). However, the state was not envisaged in God's original plan for mankind, as it places some people in a position of authority over others. At the beginning of the creation, however, Genesis tells us that man and woman lived in close fellowship with God, under His direct and sole authority. Then Thomas Paine (1737–1809), a non-Christian himself, expressed the biblical worldview when he uttered these words:

'Government even in its best state is but a necessary evil; in its worst state an intolerable one; for when we suffer, or are exposed to the same miseries *by a government*, which we might expect in a country *without government*, our calamities are heightened by reflecting that we furnish the means by which we suffer.

Government, like dress, is the badge of lost innocence; the palaces of kings are built on the ruins of the bowers of paradise.'

The understanding of civil government as a result of our sinful condition justifies the *doctrine of limitation of the state powers*. It inspired, in both Britain and America, the establishment of a constitutional order based on checks and balances between the branches of government—namely *legislative, executive and judicial*. Such division obeys the biblical revelation of God as our supreme Judge, Lawgiver and King (*Isa 33:22*). Since all human beings are born of a sinful nature, the functions of the state ought to be legally checked, because no human being can be trusted with too much power.

Because God instilled in each of us a desire for freedom, *political tyranny,* ***which has infected USA lately***, and as Lord "Fortescue" (1394–1479) explained, is the attempt on the part of civil authorities to replace natural freedom by a condition of servitude that only satisfies the 'vicious purposes' of wicked rulers or ***party***.

As Fortescue put it, the law of England provided freedom to the people only because it was fully indebted to the Holy Scriptures. Thus, he quoted from *Mark 2:27* to proclaim that the kings are called to govern for the sake of the kingdom, not the opposite. In this sense, he also remarked:

'A law is necessarily adjudged cruel if it increases servitude and diminishes freedom, for which human nature always craves. USA is a recent living example crafted to deprive its citizens the true freedom of choice, such as the hidden agenda for destruction of civil liberty to impose a socialist government under the false name of "Climate Change" or "Covid 19." Also, spreading homosexuality and introducing gender binary among innocent children in an attempt to destroy family structure. The government of USA has lately coerced other branches of the government of USA, including the FBI, the CIA, the IRS, and the DOJ to unjustly punish the opposing party and subdue the general populace.

Servitude was introduced by governmental men and women for vicious purposes. But freedom was instilled into human nature by God. Hence freedom taken away from men always desires to return, as is always the case when natural liberty is denied. Therefore, he who does not favor liberty is to be deemed impious and cruel.'

If USA were to place God in its highest order, as it was when the US constitution were established, harmonious order and justice will return to back to USA. However, by placing God's higher laws above human law, Sir Edward Coke (1552–1634) considered that the basic laws of England were not designed by the state, but 'written with the finger of God in the human heart.' Coke described the constitution of England as a 'harmonious system' sustained primarily by God's higher laws. Then he went on to declare that no statute enacted by the Parliament is valid if it does not respect God and the law. Finally, Lord Coke wisely pointed out:

'In nature, we see the infinite distinction of things proceed from some unity, as many rivers from one fountain, many arteries in the body of man from one heart, many veins from one liver, and many sinews from the brain: so, without question this admirable unity and consent in such diversity of things proceeds only from God, the fountain and founder of all good laws and constitutions.'

HOW THE IMPRESSION OF 'EVOLUTION' UNDERMINES THE RULE OF LAW

The notion that human law is always subject to God's higher laws started to be more deeply challenged in the nineteenth century. After the work of Charles Darwin, belief in evolution presupposed the non-existence of God's natural moral order as a primary source of positive law. Thus, legal positivists decided to regard the positive law of the state as a mere result of sheer force and social struggle. In brief, a product of human will.

Fig.29.7: J.R. "Rushdoony" (1916 – 2001)

But, if laws are caught up in the faith of 'evolution', laws can no longer be regarded as possessing a transcendental dignity. Then the very idea of government under law loses its philosophical foundations, and, as a result, societies start to lack a moral condition of legal culture that allows them to effectively restrain the emergence of an all-powerful state. As J.R. "Rushdoony," Figure 29.7, pointed out:

'When man is made controller of his own evolution by means of the state, the state is made into the new absolute. "Hegel," in accepting social evolution, made the state the new god of being. The followers of Hegel in absolutizing the state are Marxists, Fabians, and other socialists ... In brief, God and His transcendental law are dropped in favor of a new god, the state. Evolution thus leads not only to revolution but also to totalitarianism.

SOCIAL EVOLUTION

Social evolutionary theory, as it came to focus in Hegel, has made the state the new god of being. Biological evolutionary thinking, as it has developed since Darwin, has made revolution the great instrument of this new god and the means to establishment of this new god, the scientific socialist state.'

THE STATE BECOMES A 'GOD'

Behind every legal order there is always a god, be it God Himself or those who have control over the state machinery. The state becomes a 'god' in itself if there is no ultimate appeal to higher laws and authority. Whenever the law of the state is regarded as the only source of legality, ***civil rulers become all-powerful authorities*** over the life and liberties of the individual. For no legal protection can be reasonably afforded against tyranny, if the supremacy of God's higher laws is not made to prevail. In this way, "Douglas W. Kmiec," a law professor at the University of Notre Dame, has correctly remarked:

'Views and opinions antagonistic to God's plan, whether fashioned in legislative enactment or "spontaneously" over an extended period of time in judicial decree, are hardly immutable first principles and they have led, and continue to lead, to the defeat of our happiness.'

UNIVERSE HELD BY SUPREME LAWMAKER

The complexity of things that are held together in the universe indicates the existence of a Supreme Lawmaker. As we see the world as it really is, we must concede that its motions are directed by invariable and fixed rules of law. If there are laws sustaining the world, then who has created these laws? In this regard, as "Montesquieu" commented:

'Those who assert that a *blind fatality* might have produced the various effects we behold in this world are guilty of a very great absurdity; for can anything be more absurd than to pretend that a *blind fatality* could be productive of intelligent beings?

'God is related to the universe as Creator and Preserver. The laws by which He has created all things are those by which He preserves them. He acts according to these rulers because He knows them; He knows them because He has made them; and He made them because they are relative to His wisdom and power.

'Particular intelligent beings may have laws of their own making, but they also have some which they never made ... To say that there is nothing just or unjust but what is commanded or forbidden by positive [human] laws is the same as saying that before the describing of a circle all the radii were not equal.

'We must therefore acknowledge the existence of relations of justice antecedent to positive law, and by which they are established ... If there are intelligent beings that have received a benefit of another being, they ought to be grateful; if one intelligent being has created another intelligent being, the latter ought to continue in its original state of dependence.

'But the intelligent world is far from being so well governed as the physical one. For though the former has also its laws which of their own nature are invariable, yet it does not conform to them so exactly as the physical world. This is because on the one hand intelligent human beings are of finite nature and consequently liable to error; and, on the other, their nature requires them to be free agents. Hence, they do not steadily conform to their primitive laws; and even those of their own instituting they frequently infringe.

'Man, as a physical being, is, like other bodies, governed by invariable laws. As an intelligent being, he incessantly transgresses these laws established by God and changes even the ones which he himself has established. Then, he is left to his own direction, though he is a limited being and subject like all finite intelligences to ignorance and error. And even the imperfect knowledge he has, he loses it as a sensible creature when it is hurried away by a thousand impetuous passions. Such a being might every instant forget his Creator. For this reason, God has reminded him of his obligations by the law of [the Judaeo-Christian] religion.'

GOD'S LAW IS ABOVE THE STATE LAW

The human intellect should not be our basic reference in terms of legality because everyone is affected by a sinful nature. Then, our basic legal rights should be considered the ones revealed by God Himself through the Holy Scriptures. According to the doctrine of *natura delecta*, which means that our human nature has been damaged by the original sin, law is not so much to be based on human wisdom as on God's wisdom and sovereign will. As the Bible says, 'The foolishness of God is wiser than man's wisdom, and the weakness of God is stronger than man's strength' (*1 Cor 1:25*).

The rule of law can only subsist if civil authorities are able to respect the hierarchical prevalence of God's higher laws over the state law. Although the law of God is always perfect, for God's wisdom is always perfect, human authorities are sinful creatures who might have their minds controlled by desires of the flesh. They may be slaves of sin and rebels against God, although citizens who elect sinful people and obey their wicked rulings are slaves of sin as well.

Fig.29.8: "William Blackstone" (1723–1780) biblical understanding of the rule of law.

A basic question of the rule of law is to know which sort of authority we want as the ultimate source of power over ourselves: the authority of a loving God or the authority of a sinful human ruler. If we decide for the sinful ruler, then, as "R.J. Rushdoony" puts it, 'we have no right to complain against the rise of totalitarianism, the rise of tyranny—we have asked for it.'

To avoid tyranny, "William Blackstone" (1723–1780), Figure 29.8, once declared that no human law could be valid if it contradicted God's higher laws which maintain and regulate natural human rights to life, liberty, and property. According to Blackstone's biblical understanding of the rule of law.

'No human laws should be suffered to contradict [God's] laws ... Nay, if any human law should allow or enjoin us to commit it, we are bound to transgress that human law, or else we must offend both the natural and the divine.'

THE BIBLICAL UNDERSTANDING OF LAWFUL RESISTANCE AGAINST TYRANNY

Fig.29.9: God Explicitly Commands Civil Disobedience Against the State of Tyranny

Fig.29.10: Egyptian Midwives Refused to Obey the Pharaoh's Order to Kill Hebrew Babies. They Saved Moses' Life.

When God delegates His supreme authority to human rulers, they have no liberty to use it in order to justify tyranny. In fact, there are quite remarkable examples in the Holy Scriptures where God explicitly commands civil disobedience against the state, Figure 29.9. For example, Egyptian midwives refused to obey the Pharaoh's order to kill Hebrew babies, Figure 29.10. As the Bible says, '[they] feared God and did not do what the king of Egypt told them to do' (*Exod 1:17*). Likewise, three Hebrews did not obey Babylon's King Nebuchadnezzar, when he commanded everyone to bow down and worship his golden image (Dan 6), Figure 29.11.

Fig.29.11: *Three Hebrews Thrown in Fire as they did not obey Babylon's King Nebuchadnezzar – Jesus Appeared in the Midst*

Daniel also refused to obey a decree enacted by King Darius, which forced everyone not to pray to any god or men except to himself, Figure 29.12.

Fig.29.12: *Daniel in the Den of Lions*

In the New Testament, we have the example of the first Apostles' attitude towards the Sanhedrin, a Jewish council of priests and teachers of the law. The council ordered them not to preach in the name of Christ Jesus. However, the Book of Acts says that the Apostles refused to obey their decision, Figure 29.11, and, as the Apostle Peter boldly declared, 'We must obey God rather than human authority' (*Acts 5:29, NLT*). In fact, the zeal of the Apostles for the Lord was so great that they refused to be silenced by unfair rulers, even if such a refusal resulted in arrest and/or execution.

Fig.29.13: *Sanhedrin Council Ordered the Apostles not to Preach. Apostles Refused to Obey*

They considered themselves bound by God's Law in the first place, and kept on preaching the Gospel as if it were no legal prohibition, Figure 29.13. To be obeyed, therefore, civil authorities have firstly to obey God and the law. As "John Stott" has pointed out:

'If the state commands what God forbids, or forbids what God commands, then our plain Christian duty is to resist, not to submit, to disobey the state in order to obey God … Whenever laws are enacted which contradict God's law, civil disobedience becomes a Christian duty.'

Although the first Apostles regarded it as totally lawful to disobey ungodly legislation, today's followers of Christ like to quote from chapter 13 of Paul's letter to the Romans in order to justify their compliance with immoral rules of positive law. However, Paul argues here that we obey the civil authority because it holds 'no terror for those who do right, but for those who do wrong' (*Rom 13:3 NIV*). If the person who holds the state power abuses his or her God-given power, 'our duty is not to submit, but to resist.' According to "F.A. Schaeffer," a more accurate interpretation of this passage would clearly indicate that 'the state is to be an agent of justice, to restrain evil by punishing the wrongdoer, and to protect the good in society. When it does the reverse, it has not proper authority. It is then a usurped authority and as such it becomes lawless and is tyranny.'

God has established the state as delegated authority, not an autonomous power above the law. When we obey the state, it is not that we obey individuals who are in charge of the state machinery, but it is rather for obedience to a God-given authority who is commanded by God to promote natural principles of liberty and justice. Therefore, as Pope "John XXIII" explains in his encyclical 'Pacem in Terris':

'Since the right to command is required by the moral order and has its source in God, it follows that, if civil authorities pass laws or command anything opposed to the moral order and consequently contrary to the will of God, neither the laws made nor the authorizations granted can be binding on the consciences of the citizens, since *God has more right to be obeyed than men*. Otherwise, authority breaks down completely and results in shameful abuse.'

Because Paul also says that the Word of God is not to be bound (*2 Tim 2:9 NIV*), the right of resistance against tyranny is an important element of the rule-of-law system ordained by Him. For this reason, as "John Knox" (1513–1572) put it, to rebel against a wicked ruler can be the same as to oppose the devil himself, 'who is the one abusing from the sword and authority of God.' Knox stated that anyone who dares to rule over a nation against the law of God can be lawfully resisted, even by force if necessary. According to John Knox, if the civil ruler seems to be effectively willing to destroy the Christian foundations of the society,

'[God] hath commanded no obedience, but rather He hath approved, and greatly rewarded, all those who have opposed themselves to their ungodly commandments and blind rage.'

Fig.29.14: Color image of Trumbull's Declaration of Independence taken from Declaration_ independence.jpg. public domain - Image Curtsy - library of Congress, Prints and Photographs Division, Detroit Publishing Company Collection.

"Samuel Rutherford" (1600–1661), a Scot Presbyterian like John Knox, developed in 'Lex Rex' a consistent doctrine of lawful resistance against political tyranny. According to Rutherford, if people wish to effectively stay free from such tyranny, then they will have to preserve their inalienable right to eventually disobey unjust legislation. For him, 'A power ethical, politic, or moral, to oppress, is not from God, and is not a [lawful] power, but a licentious deviation of a [lawful] power.' And in answer to the royalists who liked to use *Romans 13* in order to condemn any form of resistance against the government, as a resistance against God Himself, Rutherford boldly proclaimed:

American 'fathers' signing the U.S. Declaration of Independence at Philadelphia, 4 July, 1776, Figure 29.14. The document states as a 'self-evident truth' that all human beings are equally endowed by God with certain unalienable rights to life, liberty, and the pursuit of happiness.

'It is a blasphemy to think or say that when a king is drinking the blood of innocents and wasting the Church of God, that God, if he were personally present, would commit these same acts of tyranny.'

Fig.29.15: *"John Locke" (1634–1704), Whose Legal and Political Ideas Provided Legal Justification to the 1688 'Glorious Revolution' in Great Britain*

"John Locke" (1634–1704), Figure 29.15, whose legal and political ideas provided legal justification to the 1688 'Glorious Revolution' in Britain, argued that lawmakers put themselves into a 'state of war' against the society whenever they endeavor to destroy our God-given 'natural' rights to life, liberty and property. For Locke, no government has the right to reduce these basic rights of the individual citizen. If so, Locke argued that people would be left 'at the common refuge that God has provided for all men against force and violence.'

The American Founding Fathers fully acknowledged the principle of lawful resistance against tyranny, and drew heavily from this in order to justify their revolutionary actions against the British government, in 1776, Figure 29.16. Written by Thomas Jefferson, the US Declaration of Independence argues that revolution is the last recourse of a free people against 'a long train of abuses and usurpations' on the part of the government. Thus, they justified their actions on the grounds that God has endowed each human being with natural rights to life, liberty and the pursuit of happiness, which are basic rights that not even the state can take away from them.

Fig.29.16: *The American Founding Fathers Fully Acknowledged the Principle of Lawful Resistance Against Tyranny - 1776 The US Declaration of Independence Written by Thomas Jefferson.*

Of course, any revolutionary uprising, as "Pope Paul VI" comments in his encyclical 'Popularum Progressio,' can only be justified in extraordinary situations 'where there is a manifest, long-standing tyranny which would do great damage to fundamental personal rights and dangerous harm to the common good of the country.' However, the recourse to violence, as a means to right the wrongs of the state against the rule of law, risks itself to produce new forms of injustice. Therefore, Pope Paul VI, Figure 29.17, also stated that revolutionary uprising can only be carried out as the last remedy against long-standing tyranny, because, as he put it, 'a real evil should not be fought against at the cost of greater misery.'

Fig.29.17: Pope Paul VI Stated that Revolutionary Uprising can only be Carried out as the Last Remedy Against Long-Standing Tyranny

THE RULE OF LAW - CHRISTIANITY - HUMAN RIGHTS

According to the Judeo-Christian worldview, human beings were created by God and, as such, have never 'acquired' their basic rights from the state. Nor are such basic rights a result of any work performed by them, but it flows directly from the nature of each human being who is always conceived in the image of a loving God (*Gen 1:26*).

According to *Genesis 1:27–28*, God created all human beings, male and female, in His own image, commanding them to fill the earth and subdue it. We found here a very special meaning for the recognition of human dignity, as the result of the relationship between God and His human creatures, which the Fall has distorted but not destroyed. From this fact it follows, for instance, that widows will not be burned on their husband's funeral pyre, as they still are in India, and that people will not be sold to slavery, as they still are in Sudan.

Every year, Freedom House, a secular organization, conducts a survey to analyze the situation of democracy and human rights across the globe. Year after year, it concludes that the most rights-based and democratic nations are the majority-Protestant ones. On the other hand, Islam and Marxism, the latter a secular religion, seem to offer the most serious obstacles for the realization of democracy and human rights. In fact, the denial of the broadest range of basic human rights comes precisely from Marxist and majority-Muslim countries. The worst violators of human rights are Libya, Saudi Arabia, Sudan, Syria, Turkmenistan, and the one-party Marxist regimes of Cuba and North Korea.

In contrast to Islam, Christianity has democratized political manners, and still is the main moral force that holds democratic values together in the Western world. It provides the strongest argument for the protection of basic human rights. For "Paul L. Maier," Professor of Ancient History at Western Michigan University, 'no other religion, philosophy, teaching, nation, movement—whatever—has so changed the world for the better as Christianity has done.'

In declaring that we all stand on equal ground before God, Christianity gives the best moral foundations for social and political equality. If Christianity is found to be true, the individual, male or female, is not only more important but incomparably more important than the social body. This helps to explain, in "Charles Colson's" opinion, 'why Christianity has always provided not only a vigorous defense of human rights but also the sturdiest bulwark against tyranny.'

Moral Relativism – Progressive Abandonment

A visible fact in these days of moral relativism is the gradual abandonment of the Christian faith and culture in the Western world. As a result, the moral foundations for the rule of law have been seriously undermined. Westerners who believe that abandonment of Christianity will serve for democracy and the rule of law are blindly ignoring that such abandonment has already brought totalitarianism and mass-murder to several Western countries, particularly to Germany and Russia. Any honest analysis of contemporary Western history would have to recognize that no effective legal protection against tyranny can, in the long run, be sustained without the higher standards of justice and morality brought into the texture of Western societies by Christianity.

Westerners who disparage their Christian heritage should get much better informed that were it not for this religion, they would not have the freedoms they enjoy today, for instance, to dishonor the very source of these freedoms, namely Christianity. Regarding the present climate of multiculturalism, it would be better for them to think much more carefully on the words uttered by a great historian, "Carlton Hayes":

'Wherever Christians' ideals have been generally accepted and their practice sincerely attempted, there is a dynamic liberty; and wherever Christianity had been ignored or rejected, persecuted or chained to the state, there is tyranny.'

THE CREATION BASIS FOR MORALITY

News bulletins every day abound with stories of moral unrest somewhere in the world. The Columbine high-school massacre in USA a few years ago focused world attention on a once-average community which suffered the outrage of brutal murders. The atrocities that resulted in the deaths of thousands in the 9-11 World Trade Center terrorist acts brought to light how wicked human beings can be towards one another.

Society's laws are being violated at an ever-increasing rate, as evidenced by the rapid expansion of police forces and proliferation of all manner of 'security' devices and services to protect life and property. Clearly, our social ethics have not kept pace with our incredible achievements in science and technology. Social reform and massive government spending on new ideas to cure society's ills will be doomed to failure in the long run as society's shift away from God-based concepts continues. Why is this?

Just as with the motorists who slow down when faced with a speeding fine, society tends to become more responsible when, as individuals, we're accountable to someone. However, the idea that as human beings we are accountable to a Creator for our actions, and will face a judgment for them, has become 'outdated'. Constant bombardment via schools and the media with the evolutionary notion that we are simply the result of a cosmic accident has persuaded many that there is no higher being in the universe. Mankind is not responsible to anyone, in this evolutionary view. There will be no accounting for our actions during or after life here on Earth.

However, according to the Bible, a day is coming when all of us will stand before God and be judged for our actions while alive on Earth. No one will escape.

What a difference it makes when people believe that we will be called upon to give an account of ourselves for every situation. 'But I say to you that every idle word, whatever men may speak, they shall give account of it in the day of judgment' (*Matthew 12:36*).

When the Nazis during WWII carried out their brutal reign of terror upon innocent Jews, would they have done so if they had truly believed that they would reap a recompense for their actions? The Nazis had been taught that evolution was true and involved the concepts of *survival of the fittest and elimination of the weak*, and they were determined to apply this in real life. The ardent evolutionist Sir Arthur Keith, although an anti-Nazi, commented on Hitler's evolutionary stance:

'The German Führer, as I have consistently maintained, is an evolutionist; he has consciously sought to make the practice of Germany conform to the theory of evolution.'

The bloodthirsty regimes of Hitler and Stalin both based their philosophies on evolutionary principles. Injustice, cruelty, and merciless brutality were backed up by the ultimate evolutionary notion that all things made themselves, and that there is no higher authority which we must obey.

Evolutionist zoology professor "Ernst Haeckel," whose fraudulent drawings of embryos continue to feature in some school textbooks[2] and whose influence laid much of the foundation for Hitler's Germany, argued in his book *Natural History of Creation*, 'the church with its morality of love and charity is an effete fraud, a perversion of the natural order'. He said this was because Christianity '... makes no distinction of race or of color; it seeks to break down all racial barriers. In this respect the hand of Christianity is against that of Nature, for are not the races of mankind the evolutionary harvest which Nature has toiled through long ages to produce? May we not say, then, that Christianity is anti-evolutionary in its aim?'

Others have written concerning prewar Germany: 'The Jews, labelled subhuman, became nonbeings. It was both legal and right to exterminate them in the collectivist and evolutionist viewpoint. They were not considered ... persons in the sight of the German government.'

How different to the Bible. God's standards are love and kindness towards one another:

'... if there is any other commandment, it is summed up in this word, "You shall love your neighbor as yourself." Love works no ill to its neighbor; therefore, love is the fulfilling of the law,' (Romans 13:9–10).

The concepts of right and wrong must have an ultimate basis from which to appeal, or else they become simply relative to the culture. This is what we see happening in today's Western world.

The Bible teaches that God the Creator made men and women for each other, and experience confirms that we are made for each other physically. This was the Creator's intention and plan. And any deviation from this is outside the created order. So, it has been until very recent times, when, backed up by evolutionary 'science,' the concept of homosexual acts being 'wrong' has been changed. Now they are promoted as neither right nor wrong, but a 'choice.' And, unless one appeals to a Creator who sets the absolute laws for life, who can say this is incorrect?

If there is no Creator who has made us and set the rules, then all our morals and ideas of what is right or wrong are simply subjective—what we ourselves decide.

When a partner cheats on another in marriage, who is to say this is wrong? What is 'wrong'? When a person murders another, or steals what belongs to someone else, on what basis is it wrong? For if evolution is true, then ultimately one person killing another is no different than a lion killing an antelope—a fact of life and survival of the fittest, for mankind has no place of honor among the animal kingdom aside from that which we attribute to ourselves. How different from the Bible, wherein our Creator says we are special and above all else that He created. 'Therefore, do not fear, you are of more value than many sparrows' (*Luke 12:7*).

Biblical Codes of Life

Evolutionists invent all sorts of complicated explanations to account for why evolution can give rise to altruistic (considerate, self-sacrificing) behavior. Kindness and love towards others are among the hallmarks of the Biblical codes of life. It is not dependent on the giver receiving any benefit from the act.

Evolution Propagation of One's Genes

Evolution is a blind concept built on chance random events, with no purpose other than, supposedly, to benefit the propagation of one's genes. It does not know the end from the beginning. It incorporates the ideas of survival of the fittest, and the weak dying out or being exterminated to make way for the strong to survive. Love is totally contrary to this. Kindness that infers no benefit upon the one giving it makes little sense if evolution is true.

Missionaries helping the poor in third world countries, Figure 29.18, are completely wasting their time if evolution is true. They are actually working against the natural order of things, by their efforts to support individuals or groups that cannot survive by themselves. Under the truth as evolutionists really 'know' it, the weak, hungry and incapacitated ought to be allowed to perish if they cannot survive under their own strength.

Fig.29.18: American Missionaries Helping the Poor in Third World Countries

Darwin's Lamented

Darwin himself lamented the foolishness of society for caring for its 'weak members' Figure 29.19, and thus allowing them to 'propagate their kind,' with 'undoubtedly bad effects.' He said that *'excepting in the case of man himself, hardly anyone is so ignorant as to allow his worst animals to breed.'*

Fig.29.19: Darwin Lamented the foolishness of Society Caring for its Weak

In a debate between two evolutionists, "Jaron Lanier," a computer scientist, said, *'There's a large group of people who simply are uncomfortable with accepting evolution because it leads to what they perceive as a moral vacuum, in which their best impulses have no basis in nature.'*

And "Richard Dawkins," who is a Professor at Oxford University and an ardent atheist, replied, *'All I can say is, That's just tough. We have to face up to the truth.'*

So, consistent evolutionists who acknowledge what their philosophy is *really* saying know that an evolutionary worldview is devoid of absolute morals.

Some evolutionists try to escape the obvious consequences of their position. They suggest that, since mankind now knows his true origins as a product of chance in an impersonal universe, we can guide our own evolution. And man is of sufficiently high moral character, they say, to set the rules by which we measure compassion, love, altruism, ethics. But what is moral character anyway? And why is one person's opinion of what is right or good any more right or relevant than another's? So, who decides what is right? If evolution is true, then there are no such things as morals. There is no objective standard by which to measure what is 'good.'

Everything becomes a matter of one's opinion, unless there is an absolute source to appeal to.

Hitler decided that those he didn't like should be exterminated, and convinced the majority that he was right. So, on what basis would an evolutionist say he was wrong? If our only appeal is to what we think as a society, think of this: had the Nazis achieved their goals of world domination, would their children have grown up believing it was wrong to kill Jews?

Does this mean those holding to evolution cannot live a moral, ethical life? Not at all, and many do. But when they see unethical behavior by others, they have no grounds on which to judge that behavior as wrong. It may be their *choice* to be faithful to a marriage partner, and to do good to others. But it may be another's choice to live a life of unfaithfulness, or to take advantage of everyone around them in order to get ahead. If evolution were true, then neither position is right or wrong. Just different choices. Therefore, society as a whole gravitates to a less moral position.

So, when we tell children that it is wrong to murder, for instance, do we tell them *why* it is wrong? Is it wrong only because society currently says it's wrong, or because our Creator has told us it is wrong?

Ultimate Cause of Immoral Behavior

The ultimate cause of immoral behavior is of course sin, and not evolutionary beliefs as such. But the more a culture is infused with the belief that there is no right and wrong, ultimately, the more we can expect to see a rejection of the authority of the Bible and its proclamation of absolute values.

God Upheld the Validity of His Law

The Christian Gospel message of Salvation is meaningless without the concept of right and wrong. God the Creator imposed the (eternal, as well as physical) death penalty for infringement of His Law. In the ultimate act of love the universe has ever known, the Creator then became a man in order to take that penalty for mankind's sin. God upheld the validity of His Law by not relaxing its demands even when God the Son would be the one having to suffer its penalty, Figure 29.20. Those who believe this can thus know total forgiveness and acceptance by God on that basis.

THE TEN COMMANDMENTS

1. You shall have no other gods before me.
2. You shall not make for yourself an idol in the form of anything.
3. You shall not misuse the name of the Lord your God.
4. Remember the Sabbath day by keeping it holy.
5. Honor your father and your mother.
6. You shall not murder.
7. You shall not commit adultery.
8. You shall not steal.
9. You shall not give false testimony against your neighbor.
10. You shall not covet anything that belongs to your neighbor.

Exodus 20: 1-17

Fig.29.20: God upheld His Laws Even When God-the-Son is the one to suffer its penalty

Inconsistent Evolutionists

Some are eager to point out that in cases like the Inquisition, and more recently clergy involved in sex abuses against children, 'the church' is hypocritical. Of course, the Bible strongly condemns these and other forms of crime, and those acting in these ways are inconsistent with what the Bible teaches.

However, evolutionists who argue this way are being inconsistent with their own philosophy. Because if there is no Creator who sets absolute rules, such as 'You shall not murder' (which is what the Nazis did to the Jews, totally consistent with their evolutionary beliefs), then the only rules that exist are what we as a society make up.

A theistic evolutionist may profess belief in a rule-giving Creator, but because of his/her *de facto* rejection of a reliable revelation, how can the rules be known for certain?

Simply put, if the Bible misleads in Genesis, how do we know when it can be trusted elsewhere, including its moral teaching? Jesus said (*John 3:12*): 'I have spoken to you about earthly things and you do not believe; so how will you believe if I speak of heavenly things?'

MORALITY AND HISTORY

From very early records we see that man has shown a high degree of culture and understanding in law and moral/societal behavior. Dating from the 17th century before Christ is the Code of Hammurabi, a Babylonian king who, according to secular historians, came to power about 1750 BC. This set of laws, governing situations such as marriage, commerce and theft is generally regarded as one of the best and earliest written codes of law for a society. The proper functioning of law depends on the existence of an ultimate authority. Speaking of a society which was crumbling because of a lack of authority, the Bible says: 'In those days there was no king in Israel: every man did that which was right in his own eyes' (*Judges 21:25*).

The Ten Commandments, Figure 29.20, are considered, even by many non-Christians, to be a foundational set of rules for moral and ethical living. But if they were written by only a man, then they are no more 'right' than someone else's opposite view. In rejecting Biblical absolutes, will modern law eventually cease from allowing criminals to be branded with 'wrong-doer' in favors of the more evolutionarily consistent concept of a 'socially-unacceptable choice'? Some evolutionists have excused even rape on the grounds that males' genes and 'less civilized' evolutionary past predispose them to such actions.

As the evolutionary foundation of ***no right or wrong*** overtakes the Biblical concept of ***personal responsibility***, we can expect people to have less defined boundaries in both young and adult life, resulting in increasing social and family problems.

CHILDREN AND BEHAVIOR

Every parent knows that young children do not have to learn selfishness or misbehavior. Such conduct is innate and this is why 'No' is one of the first words babies become familiar with. An Oxford University study reported in 1999 that even children with no religious input had '... abstract notions of a Creator', thereby paving the way for a solid foundation in the concepts of right and wrong.

Increasingly, public dismay is felt worldwide at reports of children committing violent crimes. And as personal responsibility is becoming less recognized, the concept of doing 'what is right' for its own sake is becoming less practiced.

People tend to take offence at someone else telling them what they can or can't do. The answer is, however, to recognize the truth that the Creator, not man, has set rules, such as not stealing, which apply to anyone, at anytime, anywhere.

THE BIBLE - BASIS OBJECTIVE MORALITY

Many recent articles were published in USA lamenting disturbing and troubling trend: many children in school do not think there is such a thing as a 'moral fact'—otherwise known as 'moral realism.' Traditionally in Western societies, we have believed there are many moral facts—things that are objectively right or wrong. 'Murder is wrong,' 'Helping people is good,' and 'Racism is wrong' are just a few truths that express the idea that some things are objectively right or wrong.

But where do these ideas come from? Certainly not from evolution—if murder helps me survive and reproduce more than my competitors, then it is an evolutionary good. Lions often kill the cubs in a pride they've taken over, so that the pride's resources aren't 'wasted' raising another lion's cubs. Even though some have suggested that altruism might be an evolutionary good, they can't have it both ways. This is clearly inconsistent with an evolutionary worldview since helping others means fewer resources for 'me'. Why should I help people who might be less 'fit'? And racism is one of the most evolutionarily consistent views Darwin believed. If evolution is true, then the people more closely related to me are in competition with other people supposedly less closely related to us.

The sort of moral facts that have been accepted as more or less self-evident, and which form the basis of our law codes, actually come from Scripture. Western societies traditionally held the Bible, e.g., the Ten Commandments, to be the standard for how we should act. Its teachings, and the concept of God as the moral law giver, became the most overarching influence of our civilization.

Nonetheless, today, more and more people are completely ignorant of what it teaches and the important role it played in our history. And because the Bible, and in particular its foundational book—Genesis—is increasingly marginalized, so is the Bible's authority in other areas such as its morality.

EVOLUTION HAS NO OBJECTIVE BASIS FOR MORALITY

As the Bible becomes less of an influence in our American society, people start to experience confusion regarding what used to be considered basic moral principles. The result is that many decide they can make up their own rules because there are no moral absolutes. This is exactly what the aforementioned survey found amongst young people, who, no doubt, are also ignorant about the Bible, its history and impact on society. And once the foundation of Scripture is rejected it is only a matter of time until the edifice of morality built upon that foundation crumbles. Sadly, the US government overtly supports various groups that promote immorality, abortion, gender binary and tyranny among school children of all ages, accusing concerned as well conservative parents as terrorists--mainly to gain votes and maintain tyrannical control. Also, US current government stealthily encouraging to change US constitution into Marxist society.

If 'feelings' are the only thing that determines right and wrong, very much a youth sentiment today, then why accept what anyone else has to say about what is right and wrong? Moral anarchy on a societal level is the *logical* end of such a view.

Christians aren't the only ones concerned about the logical moral outcomes of evolution. History is replete with holocaustic examples, and sadly, history has a habit of repeating itself. Former Oxford Professor Richard Dawkins is probably the world's best-known anti-creation, but even he had to admit:

"I'm a passionate Darwinian when it comes to science, when it comes to explaining the world, but I'm a passionate anti-Darwinian when it comes to morality and politics."

And Dawkins has a right to be concerned. Evolutionary beliefs underpinned actions such as the forced sterilizations of the 'feeble-minded' in America, to the killing fields of Cambodia, to the murder of Aborigines in Australia under the guise of science. The idea that we are evolved has led time after time to the idea that some of us are more evolved than others, and therefore the less-evolved or fit are expendable.

Return to God's Word

If we want our children to be able to tell right from wrong, and to have an objective basis for doing so, the answer is a wholesale ***return to God's Word*** as the authority in our lives. And people of all ages must be able to defend what they believe and trust the Bible in everything it teaches. As Jesus said, *"If I have told you earthly things and you do not believe, how will you believe if I tell you heavenly things?"* (*John 3:12*). If the Bible's history is wrong, why should we believe its morality?

UNFOLDING THE PLAN

Skeptics, liberals, and others sometimes claim that man's concept of God is something which evolved, and that the Bible is merely man's efforts to provide himself with a religious prop to explain the otherwise unexplainable or to ease the burden of life. However, nothing could be further from the truth. The Bible is a book from God about God, His glory, and His plan of salvation for sinful humanity, Figure 29.21.

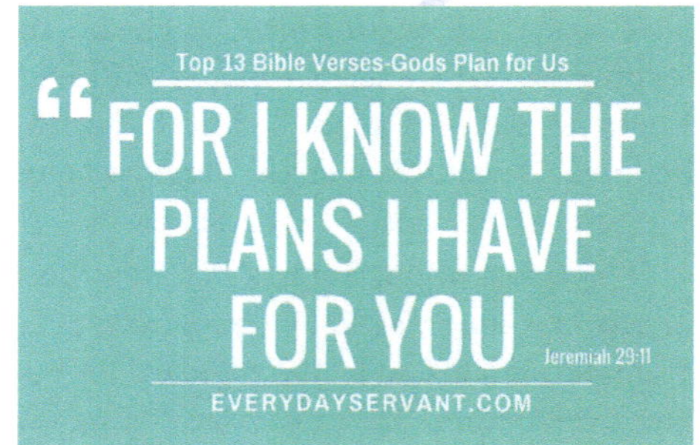

Fig.29.21: *The Bible is a Book from God about God, His glory, and His Plan for You.*

When we read the Bible, we find that God does not tell us everything about these things all at once. He gives us successive revelations. A name for this is 'progress of doctrine', which simply means that we learn more about God and His dealings with us from each book, as we read through the Bible.

This concept can be likened to the raising of a blind in a dark room. Outside the sun is shining. As the blind goes up, it does not increase the amount of light emanating from the sun, but it does let more and more light into the room. Let us see how this works out with regard to four things that God tells us about Himself in Genesis, thus affirming the crucial nature of this foundational book.

I. God is Creator

The first thing God tells us about Himself in Genesis is that He is *Creator*. The very first verse of the Bible reads: 'In the beginning God created the heaven and the earth' (*Genesis 1:1*). The rest of *chapter 1* details what God created and how He did it, namely by His Word, as well as over what time frame.

As we read through the Bible, we learn that 'the Word' is a title given to Jesus Christ (*John 1:1-14*), and that it was through Jesus that everything was created (*John 1:3; Colossians 1:16; Hebrews 1:2*).

We also learn, following the death and resurrection of the Lord Jesus, that God began a new creation—those who repent and put their faith in Jesus Christ are described as being a 'new creature' [new creation] (*2 Corinthians 5:17*).

Finally we learn that, at some time in the future, God will create a new heaven and a new earth, as prophesied in *Isaiah 65:17*, and described in Revelation chapters *21-22*.

II. God is Lawgiver

A second matter God tells us about Himself in Genesis is that He is *Lawgiver*, when He says to Adam: 'Of every tree of the garden thou mayest freely eat: But of the tree of the knowledge of good and evil, thou shalt not eat of it: for in the day that thou eatest thereof thou shalt surely die' (*Genesis 2:16–17*).

Fig.29.22: *Moses Delivering the Ten Commandments of God*

Practically this was a test of Adam's love for God, that would be shown by whether or not he obeyed this one command. In essence, however, this was the first declaration to humanity of the moral law of God and of God's right to impose His law upon mankind. The moral law of God may be defined as 'the expression of God's will, enforced by His power, for His rational beings (angels and men).'

Someone may ask, 'What right has God to impose His moral law upon mankind?'

The answer is that God is Sovereign, and whatever He does is right. The fact that God is man's Creator gives Him the *right* to be man's Lawgiver (*Ecclesiastes 7:29; Ephesians 4:24*), Figure 29.22. The fact that 'God is love' (*1 John 4:8*) means that His laws are in our best interests.

Jesus summed up God's moral law for His followers in the words: 'Thou shalt love the Lord thy God, with all thy heart, and with all thy soul, and with all thy mind … and … thou shalt love thy neighbor as thyself,' citing Deuteronomy 6:5 and *Leviticus 19:18* (*Matthew 22:37-39*).

III. God is Judge

A third matter that God tells us about Himself in Genesis is that He is *Judge*. After Adam and Eve had eaten the forbidden fruit, the record states:

- 'And the Lord God said unto the serpent, because thou hast done this, thou *art* cursed …
- Unto the woman He said, I will greatly multiply thy sorrow … And
- unto Adam He said … cursed *is* the ground for thy sake … and
- unto dust shalt thou return' (*Genesis 3:14-19*). This role of God as Judge is seen throughout the Bible.
- In *Genesis 6–9*, God judged the wickedness of the people of that day with the Flood.
- In Genesis 11, by causing the confusion of languages, God judged the people of Babel, who had refused to obey His command to fill the earth (Genesis 9:1).
- In Genesis 18-19, God judged the blatant homosexual behavior of the men of Sodom and Gomorrah—'because their sin is very grievous' (Genesis 18:20) and they 'committed abomination' (Ezekiel 16:50)—by destroying those cities and everyone in them.
- In Exodus 5-14, God judged the nation of Egypt with various plagues for its refusal to obey the Lord and let His chosen people go.
- In the Old Testament, God judged the nation of Israel for its idolatry.

- In the Gospels, the death of the Lord Jesus on the Cross is the expression of God's judgment upon mankind's sin. Christ, our substitute, paid the full penalty which we, each one, deserve.[10]
- Further on in the New Testament, God tells us that He has appointed a day in which He will judge the world in righteousness and that the judge will be the person whom He raised from the dead, namely the Lord Jesus Christ (Acts 17:31; Romans 14:10).
- Finally, in Revelation, the last universal judgment of 'the dead, small and great' is seen in what the Bible calls the 'great white throne' judgment (Revelation 20:11-15).

IV. God is Savior

A fourth matter that God tells us about Himself in Genesis is that He is *Savior.* Amid the judgment set out in Genesis chapter 3, the Lord God gave the promise of the Savior, Figure 29.23, who would come and defeat Satan, while He Himself (in the person of Jesus Christ) would suffer in the process—the so-called 'Protevangel' spoken to the serpent: 'And I will put enmity between thee and the woman, and between thy seed and her seed; it shall bruise thy head and thou shalt bruise his heel' (Genesis 3:15).

Fig.29.23: *The Wage of Sin is Death*

The role of God as Savior, in the person of Jesus Christ, is a major theme of the Bible. The Old Testament points forward to this through many 'types,' such as the sacrificial lambs, the priesthood, the tabernacle, etc. Then in the Gospels, we have the historical record of the birth, life, death, resurrection, and ascension of the Savior, Jesus Christ. This was God's fulfillment of His promise of Genesis 3:15.

Finally, after the outworking of all that was involved in the missionary task of teaching, preaching, and planting churches as exemplified in the rest of the New Testament, Revelation portrays the ongoing triumph of the Creator/Lawgiver/Judge/Savior Jesus Christ over the powers of evil and His enthronement with His redeemed people, first in Heaven, and later in Heaven-on-earth (Revelation 21). Thus, what God began at the commencement of the Bible in Genesis 1 is fulfilled, consummated, and perfected at the end of the Bible in Revelation 21-22.

STRUGGLING TO HARMONIZE SCRIPTURE WITH EVOLUTION

Atheists use the Bible's progress of doctrine to justify their belief that Christianity itself evolved in the mind of man, with doctrine becoming more sophisticated as people and their culture evolved.

A confused evolutionist speaker claimed that progress of doctrine demonstrated that evolution was a fact. 'As man's brain grew bigger and bigger,' he said, '*he gradually understood more and more of what God was trying to tell him.*'

The Bible is not just a book about 'religious ideas or emotions, it is a book about God, and His dealings with humanity and humanity's home (the universe), from the beginning to the end of time. The early chapters of Genesis are crucial to a proper understanding of these dealings. These chapters are totally reliable and may not be manipulated to make them conform to the temporary wisdom of any particular age.

To the person willing to submit to its authority, the Bible, revealing the consistent unfolding of God's plan, shows itself to be a self-authenticating, utterly trustworthy whole, from beginning to end.

GOD, MIRACLES, AND LOGIC

Why do atheists call the miracle-working God a '*magic man in the sky*'? And how do atheists account for the laws of logic?

Fig.29.24: Why do atheists call the miracle-working God a 'magic man in the sky'? And how do atheists account for the laws of logic?

God does not perform miracles arbitrarily; He uses miracles in His providential care of history in a purposeful and logistically careful way, Figure 29.24.

Nearly every skeptic we have ever encountered had the same problem with believing in the Bible. It was God performing miracles, which they in turn use to equate God with a "magic man in the sky," even though the sky is part of finite creation and God exists outside of creation.

When atheists call God imaginary, are they also calling the laws of logic and unchanging moral laws imaginary as well?

When an atheist misrepresents God as a '*magic man in the sky*', it's not an argument; it's an insult. But there's an unvoiced belief behind it that can be put into an argument: if God does miracles, God is arbitrary, and we therefore couldn't trust that science would work if God existed. But this is nonsense. God is rational. He's also *unchangeably* rational.

There are also reasons for God to sustain the world in ways that seem to us 'immovable.' For instance, God would sustain the world so we could make choices in it. But to be able to make choices, we need to be able to expect our choices to result in consistent consequences. For instance, the choice to walk requires the world to behave in consistent ways for us to be able even to put one foot in front of another. If we couldn't expect a consistent consequence when we make a particular choice, then we wouldn't be able to choose *anything*, which undermines everything about us, not least of which would be our ability to relate to God. As such, not even the Christian theist would expect God to do earth-shattering miracles all the time, and He certainly wouldn't do anything that destroys our ability to make choices—that would be inconsistent with both-- His perfect mind and perfect character.

Miracles, especially if they reveal God to us in a special way, don't automatically contradict this, and an all-knowing God would certainly know how to use miracles in the world in ways that don't destroy our ability to make choices.

But why would God bother to 'intervene' with miracles? The simplest reason is that He may wish to communicate more about himself than we can learn through general revelation. This becomes especially poignant when we realize that God is morally perfect and we are not (Romans 3:23). A morally perfect God can't simply ignore our sin, and without His special intervention to reconcile us to himself, we would all be doomed. So, does He refuse to intervene and let us all perish, or does He do something to restore us to himself?

It seems obvious that a gracious God who steps into history to offer an olive branch to His enemies is more praiseworthy than a god who just lets us all suffer. As the apostle James says: "Mercy triumphs over judgment" (James 2:13). And that of course is precisely what we claim God did in Christ's life, death, and resurrection and why would a loving God allow death and suffering?

Now, as to the question of whether an atheist can account for the laws of logic, they would say that truths like 'a rock is a rock' and 'a rock is not a car' are true regardless of whether a mind exists or not. They might parse this out in a couple of different ways. One option is to say that the laws of logic, along with numbers, truths, properties etc., are abstract objects; i.e., they are objects that have no causal power, but exist necessarily and independently of anything else. As such, the existence of God is irrelevant to the question, since these objects would exist independently of God whether or not He existed.

Another option is to say that the laws of logic are not *things* needing an explanation, since the laws of logic don't 'exist' like a chair or a bee or a star 'exists.' Rather, the 'laws of logic' are just descriptions of the essential self-consistent nature of reality. In other words, they would say that the structure of reality is inescapably logical, even if physical reality is all there is.

One way around this is an argument for God known as the argument from concepts for God's existence, which treats e.g., numbers and the laws of logic as *objects* of a special type—ideas. The argument runs like this: the laws of logic are ideas, and ideas can only exist in minds, but since the laws of logic exist necessarily, the laws of logic must be constituted in a necessary mind, such as God, therefore God must exist.

This argument rests on the notion that ideas are real objects, like chairs and stars, that can only reside in minds. But if that's not true, then the argument fails. And there is debate even among theists over whether the laws of logic are ideas, and whether ideas are really existing things. As such, not every theist would consider this argument for God sound. Nonetheless, it's an open question. So, if you choose to use this argument, just beware that the atheist may reject the argument in similar ways to how some theists would.

Another possible solution is that the theist can agree that the laws of logic are not 'things,' but are just our descriptions of the essential self-consistent nature of reality. In this case, rather than making the laws of logic *things* that are somehow dependent on God for their existence, the theist could say that 'logic' is just a description of how God's mind works; i.e., God is logical. This doesn't necessarily put theism and atheism on level footing, since there are arguments for God that argue for His necessity and don't rely on the laws of logic being actual objects—e.g. moral arguments, arguments from contingency, and ontological arguments. If these arguments are successful, they would show that while the atheistic analysis of the laws of logic in itself is coherent, saying that this is true 'even if God doesn't exist' is ultimately incoherent, since God necessarily exists. Rather, the logical structure of reality is ultimately grounded in God's logical nature.

Whether we believe logic is an object or not, the facts that God cannot fail to exist, and is himself perfectly logical, provide plausible explanatory grounds for logic that atheism simply lacks. And since God is perfectly logical, we can trust that the way He providentially orders miracles in history will also be logistically solid.

IMPOSED LAW AND DIVINE WILL

In order for science to progress, it was necessary to reject the erroneous view of nature handed down by Greek philosophers, and which dominated among the intellectual elite during much of the medieval period.

Leading historians of science acknowledge that the Christian doctrines of God and Creation played a pivotal role in this process.

The Greek view of *nature as a living organism* was replaced by the biblical view that only people and animals have souls. This led to the rejection of the Greek explanation for motion as arising from tendencies internal to objects, and its replacement with the concept of external, divinely imposed laws.

The Greek view that natural processes are governed by eternal principles binding even on the gods was replaced by the biblical view of God's omnipotence and His freedom to create as He willed. This led to the belief that the laws of nature were determined entirely by God's choice and could, therefore, only be discovered by observations.

THE HEART OF SCIENCE

At the heart of scientific enquiry is the *faith* that the world is orderly and behaves consistently from one day to the next. One might ask, however, how this belief arose. According to "Peter Harrison," formerly Professor of Science and Religion at Oxford University, it was, in a large part, "the *theologically informed assumption* that there are laws of nature, promulgated by God and discoverable by human minds (emphasis added)".

Eminent Philosopher of Science "Alfred North Whitehead" would agree. He wrote: "My explanation is that the faith in the possibility of science, generated antecedently to the development of modern scientific theory, is an unconscious derivative from medieval theology."

Prior to the late medieval period, Greek philosophy dominated among the intellectual elite. However, around the 13th century onwards, there was a reaction against this by Christian theologians.

Philosophers and historians of science have argued that this rejection of the Greek understanding of nature, particularly the teachings of Plato, Aristotle and the Stoics, and its replacement by the biblical worldview, substantially underpinned the rise of modern science.

THE GREEK VIEW OF NATURE

In the thinking of most Greek philosophers, Figure 29.24, the world was a living, divine organism. For some, even matter was understood to have god-like attributes, being self-sufficient and unchangeable, with inherent properties determining a universal world order binding even upon the gods. This divine substance governed the development of the world and dictated the movements of the heavenly bodies. Consistent with this belief, the Greek poet Hesiod (c. 700 BC) thought that the earth *generated* the mountains. In contrast, the Hebrews saw the world forming according to God's *command* (e.g., *Genesis 1*).

Fig.29.25: Greek Philosophers

Plato

In the thinking of Plato (c. 428–348 BC), Figure 29.25, true reality is found in the realm of thoughts, rather than by observing and learning from our world. Using our senses, we perceive only shadows of reality. The principle is applied generally so that everything known in our world, material or immaterial, exists as a more perfect 'ideal' or true 'form' in some higher plane.

Legal decisions made in courts are our best attempts to administer justice but are never wholly successful as true justice remains transcendent. Round plates produced by a potter or circles painted by an artist are merely imperfect representations of true roundness which can only be pictured in the mind or expressed mathematically. These 'forms' or 'ideas' are divine and explain the nature of objects.

Fig.29.26: Plato (424/3 - 348/47 BC) – Greek Philosopher

While 'forms' are eternal and immutable, objects are changeable. Hence, 'forms' are considered more certain than what is observed, and logical reasoning and analysis are understood to be more reliable than fallible observations.

According to Plato, when 'the Demiurge' (the creator) shaped the world, he was constrained to follow these preordained 'ideal' patterns, rather than being free to make it as he wished. In addition, he had to use materials he had not created himself and these tended to resist his attempts to form them. "Galen" (c. 129–216) was another influential Greek writer who rejected the Genesis account of creation because this was contrary to his understanding that the creator would be limited in his work by the nature of matter.

Instead of studying the motions of the planets and concluding from this that they follow elliptical orbits, as did Johannes Kepler (1571–1630), Plato 'reasoned' that they must follow circular paths because circular motion is most perfect, an 'ideal' form, and most befitting to the gods. Similarly, he 'deduced' that the universe must be spherical because this is the 'ideal' shape. In fact, Plato, Figure 29.26, explicitly rejected the view that astronomical observations were useful, arguing that we should "leave the starry heavens alone". In this he followed his teacher, Socrates (c. 470–399 BC), who, while regarding astronomy worthwhile in determining the time or day of the year, considered that learning the courses of the stars, or enquiring about the causes of their movements, was a waste of time. According to his pupil, Xenophon (430–350 BC), Socrates "held that speculators on the Universe and on the laws of the heavenly bodies were no better than madmen".

Aristotle

Like Plato, "Aristotle" (384–322 BC), Figure 29.27, believed the world to be as it is due to necessity, conforming to eternal, unchangeable principles which could be deduced by processes of reason. To him the world was like a huge animal which breathed, grew, and decayed. In this he again followed Plato who asked: "In the likeness of what animal did the creator make the world?" In his writings Aristotle continuously appealed to biological similes, for example, likening earthquakes to animal digestion and the motion of stars to the locomotion of quadrupeds. As did Plato, he saw the heavenly bodies as living beings.

Fig.29.27: Aristotle (384–322 BC)

In the thinking of Aristotle, physical objects are a compound of 'matter' and 'form', where 'form' unifies some matter into a single object and determines its structure, properties, and activities. Without 'form,' matter cannot even exist. Aristotle's god, however, has little power over nature, having jurisdiction over neither the matter nor the form of natural objects.

Aristotle distinguished between 'natural motion' and 'violent motion,' the former arising from the nature of an object, the latter being imposed on it. For example, the natural motion of a stone would be to fall to the ground. However, if thrown, it will for a time move in an unnatural or 'violent' way. Whereas the natural motion of terrestrial bodies is rectilinear, the natural motion of celestial bodies, due to their being made of a different substance, is circular.

The Stoics

According to the "Stoics," the material world was impregnated with reason, and objects, along with people and animals, had souls. All was part of a universal world soul, with its individual parts in sympathetic relations to one another. To the Stoics, 'natural law' was 'immanent,' i.e., inherent in the structure of things, and this explained everything from the behavior of people and animals to the movements of the heavenly bodies. 'Laws of nature' arise out of necessity, in the properties of matter, and hence knowledge of the nature of things is thought to be the key to understanding their relations to one another.

Medieval scholastics often amalgamated Greek and biblical thinking. For example, "Thomas Aquinas" (c. 1224–1274), Figure 29.28, while accepting the omnipotence of God, also saw natural law as immanent in the universe. For him, eternal law is nothing other than God. Some wrote of 'substantial forms' impregnated in nature, internal causes of processes arising from objects possessing soul-like powers. Unobservable 'occult qualities' adhered to objects like 'little ghosts,' producing effects by 'sympathy' and 'antipathy.' Sympathy, for example, was thought to explain the attraction of iron to a magnet—just as man is attracted to woman. The doctrine of *horror vacui* (abhorrence of a vacuum) was thought to explain why water rose in 'suction' or 'vacuum' pump barrels. Supposedly, this was because nature had an antipathy to empty space.

Fig.29.28: *"Thomas Aquinas" (c. 1224–1274)*

AN IMPEDIMENT TO SCIENCE

Platonic thinking was antithetical to science because it detracted from the view that the world could be understood by learning from observations. In contrast, biblical thinking pointed to this as the only way of discovering reality. The Bible teaches that God is omnipotent and was in no way constrained to create according to any prescribed pattern. For, "Whatever the Lord pleases, he does, in heaven and on Earth, in the seas and all deeps" (*Psalm 135:6*). Since He created matter *ex nihilo* (from nothing) he could endow it with whatever properties He chose. In biblical thinking, the natural order arose as a result of a historical act of creation (*Genesis 1:1ff*); nothing about it is either eternal or necessary and the Creator was not constrained to follow pre-existing 'forms.' Rather, the world is as it is, and behaves as it does, because of *divine choice*, the will of a sovereign deity. Hence, it is impossible to determine the nature of things based on reason alone. Only by studying His creation could God's design be known.

The rejection of Greek thinking by the founders of modern science is exemplified in Roger Cotes' preface to the second edition of Isaac Newton's *Philosophiae Naturalis Principia Mathematica* (*Mathematical Principles of Natural Philosophy*):

"Without all doubt this World … could arise from nothing but the perfectly free will of God directing and presiding over all. From this fountain it is that those laws, which we call the laws of Nature, have flowed; in which there appear many traces indeed of the wisest contrivance, but not the least shadow of necessity. These therefore, we must not seek from uncertain conjectures; but learn them from observations and experiments."

Newton himself, in the very first sentence of his preface, wrote of how modern thinkers, having discarded "[soulish] substantial forms and occult qualities have endeavored to subject the phenomena of nature to the laws of mathematics." A committed biblical creationist, he also rejected the Greek view that God would have been constrained in His acts of creation in any way. He wrote of God:

" … we admire him for his perfections; but we reverence and adore him on account of his dominion … and a God without dominion, providence, and final causes [i.e. design], is nothing else but Fate [i.e., necessity] and Nature."

THE WORLD—ANIMATE OR INANIMATE?

Plato taught that the cosmos created by the Demiurge was a living organism, that the world had a divine soul, and the stars and planets were gods. In a similar vein, Aristotle taught that stones fall to the ground because they have a *yearning* for the center of the universe (which he believed to be the center of the earth). Such thinking was an obstruction to science because it attributed causes of motion to motives and inner compulsions, rather than to impersonal, external forces.

In contrast, the Bible clearly distinguishes between the Creator and the creature (i.e., that which was created). God is spirit (*John 4:24*) and is a being separate from the world. There is only one God (*Isaiah 45:5*) and His creation is not divine; for God said: "Before me no god was formed, nor shall there be any after me" (*Isaiah 43:10*). Indeed, to attribute divinity to the creature is idolatry. As argued by Oration priest "Nicole Malebranche" (1638–1715), there can be only one cause which is "nothing but the will of God," For Malebranche, Greek 'forms' are nothing more than "the little gods of the heathen" introduced by the evil one to occupy the hearts which the Creator has made to belong to himself.

The cosmos is not an organism and does not have a soul, this being firmly established in the very first book of the Bible. Here only animals and people are described as 'living creatures' (Genesis 1:20, 24). The universe is not eternal and does not have any self-sustaining or self-generating powers. Rather it is the work of a single Creator upon whom it is totally dependent. Hence, objects do not have minds and desires, and are not subject to laws inherent within their natures; instead, the non-living world operates according to laws imposed on it from without. The moon gives rise to tides, not because it has some sort of friendship with the water of the oceans, but because of the impersonal law of gravity.

THE LAWGIVER

The God of the Bible is the lawgiver in both the moral and physical realms. He gave the 10 commandments to Moses (Exodus 20:3–17) and wrote the requirements of the law on the hearts of men so that they "by nature do what the law requires" (Romans 2:14–15). He is the one who gathered the waters together (Genesis 1:9) and "assigned to the sea its limit, so that the waters might not transgress his command" (Proverbs 8:29). He "made a decree for the rain and a way for the lightning of the thunder" (Job 28:26). He created the sun to govern the day and the moon to govern the night (Genesis 1:16), "commanded the morning … and caused the dawn to know its place" (Job 38:12). He created the stars to mark the seasons (Genesis 1:14), knows "the ordinances of the heavens", and established "their rule on the earth" (Job 38:33). He continually "upholds the universe by the word of his power" (Hebrews 1:3).

In the Old Testament, God's commands to nature are often expressed in legal language. For example, the Hebrew word *huq* is used in both Proverbs 8:29 and Job 28:26. Its verbal form means to 'engrave' or 'legislate' and is often used in the context of God giving moral and ritual laws. In both these verses, the 4th century Vulgate translation uses the Latin word *lex*, meaning 'law.' According to philosopher of science "Edgar Zilsel," verses such as these "were quoted through the centuries again and again, and have decidedly contributed to the formation of concepts in rising natural science." "Galileo Galilei" (1564–1642), for example, wrote that nature "never transgresses the bounds of the laws imposed to it", being a "most careful executor of the orders of God" and argued for nature's strict observance of God's commands citing, among others, Job 28:26, 38:8–11 and Psalm 104:9. According to Professor "Friedrich Steinle," for Galileo the concept of law in nature was "most intimately and inextricably connected with theological considerations concerning God's activity as legislator."

Nobel Prize winner "Melvin Calvin" also acknowledged the influence of the Bible in these matters. Referring to the necessity of conceiving of the world as orderly, he remarked:

"As I try to discern the origin of that conviction, I seem to find it in a basic notion … enunciated first in the Western world by the ancient Hebrews: namely that the universe is governed by a single God, and is not the product of the whims of many gods, each governing his own province according to his own laws. This monotheist view seems to be the historical foundation for modern science."

Drawing on his Christian theology Newton wrote:

"This most beautiful system of the sun, planets, and comets, could only proceed from the counsel and dominion of an intelligent and powerful Being. … This Being governs all things, not as the soul of the world, but as Lord over all; and on account of his dominion, he is wont to be called *Lord God or Universal Ruler*."

IMPOSED VS IMMANENT LAW

The Greek view of the cosmos as an organism drew upon an analogy between the natural world and a human being. As such it was understood to have been endowed with intelligence and life. In contrast, the Christian view was based on an analogy between the natural world and a machine. Hence, the movements of bodies were not due to their being capable of controlling themselves; nor did they arise from 'immanent laws' (i.e., those inherent in objects and in the structure of reality itself). Rather they were the result of 'imposed laws' set up by an external, omnipotent Designer. According to the Bible, this same God had created the mind of man after His own likeness (Genesis 1:26–27, 5:1–3); hence it was considered possible for us to understand His designs and describe the scientific principles by which they operated.

The Divine Will

"Francis Oakley," formerly Professor of the History of Ideas, Williams College, Massachusetts, documents how, beginning around the 13th century, European theologians rejected Greek thinking about God and nature and replaced it with biblical thinking. This began in 1277, when "Etienne Tempier," Bishop of Paris, and "Robert Kilwardby," Archbishop of Canterbury, formally condemned a list of 219 philosophical propositions as contrary to the Christian faith. These focused particularly on the teaching of Aristotle and the need to refute the view that God was in any way limited in His absolute power to do whatever He wishes.

Fig.29.29: Portrait of The Honorable Robert Boyle (1627–1691), by Kerseboom. Credit: Wellcome Collection. CC BY

This emphasis on the 'divine will' was strengthened, among others, by "William Ockham" (c. 1332), who insisted that both moral law and the whole of creation are entirely subject to God's choice. Ockham drew attention to God's ability to overrule natural law by reference to Shadrach, Meshach, and Abednego's emerging unscathed from Nebuchadnezzar's fiery furnace (Daniel 3). "Jean Buridan" (c. 1350) argued that God, "in his most free will" may have created things which did not seem reasonable to the human mind—and this, of course, is true. Who would have thought that light would sometimes behave like a wave and sometimes a particle? "Nicole

Oresme" (c. 1377) rejected Aristotle's assertion that the earth must be stationary and that the heavenly bodies must move in circular orbits. To Oresme, God would do as He pleased.

The 'father of chemistry,' "Robert Boyle" (1627–1691), Figure 29.29, argued that God could have made other worlds where the laws of nature were different.

Of great significance is that two of the great German reformers, "Martin Luther" (1483–1546) and "Philipp Melanchthon" (1497–1560), in order to demonstrate the power of God over nature, also referred to Shadrach, Meshach, and Abednego's deliverance, along with, "Francisco Suárez" (1548–1617), "William Perkins" (1558–1602), "John Preston" (1587–1628), "William Ames" (1576–1633), "Thomas Shepard" (1605–1649), "John Norton" (1606–1663), "Increase Mather" (1639–1723), and "Samuel Willard" (1640–1707). Another was the 'father of modern chemistry', "Robert Boyle" (1627–1691; Figure 29.28, who referred to this incident in no less than three of his works. Oakley argues that this is strongly supportive of his contention that "the scientific idea of divinely imposed laws of nature had its origins in a living theological tradition which went back to the last years of the thirteenth century."

Kepler argued that the failure of the Greek philosophy to birth the concept of mathematical law could be explained by Aristotle's belief that the world was eternal and that Aristotle's god did not impose order on the world. In contrast, Kepler maintained that "our faith holds that the world, which had no previous existence, was created by God in weight, measure, and number, that is in accordance with ideas co-eternal with Him." Kepler's understanding of 'eternal ideas', however, was very different to that of the Greeks. For Plato, because 'ideas' were eternal, they were also immutable and binding on the gods. For Kepler's God (the God of the Bible), the principles which He used to determine the order of nature were entirely of His own choosing.

Both Boyle and Newton argued that God could vary the laws of nature. Boyle considered it plausible that God had made other worlds where "the laws of this propagation of motion among bodies may not be the same with those that are established in our world." Similarly, Newton argued that "God is able … to vary the Laws of Nature and make Worlds of several sorts in several Parts of the Universe."

This belief that the laws governing the natural world were determined entirely by the Creator's choice led to the realization that the world order could not be deduced by *a priori* reasoning, but only empirically, through observation and experiment.

The Orderliness of Creation

- The God of the Bible is the One who "laid the foundation of the earth … determined its measurements" and "laid its cornerstone" (Job 38:4–6).
- He "gives the horse its might", and it is by His "understanding that the hawk soars and spreads his wings toward the south" (Job 39:19, 26).
- The Israelites were told to consider the stars and remember their God, who "brings out their host by number, calling them all by name; by the greatness of his might and because he is strong in power, not one is missing" (Isaiah 40:26).
- He is the One who established a "covenant with day and night and the fixed order of heaven and earth" (Jeremiah 33:25).
- "The Lord is the everlasting God, the Creator of the ends of the earth … his understanding is unsearchable" (Isaiah 40:28).
- In forming the world, God brought forth order from disorder. The original creation was "without form and void" (Genesis 1:2) and
- the first man was made from the dust of the ground (Genesis 2:7).

- Israel is said to be like clay in a potter's hand (Jeremiah 18:6); according to the apostle Paul, God is not a God of disorder but of peace (1 Corinthians 14:33).
- In the first chapter of John's gospel, the Creator is revealed to be 'the Word' (Greek *logos*), the incarnate Son of God. *Logos* also carries the sense of logic and reason.
- This *logos* is also the One whose act of redemption will one day liberate the fallen creation from its bondage to decay (Romans 8:21). Early Christian theologian Origen (c. 185–254) argued that God conferred upon His creation an intrinsic rationality and order that reflected the divine nature itself. This is in stark contrast to the polytheism of some pagan religions where the natural world might be subject to the whims of temperamental deities with conflicting interests. In such a world, almost anything could happen!

Paul Davies, acknowledges in his book, *"Mind of God"* that "the justification for what we today call the scientific approach to inquiry was the belief in a rational God whose created order could be discerned from a careful study of nature." According to "Alistair McGrath," Professor of Science and Religion at Oxford University:

"This insight is directly derived from the Christian doctrine of creation and reflects the deeply religious worldview of the medieval and Renaissance periods … . This foundational assumption of the natural sciences—that God has created an ordered world, whose ordering could be discerned by humanity, which had in turn been created 'in the image and likeness of God'—permeates the writing of the period."

THE NATURAL WORLD AS A MECHANISM

According to the Bible, God is the Creator and sustainer of the universe and, at the same time, wholly separate from it. This, together with the sense of the orderliness of the creation, led theologians and philosophers to see the natural world as designed mechanism. Discussing blood circulation in his *"Discours de la Méthode"* (*Discourse on Method*), René Descartes (1596–1650; Figure 29.30, stated that "the rules of nature are identical with the rules of mechanics" and, in his *"Le Monde"* (*The World*), he asserted "that God is immutable, and that acting always in the same manner, He produces always the same effect".

These laws, he said, are not immanent but 'imposed' on nature by God. The courses of the planets, the oceanic tides and the universe in general are regular and predictable because they are determined by the God of the Bible who is faithful and sure. Descartes' contention that the natural world is governed by an unchanging God, and hence behaves consistently from one day to the next, was an essential step in scientific progress.

Fig.29.30: The 'Father of Mathematics', René Descartes (1596–1650), wrote that "the rules of nature are identical with the rules of mechanics." Curtsy - wikipedia.org

Although Zilsel controversially argues that the concept of laws of nature arose primarily from sociological factors—for example, the politics of absolute monarchy—he acknowledges that Descartes "took over the basic idea of physical regularities and quantitative rules of operation from the superior artisans of his period. And from the Bible he took the idea of God's legislation. By combining both he created the modern concept of natural law."

French Bishop "Nicole Oresme" (c. 1320–1382) and French theologian "Pierre D'Ailly" (1350–1420) both wrote of the workings of the world as analogous to a clock. "Melanchthon" (1497–1560) referred to the "whole machine of the world" serving "perpetual laws" and insisted that God is a "most free agent, not, as the Stoics used to teach, bound by secondary causes". In his *"De Revolutionibus Orbium Coelestium"* (*On the Revolutions of the Heavenly Spheres*), published in 1543, Nicolaus Copernicus (1473–1543) wrote of the "the movements of the world machine, created for our sake by the best and most systematic Artisan of all."

THE UNIVERSE IS FRAMED BY GOD

In a work containing numerous biblical quotations, Boyle argued that "the universe being once framed by God, and the laws of motion being settled and all upheld by his incessant concourse and general providence, the phenomena of the world thus constituted ... operate upon one another according to mechanical laws." He also expressly denied the concept of immanent law, arguing that "the laws of motion, without which the present state and course of things could not be maintained, did not necessarily spring from the nature of matter, but depended upon the will of the divine author of things." According to Professor "Hooykaas," in the thinking of Boyle and his contemporaries, "Holy Scripture ... had made their science truly free."

"Michael Foster," Oxford Philosopher

"[T]he method of natural science depends upon the presuppositions which are held about nature, and the presuppositions about nature in turn upon the doctrine of God. Modern natural science could begin only when the modern presupposition about nature displaced the Greek ... but this displacement itself was possible only when the Christian conception of God had displaced the Pagan as the object ... of systematic understanding. To achieve this primary displacement was the work of Medieval Theology."

By de-deifying nature and de-personalizing motion, Christian theology emancipated science from its stagnation under Greek philosophy. It asserted that the universe is not eternal but created, and its nature and operating principles did not have to conform to any eternal, unchangeable 'forms'.

Emphasizing God's omnipotence and His freedom to create as He willed led to the view that the scientific method necessitated observations. The belief that there are laws imposed upon a world by an orderly, faithful, and immutable God caused philosophers to see the universe as a designed mechanism, rather than an eternally existing organism. This, in turn, led to the belief that the workings of God's creation could be investigated, understood, and described mathematically. All this hung on the Christian doctrine of creation, as articulated so clearly in the "Nicene Creed:" "We believe in one God, the Father Almighty, Maker of all things visible and invisible."

GOD THE CREATOR IS THE LAWGIVER

Where does knowledge come from?

From research and inquiry?

It seems an obvious question, but is it really? It is only obvious if we realize what the unstated assumptions are that go with it.

Why do men study science anyway? Why do they believe it a fruitful activity at all?

The answer must be founded on the belief that the laws of nature, which they are attempting to discover, are the same today as they were yesterday, and will be the same again tomorrow. What is the justification for this?

Fig.29.31: Michael Faraday Curtsy - Photo Wikipedia

If we look back 500 years we see a list of names of well-known philosophers and scientists: **Tycho Brahe, Figure 29.34**, Nicolas Copernicus,

Figure 29.33, Galileo Galilei, Gottfried Leibnitz, Isaac Newton, **Leonardo da Vinci, Figure 29.32**, Johannes Kepler, Antony van Leeuwenhoek, Carolus Linnaeus, Leonhard Euler, John Dalton, Christian Huygens, Robert Hooke, **Michael Faraday, Figure 29.31**, Joseph Henry, James Joule, Louis Pasteur, William Thompson (aka Lord Kelvin), James Clerk, Maxwell, John Strutt (aka Lord Rayleigh), John Ambrose Fleming, to only name a few. These men all believed in the truth of the Bible, so much so they believed that the laws governing nature that we study in the laboratory can be extrapolated to the universe and apply both in the past and the future. We know this by the very nature of the scientific endeavors they were involved in.

If they studied the cosmos, e.g., planets in the solar system, they believed that what they observed had some order to it—some law by which the planets moved. This eventually led Newton to discover the universal law of gravitation, which 'tells' a planet how to move around the Sun. But *the law* that was discovered was then used to predict the future positions of the planets and it was found to be very reliable. If they studied microbes, for example, they believed that the experiments they performed one day would be repeatable the next and the results could be consistently interpreted.

THE EXPECTATIONS BEHIND SCIENCE

Ultimately this idea has led to what we now call the 'Scientific Method'—the notion that an experiment can be repeated and the result will be the same if the conditions are identical. On top of this is the idea that a theory must make predictions that are testable and if a prediction fails then the theory is found to be wrong and must be discarded or modified. This leads to the whole subject of the philosophy of science, which is a subject on its own, but the main point to be taken from this is that scientists in both the past and the present rely implicitly on the idea.

Fig.29.32: Leonardo da Vinci - Curtsy Photo wikipedia

Explicitly, science depends on the idea that the laws of Nature are constant in time and in space. But how can we justify this? Some might justify constancy 'in time' simply by repeating the experiment and getting the same results day in and day out, but what about 'in space'?

However, this is strictly speaking the *fallacy of induction*, as the skeptic "Bertrand Russell" illustrated with his 'inductivist turkey' parable: a turkey is brought to a farm in America, and for 364 days is fed the same food every day, and goes to sleep happily that night. So, he has every reason to assume that Thanksgiving will be the same, and he will once more go to sleep happily. But his head is chopped off instead, and he ends up as Thanksgiving dinner. So, on a secular basis, there is no proof that the laws of nature will be constant *tomorrow*, just because we have observed them to be constant *so far*.

The same applies to the assumption that the laws of nature are the same in all parts of the universe. Since we can test a theory locally, on earth or in the solar system, it seems to be a fair guess for the rest of the universe. But we can't know for sure that they are not different in the core of Pluto, for example. We cannot experimentally test the laws of nature in other parts of the universe so they are only assumed to be true there also.

The same problem lies with the assumption of an orderly universe. This can't be proven: first because of the inductive fallacy, and second, because any alleged proof would have to presuppose the very order it claims to prove.

THE BIBLICAL BASIS FOR THESE ASSUMPTIONS

Fig.29.33: Nikolaus Copernicus – Curtsy Photo wikipedia

So why did science advance rapidly since the Renaissance? This has largely been from Europe and based in the Western culture. The underlying assumption that the laws of Nature are constant has been the result of Judeo-Christian thinking. As "Thomas Aquinas" put it:

'Since the principles of certain sciences—of logic, geometry, and arithmetic, for instance—are derived exclusively from the formal principles of things, upon which their essence depends, it follows that God cannot make the contraries of these principles; He cannot make the genus not to be predictable of the species, nor lines drawn from a circle's center to its circumference not to be equal, nor the three angles of a rectilinear triangle not to be equal to two right angles.'

The creator God by his very nature created laws that are reliable. The Apostle Paul tells us:

For God is not a God of disorder but of peace. (*1 Cor. 14:33*).

Excluding the occasional miracle, God Himself chooses to generally operate in conformity with His own laws—the laws of Nature. These are part of His Creation, the expression of His own orderly will. He is Lord and judge and lawgiver.

For the Lord is our judge; the Lord is *our lawgiver*; the Lord is our king; he will save us. (*Isa 33:22)*

He is the same always. Jesus Christ is the same yesterday and today and forever. (*Heb 13:8*)

If I have not established my covenant with day and night and *the fixed laws* (or 'fixed order' in ESV) *of heaven and earth*, (*Jer 33:25*)

Clearly God is speaking of the laws of nature like gravity. The Scripture says:

Thus says the Lord, who gives the sun for light by day and *the fixed order* of the moon and the stars for light by night, … (*Jer 31:35*)

God usually operates in a regular way that scientists describe as 'natural law'. He makes it clear that the continuity of these laws can be relied upon to the same extent as His covenantal word, and vice versa—i.e., totally. He says …

If this *fixed order* departs from before me, declares the Lord, then shall the offspring of Israel cease from being a nation before me forever. (*Jer 31:36*)

Thus says the Lord: If you can break my covenant with the day and my covenant with the night, *so that day and night will not come at their appointed time*, then also my covenant with David my servant may be broken, so that he shall not have a son to reign on his throne, and my covenant with the Levitical priests my ministers. As the host of heaven cannot be numbered and the sands of the sea cannot be measured, so I will multiply the offspring of David my servant, and the Levitical priests who minister to me. (*Jer 33:20–22*).

God is saying that if He doesn't keep His own laws that hold the universe together then even the Nation of Israel is in jeopardy, God's covenant is broken—the lineage of David is broken and it even follows that there may be no son of David, who will reign over the Creation, *Christ Jesus.*

No doubt God is sovereign, but He is bound by His own nature, which is non-capricious, faithful and unchanging. The 'laws' are our description of the way He normally acts in a dependable way. The Bible is the basis for the Judeo-Christian culture from which our modern scientific age has developed.

THE KORAN

On the other hand the **Koran** ' … portrays Allah as absolutely sovereign and bound by nothing. This sovereignty was so absolute that it precluded a key assumption that helped foster the development of science in Europe: … that God is good, and that His goodness is consistent. Therefore, He created the universe according to rational laws that can be discovered, making scientific investigation worthwhile.'

Fig.29.34: *Tycho Brahe - Curtsy Photo wikipedia*

"Robert Spencer," an authority on Islam, makes this very relevant comment: that in Islam the very idea of Laws of Nature would be blasphemy as it is 'a denial of Allah's freedom.' Spencer goes on to say that the very idea that Allah created the Universe with consistent and rational laws means He cannot do something, which would bind his absolute sovereignty. (In Christianity, the sovereignty of God does not preclude Him from being bound by either his covenantal word, or aspects of His divine nature. In short, God is not limited by anything *outside* Himself.)

So, in the world of Islam there is no logical connection between the order in Nature from one instant to the next. No wonder Spencer says in his book that '*Allah killed science*.' It makes much more sense that science developed in Christian Europe where Islam failed to reach. Islam was stopped by the armies of the Christian west around the 1400s just as science was starting to rapidly develop, fostered by the Christian church, which kept learning and knowledge alive.

One Muslim website states … "The Muslim says: "The creatures (includes everything) are given by God, their qualifications (including the physical laws, equations, formulas [we may be unknowing everything, the real is known by Allah]) and they are executed and *may be changed by Allah when(r)ever He wants*.". [emphasis added]

The website argues in favors of the Islamic position, but there is a clear admission and corroboration of Spencer's conclusion above. It seems to be saying that Allah can change the laws, and their relationships whenever and wherever he likes—a very capricious god, impulsive and unpredictable. The website compares the above with the view of the atheist as:

The (atheist) scientist says "The properties (including above mentioned laws and so on) of everything are intrinsic to them and absolute and not subject to change except for any specific conditions."

This would be the worldview of most secular scientists today. The atheist assumes the properties of matter or laws of nature are intrinsic to nature itself. That is the position of the materialist who believes all things, all laws, arise spontaneously from the universe itself.

The humanist worldview tries to establish itself in a rational consistent universe without a Creator. But how is that possible unless some order is impressed upon matter itself? In fact, the humanist must borrow the order from the biblical worldview before he can even think in an orderly fashion about his universe. The biblical worldview involves a universe that is rational and behaves consistently. It is the expression of an orderly Creator, whose laws reflect His own unchanging nature and the fact that, while He is able to engage in miraculous activity outside of those laws, He chooses not to do so capriciously or arbitrarily. I.e., He 'keeps' His laws Himself.

Could Adam have Appealed the Verdict?

Usually, in countries not run by a dictator, a person who has been found guilty of any offence has the right to appeal to a higher court against his or her conviction.

Genesis chapters 1–3 record that God created Adam, gave him the Garden of Eden to live in, told him he could freely eat the fruit from any tree there except from one called the Tree of the Knowledge of Good and Evil, and warned him that if he did eat from that tree the penalty would be death. This was not advice, but a command. The tree was not poisonous; it was disobedience that had death in it. Adam (and his wife, Eve) ate the forbidden fruit. God summoned them, interrogated them, and pronounced the sentence of death. So, could Adam have appealed this verdict? Figure 29.35.

Fig.29.35: Did Adam Appeal His Verdict – Curtsy stockxpert

What Adam and Eve did was not a minor matter, like for example our incurring a parking fine. Adam had been clearly given one law to obey, 'Don't eat that particular fruit.' Equally clearly, he had been told the penalty, 'If you do, you will die.' He could not claim that the law was unjust or that his situation was unreasonable.

The prohibition was a simple and fair test of his (and Eve's) obedience to and love for God, and a means by which they might acknowledge that God ruled over them. Clearly, they were being tested.

THE JUDGE

Was it appropriate for God to exercise the role of judge with respect to Adam and Eve? Figure 29.36.

Fig.29.36: Eve Attentive to the Serpent

Answer: Yes. From a human point of view, the creator of a game, say like Monopoly, has the absolute right to decide the rules and plays of the game. As Adam and Eve's Creator and Lawgiver, God also had the absolute right to be their Judge. However, regardless of any human rationalization we may envisage, Almighty God, for no other reason than that He is Almighty God, had the absolute right to be Adam's Lawgiver and to set any rules He chose (consistent with His holy, righteous and just character) for Adam and Eve to obey.

Observe that God had not sought their downfall. On the contrary, He had made obedience both easy and pleasant. He had 'created man without a sinful nature, placed him in an ideal environment, provided for all his temporal needs, endowed him with strong mental powers;'

He had given him 'work to engage his hands and his mind, provided a life-partner for him, warned him of the consequences of disobedience, and entered into personal fellowship with him.'

Christ ('the last Adam,' as He is called in *1 Corinthians 15:45*) too was tempted by Satan, but He overcame the tempter, thereby wresting from the devil 'that dominion over the whole race which he had secured by his victory over the first human pair.'

THE INTERROGATION

In most human courts, by far the most time is spent in establishing what the facts were of past events, and sometimes errors occur in evaluating the evidence given. However, God did not need the divine equivalent of a human trial, either to establish the facts, or to evaluate them, or to pronounce the penalty. Being omniscient, He knew what Adam and Eve had done the moment they did it. Not only so, but God knew what the facts would be even before they happened. Before He created them, God had known that Adam and Eve would disobey Him and fall into sin.

When God next appeared in the Garden, the consequences of sin were already in operation in Adam and Eve—shame, guilt, and fear, shown in their hiding from God, with whom they previously had had perfect fellowship. He began by calling Adam, the authority figure in the family. God's questions, 'Where are you?' etc. were not to obtain information, but were rhetorical, possibly in order to encourage Adam (and then Eve) to confess their sin.

THE DEFENSE

Was there any evidence that Adam could have put forward but didn't?

Answer: No. Adam did not deny the facts. His only defense was both frivolous and blasphemous, namely that Eve had given him the fruit, and that God was the one who had given him Eve. So, if anyone was to blame, it was one or both of them! Eve, in turn, blamed the serpent. The record shows that these were invalid excuses.

God had not allowed Satan to tempt Eve in the disguise of 'an angel of light' (*2 Corinthians 11:14*), whom she might have mistaken for a divine emissary with new instructions, but rather in the form of a serpent, 'a creature, not only far inferior to God, but far below themselves.' Thus 'they could have no excuse for allowing a mere animal to persuade them to break the commandment of God. For they had been made to have dominion over the beasts, and not to take their own law from them.'

God had given His law to Adam, before Eve had been created. This suggests that Adam had later passed it on to Eve, as she quoted the prohibition about eating the fruit to Satan (*Genesis 3:3*), albeit with the addition about not touching it. Adam might have said to her, 'So don't you even touch it.' Be that as it may, they were both required to be subject to God's authority.

THE VERDICT AND SENTENCE

With the pair finally acknowledging their guilt (*Genesis 3:12–13*), God pronounced sentence, sometimes called 'the curse' (*Genesis 3:14–19*). This was to have consequences, not only for Adam and Eve, but also for their descendants, that is, for all mankind. The original warning had been, 'Dying you shall die.'

From that moment, their human bodies began to decay, and would eventually 'return to the dust' (*Genesis 3:19*). Adam 'received into his nature the germ of death, the maturity of which produced its eventual dissolution into dust.'

Eve was told that child-bearing would involve 'pain' and that her husband would 'rule over' her. The ground was cursed, so that Adam's work from then on would involve 'painful toil.' Originally, they had been given dominion over the earth (*Genesis 1:28*), but now all mankind would be in subjection to the ground. 'Everything injurious to man in the organic, vegetable and animal creation, is the effect of the curse pronounced upon the earth for Adam's sin ...

consequently, many things in the world and nature, which in themselves and without sin would have been good for him, or at all events harmless, have become poisonous and destructive since his fall.' Finally, Adam and Eve were cast out of the Garden of Eden.

Was the Penalty Too Harsh for Such a 'Small' Offence?

Answer: No. The seriousness of an offence depends on the one offended. Offence against a fly (for example) is much less serious than one against a man. How much more then, offence against Almighty God! The motive for Adam and Eve's disobedience was not appetite, but pride—the ambition to be like God (*Genesis 3:5*). All sin is essentially rebellion against God's authority and His revealed will. The measure of God's wrath against sin is the measure of His holiness. And the measure of the penalty—death—was, and is, the measure of the enormity of the offence of rebellion against God, Figure 29.37.

Fig.29.37: Adam and Eve cast out of Paradise - after eating from the Tree of knowledge in the Garden of Eden. Curtsy - From Old Testament Stories Stock Photo

Was There a Court of Higher Authority to Which Adam Could Have Appealed Against this Sentence?

Answer: No. Almighty God, omnipotent, omniscient and omnipresent, is also the Most High God. He is also 'the only true God' (*John 17:3*). There is no higher authority than He.

What of God's Mercy?

God's mercy can be seen in several aspects of this judgment:

1. Knowing what would happen, God chose His Son, Jesus Christ, 'before the foundation of the world' to be the means by which we could be redeemed from the results of Adam's and our own sin, and restored to Himself. (*1 Peter 1:18–20; Revelation 13:8; Ephesians 1:4*).
2. God promised Adam and Eve, and us, that the seed of the woman (the virginally-conceived Jesus Christ) would 'bruise the head' of the serpent, i.e., cause the ultimate downfall of Satan (*Genesis 3:15*).

3. This was accomplished by Christ through His death on the cross for our sins, and His Resurrection. Now, those who repent of their sins and have faith towards God will, after death, be united with God and attain that state of holiness and fellowship with God which Adam and Eve lost for us in the Garden of Eden.

Relevance for Us

The Bible tells us that we too must all one day appear before God to be judged. 'For it is appointed unto men once to die, but after this the judgment' (Hebrews 9:27). God has told us that He will perform this work of judgment through the Lord Jesus Christ. 'For He has set a day when He will judge the world with justice by the man He has appointed. He has given proof of this to all men by raising Him from the dead' (Acts 17:31).

Christians are those who, in this life, have already pleaded 'Guilty' to the charge of rebellion against God's authority. However, they are able to say, 'My case has already been heard. On the cross, Jesus Christ was my substitute (Isaiah 53:6). He accepted my guilt as well as the responsibility for all my sins, and He paid in full the penalty that I deserve.' He is now our 'advocate' or defense lawyer (1 John 2:1). Believers have been pardoned eternally, and will face no retrial. For them to be judged for sin again (their own or any ancestor's) would be double jeopardy. Instead of a judgment of condemnation, Christians will face an assessment of reward.

The unsaved will be judged by God's record of their lives and their relationship to Him (Revelation 20:12, 15). The opportunity to obtain mercy will be over, and there will be no higher authority to whom anyone can appeal. For this reason, unbelievers would do well to heed the words of 2 Corinthians 6:2, 'Now is the day of God's favors, now is the day of salvation.'

APPENDIX

i. McBrayer, J.P., Why our children don't think there are moral facts, 2 March 2015; opinionator.blogs.nytimes.com/2015/03/02/why-our-children-dont-think-there-are-moral-

ii. We learn many other things about God as we read the Bible, e.g. that God is self-existent and eternal; that He is merciful, righteous, and just; that He is omnipresent, omniscient, and omnipotent; that He is good, He is truth, and He is love. However, the most wonderful thing of all in the Bible, as far as we human beings are concerned, is that God in the person of the Lord Jesus Christ is Saviour of all those who repent of their sin and put their trust in Him.

iii. God cannot forgive sin merely because someone repents, because sin carries the death penalty. Being perfectly just, God can forgive only when this penalty is first paid. The Christian Gospel is that the Lord Jesus paid the repentant sinner's penalty when He died upon the Cross. The resurrection is God's affirmation of this (1 Corinthians 15:1–4).

iv. The Bible teaches that Christian believers are not under the curse of the law (Galatians 3:13; Romans 7:4; Ephesians 2:14ff). This does not mean that the Ten Commandments have no relevance today. They instruct us as to what the will of the Lord is, but are not precepts that we are to try to keep in order to become righteous. A society which repudiates them quickly sinks into moral degradation and anarchy. Christians have received the adoption of sons, and with that adoption comes the mind of the Spirit (Galatians 4:5ff), through whom we produce the fruit of the Spirit (Galatians 5:22ff; Ephesians 5:9).

v. Note that the term 'law' has several meanings in the Bible. It can refer to the revealed will of God with respect to human conduct. It can refer to the law of Moses, including the Ten Commandments, which were the terms of God's covenant relationship with His chosen people, Israel (Exodus 34:27–28; Deuteronomy 9:9-11), along with the Levitical ceremonial legislation, laws of the priesthood, and laws of purity. This covenant ended with the coming of Christ and the establishment of the New Covenant of grace under which the Christian has imparted to him all the grace he could ever need (Romans 5:1; 8:1; Colossians 2:10). In the New Testament, the Apostle Paul tells us that the law was given to reveal the sinfulness of sin (Romans 5:13; 7:7), to show the holiness of God (Romans 7:12), and to bring us to Christ (Galatians 3:24). The law served to prepare those under it for the coming of Christ.

vi. Jesus is indeed the seed of a woman, as He had no human father (Matthew 1:18–25; Luke 1:26–35)

vii. We are here discussing moral law, not physical or natural law, which relates to the material universe. This also emanates from God, and is something which He sometimes supplements by miracles.

viii. Of course, this was always God's plan and purpose (Ephesians 1:4).

ix. Note that the Hebrew word used for God in Genesis 1 is *elohim,* which is a plural form meaning 'more than two.' The doctrine of the Trinity is thus intimated in the very first verse; it is more clearly set forth in the Gospels.

x. Arguments from contingency ask the question 'why is there something rather than nothing?', and conclude that God is the naturally necessary being that provides the ultimate reason why anything at all exists.

xi. Knowledge = acquaintance with facts, truths, or principles, as from study or investigation; general erudition: knowledge of many things. First definition from <dictionary.reference.com/browse/knowledge>

xii. Matthew 5:11–12; Luke 6:22–23; 2 Timothy 4:8; James 1:12.

xiii. Matthew 6:3–4, 20; 10:42; 25:34–40; Luke 19:11–27 cf. 1 Corinthians 3:11–15; 2 Corinthians 5:10; 9:6.

xiv. By means of a supernatural miracle Satan used the serpent to tempt Eve to eat the fruit (Genesis 3:1–6; Revelation 12:9,20:2).

xv. Translated in KJV, NIV, NASB, etc. as 'surely die' (Genesis 2:17).

xvi. Hebrew: *El-Shaddai*, e.g. Genesis 17:1 'I am the Almighty God … '.

xvii. Hebrew: *El-Elyon*, e.g. Psalm 78:35 'God Most High …'.

xviii. Drawing from Isaiah 53:10; John 5:24; Romans 8:1; 2 Corinthians 5:21; Galatians 3:13; Hebrews 10:10–14; 1 Peter 2:24 .

xix. A legal term meaning that a person is immune from being judged twice for the same offence.

xx. Cf. 'If you love Me, keep My commandments' (John 14:15).

xxi. Cf. Genesis 22:1–18, where God tested the faith of Abraham; and Exodus 16:4–5, where, concerning the instructions to the Israelites about gathering manna in the wilderness, *God* says: 'I will test them and see whether they will follow my instructions.'

xxii. Thiessen, H.C, *Lectures in Systematic Theology*, Wm.B. Eerdmans Publishing Co., Grand Rapids, USA, revised edition, p. 181, 1979.

xxiii. 'in every way' just as we are—yet was 'without sin' (Hebrews 4:15).

REFERENCES

1. Frothingham, R., *The Rise of the Republic of the United States*, Brown, Boston, MA, p. 6, 1910.
2. Constant, B., *De La Liberté Des Anciens Comparée à celle des Modernes*; from: 'Écrits Politiques', Folio, Paris, pp. 591–619, 1997.
3. Fustel De Coulanges, N.D., *The Ancient City : A Classic Study of the Religious and Civil Institutions of Ancient Greece and Rome*, Doubleday Anchor, New York, p. 223, 1955.
4. St Augustine; *The City of God*, Book III, par. 28, Cambridge University Press, Cambrdge, UK, 1998.

5. Tamanaha, B.Z., *On The Rule of Law: History, Politics, Theory*, Cambridge University Press, Cambridge, UK, p. 23, 2004.
6. Batten, D. (Ed.), *The Answers Book*, Creation Ministries International, Brisbane, p. 7, 1999.
7. Bracton, H., *On the Laws and Customs of England*, Vol. II, Harvard University Press, Cambridge, MA, p. 25, 1968.
8. Spurgeon, C.H., Joy Born at Bethlehem; in: Water, M. (Ed.), *Multi New Testament Commentary*, John Hunt, London, p. 195, 1871.
9. Stott, J., *Christian Basics: An Invitation to Discipleship*, Monarch Books, London, p. 79, 2003.
10. Adamthwaite, M., Civil Government, *Salt Shakers* 9:3, June 2003.
11. Paine, T., Of the origin and design of government; in: Boaz, D. (Ed.), *The Libertarian Reader: Classic and Contemporary Writings from Lao-Tzu to Milton Friedman*, Free Press, New York, p. 7, 1997.
12. Fortescue, Sir J., *De Laudibus Legum Anglie*, translated with Introduction by Chrimes, S.B., Cambridge University Press, Cambridge, UK, Chapter XLII, p. 105, 1949.
13. 7 Co. Rep. I, 77 Eng. Rep. 277 (K.B. 1960). Apud: Wu, John C.H.; *Fountain of Justice: A Study in the Natural Law*, Sheed and Ward, New York, p. 91, 1955.
14. Coke, Sir E., Third Reports, 3. Apud: Sandoz; Elliz; *The Roots of Liberty—Introduction*; in: *The Roots of Liberty: Magna Carta, Ancient Constitution, and Anglo-American Tradition of the Rule of Law*, University of Missouri Press, Columbia, MO, p. 137, 1993.
15. See Noebel, D., *The Battle for Truth*, Harvest House Publishers, Eugene, OR, p. 232, 2001.
16. Rushdoony, R.J., *Law and Liberty*, Ross House Books, Vallecito, CA, p. 33, 1984.
17. Kmiec, D.W., Liberty misconceived: Hayek's incomplete theory of the relationship between natural and customary law; in: Ratnapala, S. and Moens, G.A. (Eds.), *Jurisprudence of Liberty*, Butterworths, Adelaide, SA, Australia, p. 145, 1996.
18. De Montesquieu, C., *The Spirit of Laws*, Book I, Chapter 1, 1748.
19. Blackstone, W., *The Sovereignty of the Law—Selections from Blackstone's Commentaries on the Laws of England*, McMillan Publishers, London, pp. 58–59, 1973.
20. Stott, J., *The Message of Romans: God's Good News for the World*, Inter-Varsity Press, London, p. 342, 1994.
21. Schaeffer, F.A., *A Christian Manifesto*, Crossway, Westchester, p. 91, 1988.
22. *Pacem in Terris*, Encyclical letter of Pope John XXIII, Paragraph 51, 1963.
23. Knox, J., *On Rebellion*, Cambridge University Press, Cambridge, p. 192, 1994.
24. Rutherford, S., 'Lex Rex', or The Law and the Prince; in: *The Presbyterian Armoury*, vol. 3, p. 34, 1846.
25. Locke, J., *Second Treatise on Civil Government*, Sec. 222; From Locke, J., *Political Writings*, Penguin Books, London, p. 374, 1993.
26. *Popularum Progressio*, Encyclical letter of Pope Paul VI, Paragraph 31, 1967.
27. Colson, C. and Pearcey, N., *How Now Shall We Live?* Tyndale, Wheaton, IL, p. 131, 1999.
28. Karatnycky, A., Piano, A. and Puddington, A. (Eds.), *Freedom in the World 2003: The Annual Survey of Political Rights and Civil Liberties*, Freedom House, New York, pp. 11–12, 2003.
29. Maier, P.L., foreword to: Schmidt, A.J., *How Christianity Changed the World*, Zondervan, Grand Rapids, MI, p. 9, 2004.
30. Schmidt, A.J., *How Christianity Changed the World*, Zondervan, Grand Rapids, MI, p. 13, 2004.
31. Hayes, C.J.H., *Christianity and Western Civilization*, Stanford University Press, Stanford, CA, p. 21, 1954.

32. Keith, A., *Evolution and Ethics,* G.P. Putnam's Sons, New York, p. 230, 1947.
33. Grigg, R., Ernst Haeckel: evangelist for evolution and apostle of deceit, *Creation* 18(2):33-36, 1996; Fraud rediscovered, *Creation* 20(2):49-51, 1998.
34. Milner, R., *The Encyclopedia of Evolution,* Facts on File, New York, p. 206, 1990.
35. Ref. 1, p. 72.
36. Whitehead, J., *The Stealing of America*, Crossway Books, Illinois, USA, p. 15, 1983.
37. Darwin, C., *The descent of man,* John Murray, London, p. 134, 1887.
38. Evolution: The dissent of Darwin, *Psychology Today,* January/February, p. 62, 1997
39. The Descent of Man—Episode 1: The Moral Animal, broadcast on *The Science Show on the ABC Radio National*, 22 January 2000; abc.net.au/science/descent/trans1.htm.
40. E.g. Sigmund Freud, 'Like an idealized father, God is the projection of childish wishes for an omnipotent protector.' Cited from *Encyclopaedia Britannica,* 19:570, 1992.
41. See also Grigg, R.M., Creation—how did God do it?, *Creation* 13(2):36-38, 1991.
42. See also Grigg, R.M., How long were the days of Genesis 1? *Creation* 19(1):23-25, 1997.
43. See also Grigg, R.M., Why did God impose the death penalty for sin? *Creation* 15(1):32–34, 1993.
44. Sarfati, J., Why does science work at all? *Creation* 31(3):12–14 June 2009.
45. Harrison, P., *God's, Man's and Nature's: Laws of nature, moral order, and the intelligibility of the cosmos*, ISSN-2045-5577, 2011; lse.ac.uk.
46. Whitehead, A.N., *Science and the Modern World*, The Free Press, New York, paperback edition, p. 13, 1967; first published 1925.
47. Stark. R., *For the Glory of God: How monotheism led to reformations, science, witch-hunts and the end of slavery*, Princeton University Press, Princeton, NJ, chap. 2, 2003.
48. Hooykaas, R., *Religion and the Rise of Modern Science*, Scottish Academic Press, Edinburgh, 1977, first printed 1972.
49. Jaki, S.L., *Science and Creation*, Scottish Academic Press, Edinburgh, UK, 1986.
50. Foster, M.B., The Christian doctrine of creation and the rise of modern natural science, *Mind* **43**(172):446–468, October 1934.
51. Glover, W.B., *Biblical Origins of Modern Secular Culture: An essay in the interpretation of western history*, Mercer University Press, Macon, GA, chap. iv, 1984.
52. Hetherington, S.F., *Planetary Motions: A historical perspective*, Greenwood Press, Westport, CT, p. 19, 2006.
53. Hammerton J.A. (Ed.), *Illustrated Encyclopaedia of World History*, Mittal Publications, Daryaganj, New Delhi, vol. 3, p. 1479.
54. Aristotle, *Physics Book II*, c. 350 BC; classics.mit.edu.
55. Aristotle, *On the Heavens*, c. 350 BC; classics.mit.edu.
56. Oakley, F., Christian theology and the Newtonian science: the rise of the concept of the laws of nature, *Church History* 30(4):433–457, December 1961.
57. Zilsel, E., The Social Origins of Modern Science, Springer Science, Dordrecht, the Netherlands, p. 176, 2003; first published c. 1940.
58. Simonyi, K., *A Cultural History of Physics*, CRC Press, Boca Raton, FL, pp. 233–234, English Translation 2012.

59. Newton, I., *The Mathematical Principles of Natural Philosophy*, Book III, 1687; translated by Motte, A., Daniel Adee, New York, p. 506, 1846; archive.org.
60. Zilsel, E., The Genesis of the concept of physical law, *Philosophical Review* 51(3):245–279, 1942.
61. Steinle, F., The amalgamation of a concept—laws of nature in the new sciences; in: Weinert, F. (Ed.), *Laws of Nature: Essays on the philosophical, scientific and historical dimensions*, Walter de Greyter, New York, pp. 316–368, 1995. Return to text.
62. Steinle, ref. 24, p. 328.
63. Calvin, M., *Chemical Evolution: Molecular evolution towards the origin of living systems on the earth and elsewhere*, Oxford University Press, Oxford, UK, p. 258, 1969.
64. Collingwood, R.G., *The Idea of Nature*, Martino Publishing, pp. 5–8, 2014; first published 1945.
65. Harrison, P., The development of the concept of laws of nature; in Watts, F. (Ed.), *Creation: Law and probability*, Ashgate, UK, 2008. Return to text.
66. The phrase, "weight, measure, and number" is found in chapter 11 of the Book of Wisdom.
67. *The Works of the Honourable Robert Boyle*, vol. 5, Printed for W. Johnston *et al.*, London, pp. 139–140, 1722.
68. Newton, I., *Opticks: A treatise on the reflections, refractions, inflections and colours of light*, Dover Publications, New York, pp. 403–404, 1952.
69. McGrath, A.E., *The Science of God: An introduction to scientific theology*, T &T Clark, London, p. 51, 2004.
70. Davies, P., *The Mind of God: The scientific basis for a rational world*, Simon & Schuster, New York, p. 77, 1992.
71. "quelles sont les Lois que Dieu lui [nature] a imposées", Descartes, R., *Le Monde*, Theodore Girard, Paris, p. 78, 1664.
72. Jaeger, L., *What the Heavens Declare: Science in the light of creation*, Cascade Books, Wipf and Stock Publishers, Eugene, OR, p. 40, 2012.
73. *The Works of the Honourable Robert Boyle*, vol. 4, Printed for W. Johnston *et al.*, London, pp. 68–69, 1722.
74. Foster, ref. 7, p. 465. See also Foster, M.B., Christian theology and modern science of nature (I), *Mind* 44(176):439–466, October 1935; and Foster, M.B., Christian theology and modern science of nature (II), *Mind* 45(177):1–27, January 1936.
75. St Thomas Aquinas, *Summa Contra Gentiles*, Book Two: Creation, Chap 25, section 14. Translated by James F. Anderson. (Notre Dame, IN: University of Notre Dame Press, 1975).
76. Robert Spencer, *The Politically Incorrect Guide to Islam (and the Crusades)* (Regnery Publishing, Inc., Washington, DC, 2005) page 96.
77. James V. Schall, *War-Time Clarifications: Who Is Our Enemy?* 2001; Quoted in ref 3.
78. www.geocities.com/athens/troy/7568/index.html accessed 26 February 2008.
79. Keil, C. and Delitzsch, D., *Biblical Commentary on the Old Testament*, Vol. 1, The Pentateuch, trans. from the German by James Martin, Wm. B. Eerdmans Publishing Co., Grand Rapids, USA, p. 93, 1968.

30

THE MYSTERY OF 6-DAYS CREATION

PROMINENT SCIENTISTS BELIEVE IN CREATION

Almost every person at one time or another asks the question, "Where did life come from?" Bound up with the answer is the additional question, "What is the purpose of life on earth?" Essentially two viewpoints exist on this question:

1. the *atheist* position, which concludes that life came about through change, time and a large number of fortuitous events; and
2. the *creationist* position, which teaches that every living organism type was created by a creator which most people call God.

Christianity has, since its inception, taught that life was created by God for a specific purpose. "You (God) created all things, and because of your will they existed and were created" (*Rev. 4:11*).

Evolutionary naturalism, often called atheism, teaches that life began by the random collision of enough atoms to form complex molecules that produced accurate copies of themselves. These hypothetical molecules eventually evolved into cells and, in billions of years, evolved into all life extant today. The key to this molecule-to-human evolution was mutations (genetic copy errors) and natural selection (the selection of favorable mutations that alter the animal or plant so that they are more apt to survive).

THE NECESSITIES FOR LIFE

The concept of this chapter is that the origin of life could not have occurred by a gradual process but must have been instantaneous. The reason this must be true is simple. Every machine must have a certain minimum number of parts for it to function, and if one part below this minimum is removed, the machine will cease to function. The example "Behe" uses is a common spring mousetrap which requires ten parts to function. The trap will no longer function if just *one* part is removed. No one has been able to show this concept to be erroneous only that under certain conditions a certain machine can operate with *one fewer part*.

Many of these "one fewer part" examples, though, are misleading. Ruse (p. 28, 1993) observes that a mousetrap can be fastened to the floor, thereby eliminating the base, he claims. In fact, it only uses a *different* base (the floor); a base is still necessary. Further, the mousetrap parts are useless without the intelligence to assemble them into a functioning unit. A trap is also useless without the bait, the knowledge and ability necessary to use the trap, and the existence of a mouse with enough intelligence to seek the bait but lacking in the experience and intelligence to avoid the trap.

A simple mousetrap system is much more complex than it first appears.

The irreducible complexity argument can be extended to the creation process which produced life. The concept argues that both an organism and its parts, including organs, organelles, cells, or even its protein, cannot function below a certain minimum number of parts. In biological organisms the smallest unit of life is the cell, and the number of parts it contains at the subatomic level is usually much larger than a trillion. As Hickman documented:

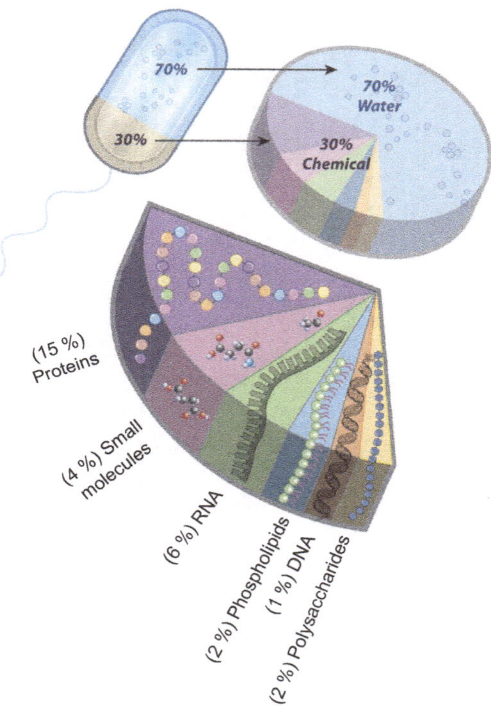

Fig.30.1: Cells are the Fabric of Life

Cells are the fabric of life, Figure 30.1. Even the most primitive cells are enormously complex structures that form the basic units of all living matter. All tissues and organs are composed of cells. In a human an estimated ~ 60-100 trillion cells interact, each performing its specialized role in an organized community. In single-celled organisms all the functions of life are performed within the confines of one microscopic package. There is no life without cells (Hickman, p. 43, 1997).

Even most bacteria require several thousand genes to carry out the functions necessary for life. *E. coli* has about 4,639,221 nucleotide base pairs, which code for 4,288 genes, each one of which produces an enormously complex protein machine. The simplest species of bacteria, Chlamydia and Rickettsia, are the smallest living things known, Figure 30.2. Only a few hundred atoms across, they are smaller than the largest virus and have about half as much DNA as do other species of bacteria. Although they are about as small as it is possible to be and still be living, these two forms of life still require millions of atomic parts (Trefil, p. 28, 1992). Many of the smaller bacteria, such as *Mycoplasma genitalium*, which has 452 genes, are parasite-like viruses and can only live with the help of more complex organisms. For this reason, when researching the minimum requirements for life, the example of *E. coli* is more realistic.

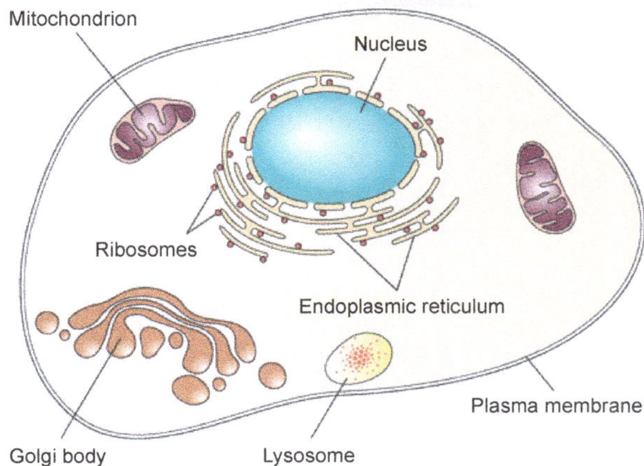

Fig.30.2: The Basic Structural and Functional Unit of Life: The Cell - The cell is structurally and functionally complex.

If the simplest form of life requires millions of parts at the atomic level, higher life-forms require trillions. All of the many macromolecules necessary for life are constructed of atoms, which are composed of even smaller parts. That life requires a certain minimum number of parts is well documented, and the only debate is *how many* millions of functionally integrated parts are necessary, not the fact that a minimum number must exist for life to live. All viruses are *below* the complexity level needed for life, and for this reason they must live as parasites that require complex cells in order to reproduce. "Trefil" noted that the question of where the viruses come from is an "enduring mystery" in evolution. They consist primarily of only a DNA molecule and a protein coat and

... don't reproduce in the normal way, [therefore] it's hard to see how they could have gotten started. One theory: they are parasites who, over a long period of time, have lost the ability to reproduce independently. ... Viruses are among the smallest of "living" things. A typical virus, like the one that causes ordinary influenza, may be no more than a thousand atoms across. This is in comparison with cells which may be hundreds or even thousands of times that size. Its small size is one reason that it is so easy for a virus to spread from one host to another—it's hard to filter out anything that small ("Trefil," p. 9, 1992).

Oversimplified, life depends on a complex arrangement of three classes of molecules:

- *DNA*, which stores the cell's master plans;
- *RNA*, which transports a copy of the needed information contained in the DNA to the protein assembly station; and
- *proteins*, which make up everything from the ribosomes to the enzymes.

Further, chaperons and many other assembly tools are needed to ensure that the protein is properly assembled. All of these parts are necessary and must exist as a properly assembled and integrated unit. DNA is useless without both RNA and proteins, although some types of bacteria can combine the functions of the basic required parts, Figure 30.3.

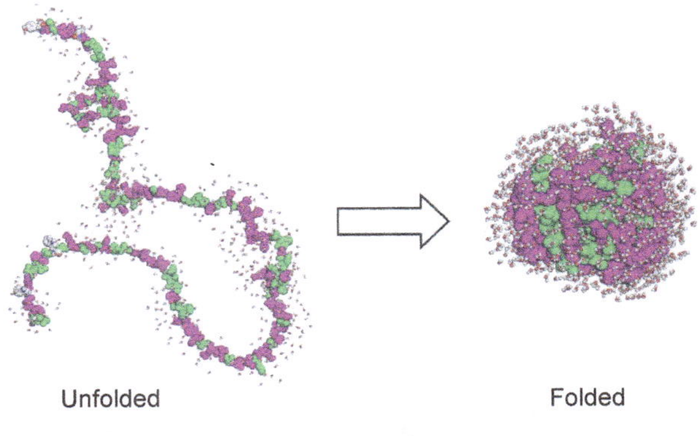

Fig.30.3: Protein Folding

The problem for evolution caused by the enormous complexity required for life is quite well recognized, and none of the proposals to overcome it are even remotely satisfactory ("Spetner," 1997). These proposals include the theory of panspermia advanced by Nobel Laureate "Francis Crick." Panspermia is the hypothesis that the earth was seeded by life from other planets (Crick, 1981), Figure 30.4. This solution, though, only moves the problem elsewhere.

Fig.30.4: *The Structure of DNA - Genetics - The Structure of DNA*

Naturalism must account for both the parts necessary for life and their proper assembly. For life to persist, living creatures must have a means of taking in and biochemically processing food.

Life also requires oxygen, which must be distributed to all tissues, or for single-celled life, oxygen must effectively and safely be moved around inside the cell membrane to where it is needed, without damaging the cell. Without complex mechanisms to achieve these tasks, life cannot exist. The parts could not evolve separately and could not even exist independently for very long, because they would break down in the environment without protection (Overman, 1997).

Even if they existed, the many parts needed for life could not sit idle waiting for the other parts to evolve, because the existing ones would usually deteriorate very quickly from the effects of dehydration, oxidation, and the action of bacteria or other pathogens. For this reason, only an instantaneous creation of all the necessary parts as a functioning unit can produce life. No compelling evidence has ever been presented to disprove this conclusion, and much evidence exists for the instantaneous creation requirement, such as the discovery that most nucleotides degrade rather fast at the temperatures scientists conclude existed on the early earth (Irion, 1998).

The problem is that the half-lives of many of the basic building blocks of life "are too short to allow for the adequate accumulation of these compounds. ... Therefore, unless the origin of life took place extremely rapidly (<100 years) ... a high temperature origin of life ... cannot involve adenine, uracil, guanine, or cytosine" ("Levy and Miller," p. 7,933, 1998). This finding is a major setback for abiogenesis, because high temperature (80°–100°C) origin of life is the only feasible model left (Levy and Miller, 1998). Creationists have only begun to exploit this huge stumbling block to Darwinism.

The simplest eukaryote life form is yeast, Figure 30.5. Most eukaryotes are much more complex than yeast, and a fertilized egg, called a zygote, is the minimum complexity possible for all multicell life-forms. Further, the development of an organism from a zygote does not provide evidence of evolution, because a zygote cannot exist as an independent unit, but is dependent on a complex designed support system, such as a womb or an egg. A complex life system designed to produce the gametes first exists, and the zygote is only part of a series of stages designed to allow it to fulfill its potential.

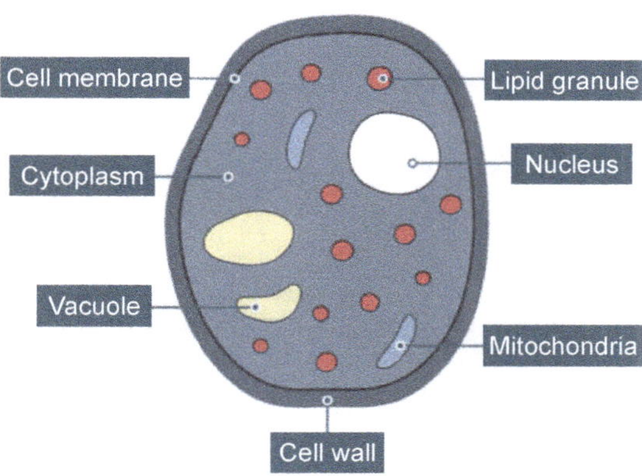

Fig.30.5: Saccharomyces cerevisiae (also known as "Baker's Yeast" or "Brewer's Yeast") is a unicellular fungus responsible for alcohol production and bread formation

An organ or an organism cannot function, nor will it be selected, until it is minimally functional. At this level it must be both enormously complex and dependent on many other parts of the system ("Behe," 1996). A gamete contains all the information needed to develop into a complete organism. When the organism is first developing, all its cells are totipotent, meaning that each cell can develop into any one of the over 200 cell types needed for an adult human to live, including epithelial, muscle, blood and other cell types.

Evolutionists once argued that all life could develop from some hypothetical first cell, because even today all new life develops from a single cell, but we now realize that a cell can develop into a complex organism only because all of the parts and instructions are in the original cell produced from conception.

The human mother passes not only 23 chromosomes but also an entire cell to her offspring, which includes all the organelles needed for life. A cell can come only from a functioning cell and cannot be built up piecemeal, because all the major organelles must have been created and assembled instantaneously for the cell to exist ("Overman," 1997).

Cells require all their millions of necessary parts to remain alive, just as a mammal must have lung, liver, heart and other organs to live. All of the millions of cell parts are required to carry out the complex biochemical business necessary for life. This business requires manufacturing and processing of proteins, and storing of genetic information to be passed on to the next generation. "Trefil" called the evolution of prokaryotes (cells without organelles) into eukaryotes (cells with organelles and other structures lacking in prokaryotes) an "enduring mystery of evolution" because of the lack of evidence of the evolution of organelles, and the total lack of plausible links between eukaryotes and prokaryotes.

The differences between prokaryotic and eukaryotic cells are striking, to say the least. But if the latter evolved from the former, why are there no intermediate stages between the two? Why, for example, are there no cells with loose DNA and organelles? If the evolutionary line really went from prokaryotes to eukaryotes, and we have many living samples of each, why did none of the intermediate stages survive? ("Trefil," p. 104, 1992).

THE MYSTERY OF LIGHT CAME TO EXISTENCE

On the first day of creation, God said, "Let there be light" (Genesis 1:3), and light appeared as a thing separate from darkness, Figure 30.6.

The phrase ***let there be light*** could be confusing to some modern English speakers who are used to using the word *let* in the context of permission, as in "Let me out of this box" or "Let me have the last cookie." Some might wonder whom is God speaking to. Was there some cosmic jailer who was keeping the light under lock and key?

Fig.30.6: *God Said: Let There Be Light*

GOD SPEAKING INTO THE VOID

The phrase *let there be light* is a translation of the Hebrew phrase ***yehi*** *or*, which was translated "***fiat lux***" in Latin. A literal translation would be a command, something like "***Light, exist***." God is speaking into the void and commanding light to come into being. The Bible tells us that God created the heavens and the earth and everything else that exists by simply speaking them into existence (Genesis 1).

His personality, power, creativity, and beauty were expressed in creation the same way an artist's personality and personal attributes are expressed through art or music. The idea of light, existing first in God's mind, was given form by the words "Let there be light" or "Let light exist."

CREATIVE POWER OF GOD'S VOICE

The reality of the creative power of God's voice has important spiritual implications that go well beyond the creation account itself. Light is often used as a metaphor in the Bible, and the word *illumination* ("divine enlightenment of the human heart with truth") has to do with bringing things into the light.

Spiritual illumination is a kind of "creation" that occurs in a human heart. "God, who said, 'Let light shine out of darkness,' made his light shine in our hearts to give us the light of the knowledge of God's glory displayed in the face of Christ" (2 Corinthians 4:6). Jesus Himself is "the light of the world" (John 8:12).

When God said, "Let there be light," at the creation, and light appeared, it showed God's creative power and absolute control. The physical light that God made on the first day of creation is a wonderful picture of what He does in every heart that trusts in Christ, the True Light. There is no need to walk in the darkness of sin and death; in Christ, we "will never walk in darkness, but will have the light of life" (John 8:12).

HOW COULD THERE BE LIGHT ON THE FIRST DAY OF CREATION

Sun was not Created until the 4th day

The question of how there could be light on the first day of Creation when the sun was not created until the fourth day is a common one, Figure 30.7.

Fig.30.7: 4th Day of Creation

Genesis 1:3-5 declares, "And God said, ***'Let there be light,'*** and there was light. God saw that the light was good, and He separated the light from the darkness. God called the light ***'day,'*** and the darkness He called ***'night.'*** And there was evening, and there was morning — ***the first day.***"

A few verses later we are informed, "And God said, 'Let there be lights in the expanse of the sky to separate the day from the night, and let them serve as signs to mark seasons and days and years, and let them be lights in the expanse of the sky to give light on the earth.' And it was so. God made two great lights — the greater light to govern the day and the lesser light to govern the night. He also made the stars. God set them in the expanse of the sky to give light on the earth, to govern the day and the night, and to separate light from darkness. And God saw that it was good. And there was evening, and there was morning — the fourth day" (Genesis 1:14-19).

How can this be? How could there be light, mornings and evenings on the first, second, and third days if the sun, moon, and stars were not created until the fourth day?

This is only a problem if we fail to take into account an infinite and omnipotent God. God does not need the sun, moon, and stars to provide light. **God is light!** First John 1:5 declares, "This is the message we have heard from him and declare to you: ***God is light***; in Him there is no darkness at all; Figure 30.8."

God Himself was the light for the first three days of Creation, just as He will be in the new heavens and new earth, "There will be no more night. They will not need the light of a lamp or the light of the sun, for the Lord God will give them light. And they will reign for ever and ever" (Revelation 22:5). Until He created the sun, moon, and stars, God miraculously provided light during the "day" and may have done so during the "night" as well (Genesis 1:14).

Fig.30.8: Transfiguration of Jesus to Glorious Light While Speaking with Moses and Elijah

Jesus said, "*I am the light of the world. Whoever follows me will never walk in darkness, but will have the light of life*" (John 8:12). Much more important than the light of day and night is the Light who provides eternal life to all who believe in Him. Those who do not believe in Him will be doomed to "outer darkness where there is weeping and gnashing of teeth" (Matthew 8:12).

THE LIGHT OF STARS BILLIONS OF LIGHT YEARS AWAY

Earth is only Thousands of Years Old

A light-year is the maximum distance that light can travel in one year in the vacuum of space. Consequently, it takes billions of years for light to travel billions of light-years through space. From our vantage point here on Earth we can see light from stars that are billions of light-years away. It is reasonable, therefore, to assume that our universe is at least billions of years old—old enough to give the light from these stars enough time to reach our planet billions of light-years away, Figure 30.9!

Fig.30.9: The Light of Stars Billions of Light Years Away from Earth

This reasonable assumption ***contradicts the Young Earth (YE)*** perspective, which claims that the ***universe is less than 10,000 years old***. If there was not a strong scientific case for the YE perspective, this contradiction would not merit a second thought. The growing body of evidence supporting the **YE** view is substantial enough, however, to warrant a thoughtful investigation into whether or not this apparent contradiction can be resolved reasonably. And so, we ask the question: ***How can the light of stars billions of light-years away reach the earth in only a few thousand years?***

GRAVITATIONAL TIME DILATION

According to Albert Einstein, space is not the empty "nothingness" that most of us perceive it to be. It is filled with what Einstein called ether. Dictionary.com defines ether as "an all-pervading, infinitely elastic, massless medium." Everything that exists within the bounds of our universe does so within this massless medium.

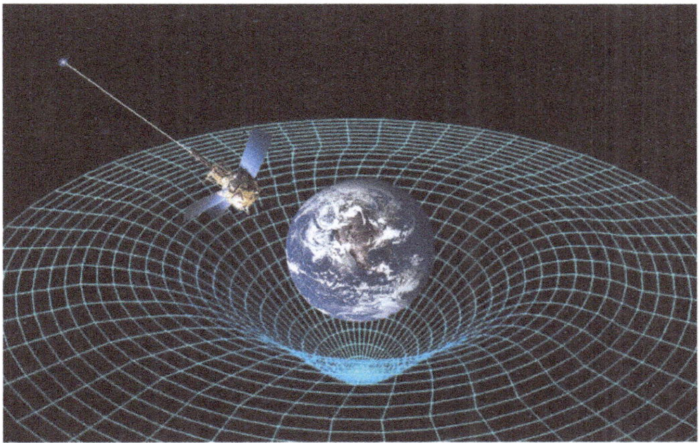

Fig.30.10: Gravitational Time Dilation

As dictionary.com notes, ether is infinitely elastic. ***It can be stretched and distorted***. In order to visualize this, imagine a tightly stretched cloth, Figure 30.10. This is ether. Now imagine dropping a heavy ball (like a bowling ball) onto the cloth, right in the middle. This would cause the cloth to sag in the middle. The heavy ball represents dense matter, like our planet. Einstein believed that matter causes space to sag, similar to how the heavy ball causes the stretched cloth to sag. These sags in space are known as gravity wells.

Now, if we placed smaller, lighter balls (like marbles) onto the cloth along with the heavy ball, they would roll toward the center, into the sag caused by the heavy ball. Moreover, they would contribute to the overall sagging of the cloth, even if only slightly. This motion towards the center represents ***gravity***. According to Einstein's view of gravity, if smaller, lighter forms of matter are close enough, they can be drawn into the gravity wells of larger, denser forms of matter. While they each create their own sag in space, some gravity wells are deeper and more influential than others (that is, they generate a stronger gravitational force). One thing they all have in common: ***they distort time***.

In the 1960s, physicists "Robert Pound" and "Glen Rebka" experimentally confirmed a theoretical consequence of Einstein's Theories of Relativity called the ***Gravitational Time Dilation Effect (GTDE).*** Pound and Rebka were able to demonstrate that time passes more slowly for objects the farther they travel into a gravity well. For example, Global Positioning System (GPS) satellites are farther away from the earth than objects on the planet's surface and are therefore less immersed in the gravity well caused by Earth's mass. The result is that time passes a little more quickly for our GPS satellites than it does for us here on the surface, since we are deeper inside of the earth's gravity well. Atomic clocks aboard the satellites and here on Earth have been used to detect and measure this difference in the rate of time's passage.

Likewise, an atomic clock in Greenwich, England (at sea level), Figure 30.11, records a slower rate of time than the atomic clock in Boulder, Colorado (at 5,430 feet above sea level). At these relatively small altitudinal differences, the measurable effect is minor. The effect across the greater cosmos can be much more dramatic. The deeper a gravity well, the stronger the GTDE. In fact, according to General Relativity, time actually stands still at the boundary of a black hole—an area known to scientists as an "***event horizon***," where gravity is so intense that even light cannot escape (hence the name "black hole").

Fig.30.11: *Atomic Clock - The World's Most Accurate Clock between Earth and the Cosmos*

Now, let's set aside the GTDE for a moment and consider another important astronomical phenomenon: stellar redshifts. Redshifts are a Doppler effect phenomenon whereby radiational wavelengths (like those of starlight) lengthen as they move farther away from an observer. The general consensus among astronomers is that observed stellar redshifts indicate that the universe is expanding (***Hubble's Law***). By extrapolating this expansion backwards, it becomes apparent that the primordial universe was somewhat denser, more compact than it is today.

In a bounded universe wherein matter has a center and an edge, the material compression as described above would serve to deepen the gravity well caused by the combined mass of the universe. This would intensify the GTDE, causing time to pass much more slowly near the center of the universe (deeper in the well) than near its edge (nearer the surface of the well).

The implication is paradoxical: even if the entire universe was created all at once in the beginning (and should therefore be the same age), ***some parts can be substantially younger than others due to the relativistic nature of time***. Light could travel billions of light-years over billions of years in some parts of the universe in what we ***on Earth would perceive to be a much shorter period of time***, Figure 30.12. As the universe expands and matter spreads out across space, the universal gravity well would gradually even out, lessening the rate of time difference across the universe.

Many astrophysicists and astronomers reject the idea of a bounded universe with our galaxy, the Milky Way, near or at its center. But this is a philosophical presupposition, not a scientific conclusion founded upon empirical data. As world-renowned astrophysicist "Dr. George F. R. Ellis" candidly explained, *"People need to be aware that there is a range of models that could explain the observations. For instance, I can construct you a spherically symmetrical universe with Earth at its center, and you cannot disprove it based on observations... you can only exclude it on philosophical grounds. In my view there is absolutely nothing wrong in that. What I want to bring into the open is the fact that we are using philosophical criteria in choosing our models. A lot of cosmology tries to hide that."* (W. Wayt Gibbs, "Profile: George F. R. Ellis," *Scientific American*, October 1995, Vol. 273, No.4, p. 55).

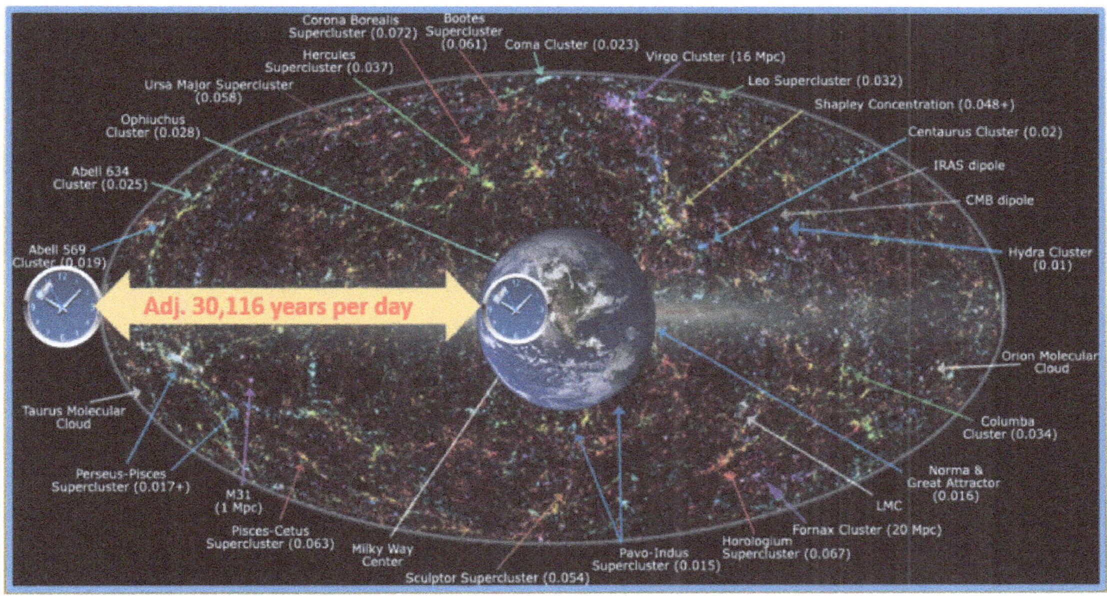

Fig.30.12: Light could travel billions of light-years over billions of years in some parts of the universe in what we on Earth would perceive to be a much shorter period of time

In summary, the Gravitational Time Dilation Effect is a theoretical solution to the YE problem of distant starlight which, amazingly, reconciles evidence for a young Earth with evidence for an old universe. Many astrophysicists and astronomers reject one of the major foundational suppositions upon which the GTDE explanation rests (a bounded universe with the Milky Way at or near the center), not because of the observable data but because of their philosophical perspectives.

WHAT HAPPENED ON EACH OF THE DAYS OF CREATION?

The creation account is found in Genesis 1—2. Most of God's creative work was done by speaking, an indication of the power and authority of His Word. Let us look at each day of God's creative work:

Creation Day 1 (Genesis 1:1–5). Figure 30.13

God creates the heavens and the earth. "The heavens" refers to everything beyond the earth, outer space. The earth is made but not formed in any specific way, although water is present. God then speaks light into existence. He then separates the light from the dark and names the light "day" and the dark "night."

Fig.30.13: Day 1 Creation

Creation Day 2 (Genesis 1:6–8), Figure 30.14

Fig.30.14: Day 2 Creation

God creates the sky. The sky forms a barrier between water upon the surface and the moisture in the air. At this point earth has an atmosphere.

Creation Day 3 (Genesis 1:9–13), Figure 30.15

Fig.30.15: Day 3 Creation

God creates dry land. Continents and islands rise above the water. The large bodies of water are named "seas" and the ground is named "land." God declares that all this is good.

God creates all plant life. He creates this life to be self-sustaining: plants can reproduce. The plants are created in great diversity (many "kinds"). The land is green and teeming with plant life. God declares that this work is also good.

Creation Day 4 (Genesis 1:14–19), Figure 30.16

Fig.30.16: Day 4 Creation

God creates all the stars and heavenly bodies. The movement of these will help man track time. Two great heavenly bodies are made in relation to the earth. The first is the sun, which is the primary source of light, and the moon, which reflects the light of the sun. The movement of these bodies will distinguish day from night. This work is also declared to be good by God.

Creation Day 5 (Genesis 1:20–23), Figure 30.17

Fig.30.17: Day 5 Creation

God creates all life that lives in the water, in all of its marvelous diversity. God also makes all the birds. The language of the passage allows that this may be the time God made flying insects as well; if not, they are made on Day 6. All these creatures have the ability to perpetuate their species by reproduction. The creatures made on Day 5 are the first creatures blessed by God. God declares this work good.

Creation Day 6 (Genesis 1:24–31), Figure 30.18

Fig.30.18: Day 6 Creation

God creates all the creatures that live on dry land. This includes every type of creature not included on previous days. God also creates man. God declares this work good.

When God was creating man, He took counsel with Himself. "God said, 'Let us make man in our image, in our likeness'" (Genesis 1:26). This is not an explicit revelation of the Trinity but is part of the foundation for such, as God reveals an "us" within the Godhead. God makes mankind in His own image, and thus mankind is special above all other creatures. He makes them ***male and female*** and places them in authority over the earth and over all the other creatures. God blesses them and commands them to reproduce, fill the earth, and subdue it (bring it under the rightful stewardship of mankind as authorized by God). God announces that humans and all other creatures are to eat plants alone. God will not rescind this dietary restriction until Genesis 9:3–4.

God's creative work is complete at the end of the sixth day. The entire universe in all its beauty and perfection was fully formed in these six periods labeled as "days." At the completion of His creation, God announces that it is "very good" (Genesis 1:31).

Creation Day 7 (Genesis 2:1–3), Figure 30.19

God rests. This in no way indicates He was weary from His creative efforts; rather, it denotes that the creation is complete. He stops creating. Further, God is establishing a pattern of one day in seven to rest. The keeping of this day will eventually be a distinguishing trait of God's chosen people, Israel (Exodus 20:8–11).

Fig.30.19: Day 7 – God Rested!

Many Christians interpret these "days" of creation as literal, 24-hour periods, a position called Young-Earth Creationism. It should be noted that certain other interpretations of these "days" suggest they were indeterminate periods of time. The Day-Age Theory and Historical Creationism are two theories that interpret the biblical data in a way that allows for an older earth. Regardless, the events and accomplishments of each "day" are the same.

THE UNIVERSE DESIGNED TO CONTAIN HUMAN LIFE

This view is also reflected in the observation that the universe appears to be designed specifically to contain human life, and functions as a unit to allow and support life ("Overman," 1997).

Creation of Humans

The problems of an instantaneous creation are best illustrated by the first man, Adam. If created as a mature adult, Figure 30.20, Adam would appear to be about, say, 30 years of age when he was only one day old. If Adam were examined medically, much scientific evidence in support of a 30-year age estimate would be found. Most medical tests completed on such a man would conclude he was and would have to be treated medically as if he, in fact, were at the prime of his life, even though only a day old.

Fig.30.20: God Created Man

This does not imply that God is deceptive, but only that to exist as a living organism, the human body *had* to be created fully formed. If his blood was not already circulating when Adam was created, the few minutes that it would take to prime the system and for blood to circulate to the brain could cause major cell death or damage. All of Adam's organs, including his heart, lungs, kidneys, and brain, must have been *functioning simultaneously* as a unit the second he was created. In other words, God created Adam as a mature man.

Although the physician who completed a physical on Adam a day after he was created would have had to conclude from development measures, such as bone-to-cartilage ratios, that Adam was 30 years old, some evidence for youth might have been found—in a one-day-old Adam, we might not have found certain effects of ageing, such as brain cell changes, which exist in the average 30-year-old today.

This, though, might have been because he was perfect, but this does not rule out the fact that some evidence, such as tissue culture examination of his cells, might have existed to prove he was in fact one week old.

Likewise, because the universe is enormously interrelated, the Creator could not have created the earth alone, but must have created the entire heavens and earth as a functioning unit, Figure 30.21. And as God likewise created the universe for a reason (such as a support system for the earth), and must have created Adam with blood *moving* in his veins, it is likewise a logical inference that the stars were created moving in their orbits and with their light in transit. Although this belief may not currently be provable, it may nonetheless be the most reasonable of the few possibilities that now exist.

This view is more viable than it may first appear. Nobel Laureate "George Wald" even stated that he believed that the universe was designed for life. In a recent interview he stated that he has concluded that the evidence is clearly obvious, because the elements carbon, hydrogen, oxygen and nitrogen "have unique properties that fit the job and are not shared by any element in the periodic system" (interview in Levy, p. 12, 1998).

Creating the universe in parts would not be unlike creating a liver and waiting a *few* days before creating a brain, then several more weeks before creating a femur bone until the body was eventually complete. No other method appears to exist to produce life other than creating instantaneously a fully functioning complete organism. This does not preclude that changes may have occurred since that time, only that a certain level of complexity must have existed for both an organism and a universe to exist.

Fig.30.21: God Created the Universe

Genetic drift, mutations and the shuffling of the gene pool can bring about only minor changes in life, changes creationists label variations within the Genesis kinds. Some creationists believe that these changes have historically been relatively significant, such as a pair of cat-kind animals producing by genetic recombination all cats existing today, including lions, tigers and cheetahs. Anatomical comparisons support this and relatively minute differences exist, for example, between tigers and lions, at least compared with other animal kinds. Also, hybrids of animals (such as tigons and ligers) have been produced to show the closeness of many animals.

The comparing of the creation of a human body with the creation of the universe has been supported by recent findings. Research has revealed that the universe is extraordinarily organized:

- our earth is organized into a solar system, which is part of a highly organized group of stars called a *galaxy*
- a *galaxy*, that is part of a highly organized family of galaxies called *clusters* which,
- *clusters,* in turn, are organized into an enormous group of clusters
- group of clusters called *superclusters.*

LIFE AND INFORMATION

One of the most compelling types of evidence in support of the instantaneous creation worldview is the daily observation that information does not come about by chance and, if left to itself, *disorder* usually soon results. Archeologists are normally easily able to discern if an object found in their field research digs was produced by humans or by natural events such as wind or rain. The criteria they use to do this is the *degree of information* the object contains ("Yockey," 1992).

Complexity and information are compelling evidence that some outside intelligent agency (which in the case of an archeologist's findings was another human) has applied design skills and intelligence to the natural world, adding a higher level of information and order on top of that which naturally exists in the nonliving world such as rocks.

Both plant and animal kingdoms manifest enormous complexity and information in their genetic codes, but this order and information *preexists* in the animal or plant and was inherited and passed on through reproduction. Except for the living world and the "world" made by humans, *the natural world operates according to preexisting physical laws and previous events*. The living world, which scientists are only now beginning to understand, represents a level of design complexity based on information existing in the genetic code which is not found anywhere in the nonliving world except that created by humans.

Hence the rationale for the belief that the living world could not come from the nonliving world. As Nobel laureate research molecular biologist "Komfield" stated in a now-famous interview that occurred over 36 years ago:

While laboring among the intricacies and definitely minute particles in a laboratory, I frequently have been overwhelmed by a sense of the infinite wisdom of God ... one is rather amazed that a mechanism of such intricacy could ever function properly at all ... the simplest man-made mechanism requires a planner and a maker; how a mechanism ten times more involved and intricate can be conceived as self-constructed and self-developed is completely beyond me (Komfield, p. 16, 1962).

In other words, the enormous amount of genetic information that is translated into the complexity that is evident everywhere in the living world is far beyond that found in both the nonliving and human-manufactured world. Products produced by the nonliving world (such as smooth stones polished by moving water) could never produce either plant or animal life because all life is based on information, and the parts produced by that information must be assembled according to a designed plan in an environment such as a certain ecosystem that supports life.

MATHEMATICAL PROOF FOR THE DESIGNER NECESSITY

That a complex structure such as a living organism could be formed by chance without intelligent input has never been demonstrated in the lab or anywhere else. Given enough time, the naturalistic worldview reasons, anything is at least possible. The problem with this view is that the degree of information and complexity required for living organisms to be able to "live" is such that, aside from deliberate intelligent design, from what we know now, no matter what the conditions, time alone will not allow for the naturalistic construction of life. Evolutionist "Stephen Jay Gould" stated that even if evolutionary history on earth repeated itself a *million times*, he doubts whether anything like *Homo sapiens* would ever develop again (Gould, 1989; also, "Kayzer," p. 86, 1997).

Many researchers have concluded that the probability of life arising by chance is so remote that we have to label it an impossibility. For example, "Hoyle" (1983) notes that the probability of drawing either ten white or ten black balls out of a large box full of balls that contains equal numbers of black and white balls is five times out of one million! If we increase the number to 100 and draw sets of 100 balls, the probability of drawing 100 black or 100 white balls in succession is now so low as to be for all practical purposes impossible.

To illustrate this concept as applied in biology, an ordered structure of just ***206 parts*** will be examined. This is not a large number—the adult human skeleton, for example, contains on the average 206 separate bones, all assembled together in a perfectly integrated functioning whole.

And all body systems—even our cells' organelles—are far more complex than this.

To determine the possible number of different ways 206 parts could be connected, consider a system of one part which can be lined up in only one way (1 × 1); or a system of two parts in two ways (1 × 2) or 1, 2 and 2, 1; a system of three parts, which can be aligned in ***six ways*** (1 × 2 × 3), or ***1, 2, 3; 2, 3, 1; 2, 1, 3; 1, 3, 2; 3, 1, 2; 3, 2, 1***; one of four parts in 24 ways (1 × 2 × 3 × 4), and so on. Thus, a system of 206 parts could be aligned in 1 × 2 × 3 … 206 different ways, equal to 1 × 2 × 3 … × 206. This number is called "***206 factorial***" and is written "***206!***".

The value ***206!*** is an enormously large number, approximately ***10^{388}***, which is a "1" followed by 388 zeros, or:

10,000,000,000,000,000,000,000,000,000,000,
000,000,000,000,000,000,000,000,000,000,
000,000,000,000,000,000,000,000,000,000,
000,000,000,000,000,000,000,000,000,000,
000,000,000,000,000,000,000,000,000,000,
000,000,000,000,000,000,000,000,000,000,
000,000,000,000,000,000,000,000,000,000,
000,000,000,000,000,000,000,000,000,000,
000,000,000,000,000,000,000,000,000,000,
000,000,000,000,000,000,000,000,000,000,
000,000,000,000,000,000,000,000,000,000,
000,000,000,000,000,000,000,000,000,000,
000,000,000,000,000,000,000,000,000,000

Achievement of *only* the correct general *position* required (ignoring for now where the bones came from, their upside-down or right-side-up placement, their alignment, the origin of the tendons, ligaments, and other supporting structures) for all ***206 parts*** will occur only once out of 10^{388} random assortments. This means one chance out of 10^{388} exists of the correct order being selected on the first trial, and each and every other trial afterward, given all the bones as they presently exist in our body.

If one new trial could be completed each second for every single second available in all of the estimated evolutionary view of astronomic time (about ***10 to 20 billion years***), using the most conservative estimate gives us ***10^{18} seconds***; the chances that the correct general position will be obtained by random is *less than once in **10 billion years***. This will produce a probability of only one out of ***$10^{(388-18)}$*** or one in ***10^{370}***.

If each part is only the size of an electron, one of the smallest known particles in the universe, and the entire known universe were solidly packed with sets of bones, this area conservatively estimated at ***100 billion cubic light years*** could contain only about ***10^{130}*** sets of ***206 parts*** each. What is the possibility that *just one* of these ***10^{130} sets***, each arranging their members by chance, will achieve the correct alignment just *once* in ten billion years? Suppose also that we invent a machine capable of making not one trial per second, but a billion-billion different trials each second on every single one of the 10^{130} sets. The maximum number of possible trials that anyone could possibly conceive being made with this type of situation would permit a total of ***10^{166} trials*** ($10^{130} \times 10^{18} \times 10^{18}$). Even given these odds, the chance that one of these ***10^{166} trials*** would produce the correct result is only one out of ***10^{388}***, or only one in ***10^{222} trials*** for all sets.

Further, all the parts must both first exist and be instantaneously assembled properly in order for the organism to function. For all practical purposes, a zero possibility exists that the correct general position of only ***206 parts*** could be obtained simultaneously by chance and the average human has about ***75 trillion cells***!

The human cerebral cortex alone contains over 10 billion cells, all arranged in the proper order, and each of these cells is itself *infinitely complex* from a human standpoint. Each of the cells in the human body consists of multi-thousands of basic parts such as organelles and multi-millions of complex proteins and other parts, all of which must be assembled both correctly and instantaneously as a unit in order to function. This required balance and assembly must be maintained even during cell division.

This illustration indicates that the argument commonly used by evolutionists "given enough time, anything is possible" is wanting. Evolutionary naturalism claims that the bone system happened as a result of time, luck, and "natural" forces, the last element actually holding the status of a god.

Time, the chief escape that naturalism must rely on to support its theory, is thus a false god. Complex ordered structures of any kind (of which billions must exist in the body for it to work) cannot happen except by design and intelligence, and they must have occurred simultaneously for the unit to function. ***Scientists recognize this problem***, and this is why "Stephen Jay Gould" concluded that humans are a glorious evolutionary accident which required 60 trillion contingent events ("Gould," 1989, also "Kayzer," p. 92, 1997).

Of course, the naturalistic evolution assumption does not propose that the parts of life resulted from an assembly of bones, but instead proposes that an extended series of stepwise coincidences gave rise to life and the world as we know it. In other words, the first coincidence led to a second coincidence, which led to a third coincidence, which eventually led to coincidence "***i***," which eventually led up to the present situation, "***N.***" Evolutionists have not even been able to posit a mechanistic "first" coincidence, only the assumption that each step must have had a survival advantage and only by this means could evolution from simple to complex have occurred. Each coincidence "i" is assumed to be dependent upon prior steps and to have an associated dependent probability "Pi." The resultant probability estimate for the occurrence of evolutionary naturalism is calculated as the product series, given the following:

- *N* the number of stepwise coincidences in the evolutionary process
- *i* = the index for each coincidence: i = 1,2,3 ...
- *Pi* the evaluated dependent probability for the i'th coincidence
- *PE* = the product probability that everything evolved by naturalism.

Innumerable steps are postulated to exist in the evolutionary sequence, therefore N is very large (i.e., *N* ...). All values of *Pi* are less than or equal to one, with most of them much smaller than 1. The greater the proposed leap in step *i*, the smaller the associated probability Pi 1, and a property of product series where *N* is very large and most terms are significantly less than one quickly converges very close to zero.

The conclusion of this calculation is that the probability of naturalistic evolution is essentially zero. Sir "Fred Hoyle" (1982) calculated "the chance of a random shuffling of amino acids producing a workable set of enzymes" to be less than ***$10^{40,000}$***, and the famous unrealistically optimistic "Green Band" equation gives the chance of finding life on another planet in the order of only one in ***10^{30}***.

These probabilities argue that the chance distribution of molecules could never lead to the conditions favorable for the spontaneous development of life. The reasoning that leads us to this conclusion is that living molecules contain a large number of elements which must be instantly assembled in a certain order for life.

The probability of the required order in a single basic protein molecule arising purely from chance is estimated at $\mathbf{10^{43}}$ (Overman, 1997). Since thousands of complex protein molecules are required to build a simple cell, probability moves chance arrangements of these molecules outside the realm of possibility.

The smallest proteins have an atomic mass of 100,000 or more atomic mass units (AMU), which is equal to 100,000 hydrogen atoms ("Branden and Tooze," 1991). And this calculation evaluates only the necessary *order* of parts, *not a functional* arrangement, i.e., one that works. Even if the gears of a clock are arranged in the correct *order*, the clock will not function properly until the gears are properly meshed, spaced, adjusted, the tolerances are correct, and the system is properly secured.

A problem with understanding the concept "life" is that although we now have identified many of the chemicals which are necessary, researchers do not yet know all of the factors necessary for life "to live." Further, even assembling the proper chemicals together does not produce life.

The proper arrangement of amino acids to form protein molecules is only one small requirement for life. Most animals are constructed of millions of cells, and the cell itself is far more complicated than the most complex machine ever manufactured by humans.

The famous illustration "the probability of life originating from accident is comparable to the probability of the *unabridged dictionary resulting from an explosion in a print shop*" argues that information and complex systems *cannot come about by chance*, but can only be the product of an intelligent designer.

Books likewise do not come about by chance, but are the product of both reasoning and intelligence (although some books may cause us to wonder about the author, but this is another problem!). Even Darwin admitted in his writings that it was extremely difficult, or impossible, to conceive that this immense and wonderful universe, including humans with our capacity of looking far backward and far into the future, was the result of blind chance.

LIFE FROM NONLIFE?

An important part of the question, "Where did life come from?" is the issue of spontaneous generation, the concept that life could produce itself if the proper circumstances existed ("Lewis," 1997). This idea is no longer accepted as possible by secular scientists except only for the beginning of life, when some believe the first living organism somehow spontaneously generated itself once or, at most, a few times, and every living thing thereafter evolved from this "first" life.

This principle of science that life only comes from life is called the "law of biogenesis." The term is from the Greek word's *bios* (meaning life) and *genesis* (meaning birth, source or creation), and means that *living organisms are produced only by other living organisms*. Biologists know only that *all life derives from preceding life*, and that the parent organism's offspring are always of the *same* kind.

The idea that *life can come from nonlife is called abiogenesis*, which is assumed by evolutionists to have occurred only once or a few times at most in earth history. This conclusion is not a result of evidence, but is obtained because the *current dominant worldview in Western science, naturalism (atheism), requires* a chance spontaneous origin of life.

The naturalistic view requires a set of unknown conditions to have existed in the distant past that operated to produce the first "living" thing. These unknown forces do not operate today to *produce flies from decaying meat* or *bees from dead carcasses*, as once believed. Scientists have demonstrated that the belief that "life" could come from "nonlife," even if millions of years were available, is untenable (Overman, 1997).

Darwinism demands a nontheistic explanation and therefore is forced to put much displaced faith in an unprovable "one-time" event that they reason must have occurred because life is here. Hoyle, in a review of the literature, concluded:

There is not a shred of objective evidence to support the hypothesis that life began in an organic soup here on the earth. Indeed, Francis Crick, who shared a Nobel prize for the discovery of the structure of DNA, is one biophysicist who finds this theory unconvincing. So why do biologists indulge in *unsubstantiated fantasies* in order to deny what is so patently obvious, that the 200,000 amino acid chains, and hence life, did not appear by chance?

The answer lies in a theory developed over a century ago, which sought to explain the development of life as an inevitable product of the purely local natural processes. Its author, Charles Darwin, hesitated to challenge the church's doctrine on the creation, and publicly at least did not trace the implications of his ideas back to their bearing on the origin of life. However, he privately suggested that life itself may have been produced in "some warm little pond," and to this day his followers have sought to explain *the origin of terrestrial life* in terms of a process of chemical evolution from the primordial soup. But, as we have seen, this [theory] simply does not fit the facts (Hoyle, 1983, p. 23).

This conclusion is not unique to Hoyle but common to thinkers not blinded by dogmatic naturalism. Einstein argued that the "scientist's religious feelings *take the form of rapturous amazement* at the harmony of natural law, which *reveals an intelligence of such superiority* that, compared with it, all the systematic thinking and acting of human beings is an utterly insignificant reflection" ("Einstein," p. 29, 1949; emphasis mine).

Scientists once argued that life was relatively simple, could spontaneously generate, and regularly did so. They now realize the human cell is the most complex machine known in the universe, far more complex than the most expensive computer. This realization has forced many persons to conclude ***life could not have evolved, but must have been created instantaneously as a fully functioning unit***.

All of the extant evidence reveals that there is nothing living on earth, either animal or plant, that did not receive its life from previous life, its sexual or asexual parent. Since the law of biogenesis states that ***life proceeds only from preexisting life***, various forms of preexisting life must have been parents of all living organisms. And since life cannot create itself, the source of life must be God: "O Lord .... For with you is the fountain of life" (*Ps. 36:6–9*).

In the words of the well-known scientist, "Robert Jastrow," "For the scientist who has lived by his faith in the power of reason, the story [of the quest for the answers about the origin of life and the universe] ends like a bad dream. He has scaled the mountains of ignorance; ***he is about to conquer the highest peak; as he pulls himself over the final rock, he is greeted by a band of theologians who have been sitting there for centuries***" ("Jastrow, p." 116, 1978).

Should Genesis be taken literally?

Fig.30.22: A Babylonian Tablet Fragment Found at Nippur - an ancient Babylonian site in the same general location that Abraham came from. The area outlined in black is a record about the Flood. Curtsy Photo Dr Clifford Wilson

There are more than 300 known records of the Flood world-wide, with about 30 of them in writing. Some are remarkably close in their details to the original—the biblical account, Figure 30.22.

Creationists are often accused of believing that the whole Bible should be taken literally. This is not so! Rather, the key to a correct understanding of any part of the Bible is to ascertain the intention of the author of the portion or book under discussion. This is not as difficult as it may seem, as the Bible obviously contains:

- ***Poetry***—as in the Psalms, where the repetition or parallelism of ideas is in accordance with Hebrew ideas of poetry, without the rhyme (parallelism of sound) and metre (parallelism of time) that are important parts of traditional English poetry. This, by the way, is the reason why the Psalms can be translated into other languages and still retain most of their literary appeal and poetic piquancy, while the elements of rhyme and metre are usually lost when traditional Western poetry is translated into other languages.
- ***Parables***—as in many of the sayings of Jesus, such as the parable of the Sower (*Matthew 13:3–23*), which Jesus Himself clearly states to be a parable and about which He gives meanings for the various items, such as the seed and the soil.
- ***Prophecy***—as in the books of the last section of the Old Testament (Isaiah to Malachi).
- ***Letters***—as in the New Testament epistles written by Paul, Peter, John, and others.
- ***Biography***—as in the Gospels.
- ***Autobiography/testimony***—as in the book of Acts where the author, Luke, after narrating the Apostle Paul's conversion on the road to Damascus as a historical fact (*Acts 9:1–19*), then describes two further occasions when Paul included this conversion experience as part of his own personal testimony (*Acts 22:1–21; 26:1–22*).
- ***Authentic historical facts***—as in the books of 1 and 2 Kings, etc.

Therefore, the author's intention with respect to any book of the Bible is usually quite clear from the style and the content. Who then was the author of Genesis, and what intention is revealed by his style and the content of what he wrote?

THE AUTHOR OF THE BIBLE

The Lord Jesus Himself and the Gospel writers said that the Law was given by Moses (*Mark 10:3; Luke 24:27; John 1:17*), and the uniform tradition of the Jewish scribes and early Christian fathers, and the conclusion of conservative scholars to the present day, is that Genesis was written by ***Moses***. This does not preclude the possibility that Moses had access to patriarchal records, preserved by being written on clay tablets and handed down from father to son via the line of Adam–Seth–Noah–Shem–Abraham–Isaac–Jacob, etc., as there are 11 verses in Genesis which read, 'These are the generations [Hebrew: *toledoth* = 'origins' or by extension 'record of the origins'] of ... .' As these statements all come after the events they describe, and the events recorded in each division all took place before rather than after the death of the individuals so named, they may very well be subscripts or closing signatures, i.e. colophons, rather than superscripts or headings. If this is so, the most likely explanation of them is that Adam, Noah, Shem, and the others each wrote down an account of the events which occurred in his lifetime, and Moses, under the guidance of the Holy Spirit, selected and compiled these, along with his own comments, into the book we now know as Genesis.

Chapters 12–50 of Genesis were very clearly written as authentic history, as they describe the lives of Abraham, Isaac, Jacob, and his 12 sons who were the ancestral heads of the 12 tribes of Israel. The Jewish people, from earliest biblical times to the present day, have always regarded this portion of Genesis as the true record of their nation's history.

Accordingly, what about the first 11 chapters of Genesis, which are our main concern, as these are the ones that have incurred the most criticism from modern scholars, scientists, and skeptics?

THE NATURE OF GENESIS CHAPTER 1–CHAPTER 11

Are any of these chapters '*Poetry?*'

Providing an answer this question, we need to examine in depth just what is involved in the parallelism of ideas that constitutes Hebrew Poetry.

Let us consider *Psalm 1:1*, which reads as follows:

'Blessed is the man that walks not in the counsel of the wicked, nor stands in the way of sinners, nor sits in the seat of scoffers.'

Here we see triple parallelism in the nouns and verbs used (reading downwards in the following scheme), **Table 30.1:**

Table 30.1: Parallelism

walks	counsel	wicked
stands	way	sinners
sits	seat	scoffers

As well as this overt parallelism, there is also a covert or subtle progression of meaning. In the first column,

- 'walks' suggests short-term acquaintance,
- 'stands' implies readiness to discuss, and
- 'sits' speaks of long-term involvement.

In the second column,

- 'counsel' betokens general advice,
- 'way' indicates a chosen course of action, and
- 'seat' signifies a set condition of mind.

In the third column,

- 'wicked' describes the ungodly,
- 'sinner' characterizes the actively wicked, and
- 'scoffers' portrays the contemptuously wicked.

Other types of Hebrew poetry include contrastive parallelism, as in *Proverbs 27:6,*

'Faithful are the wounds of a friend, but the kisses of an enemy are deceitful',

and completive parallelism, as in *Psalm 46:1*,

'God is our refuge and strength, a very present help in time of need.'

And so, we return to our question. Are any of the first 11 chapters of Genesis poetry?

Answer: No, because these chapters do not contain information or invocation in any of the forms of Hebrew poetry, in either overt or covert form, and because Hebrew scholars of substance are agreed that this is so.

Observe: There certainly is repetition in Genesis chapter 1, e.g.,

'And God said …' occurs 10 times;

- 'and God saw that it was good/very good' seven times;

- 'after his/their kind' 10 times;
- 'And the evening and the morning were the ... day' six times.

However, these repetitions have none of the poetic forms discussed above; rather they are statements of fact and thus a record of what happened, and possibly for emphasis—to indicate the importance of the words repeated.

Are any of these Chapters Parables?

No, because when Jesus told a parable He either said it was a parable, or He introduced it with a simile, so making it plain to the hearers that it was a parable, as on the many occasions when He said,

'The kingdom of heaven is like ... .'

No such claim is made or style used by the author of *Genesis 1–11*.

Are Any of These Chapters *Prophecy*?

Not in their full context, although two promises of God are prophetic in the sense that their fulfilment would be seen in the future. One of these is *Genesis 3:15*, which was the pronouncement by God to the serpent (Satan) in metaphorical form:

'And I will put enmity Between you and the woman, and between your seed and her seed; He shall bruise you on the head, And you shall bruise him on the heel.' (NASB).

Many have interpreted the 'seed' in this verse as the Messiah, including most evangelicals and even the Jewish Targums hence the Talmudic expression 'heels of the Messiah'. The Messiah would suffer wounds to His feet (on the Cross), but would completely destroy Satan's power. This verse also hints at the *virginal conception*, as the Messiah is called the seed of the woman, contrary to the normal biblical practice of naming the father rather than the mother of a child (cf. Genesis chapters 5 and 11, 1 Chronicles chapters 1–9, Matthew chapter 1, Luke *3:23–38*).

The other is *Genesis 8:21–22 and 9:11–17*,

'And the LORD said in His heart, I will not again curse the ground any more for man's sake ... and the waters shall no more become a flood to destroy all flesh.'

Are Any of These Chapters *Letters, Biography, or Autobiography/ Personal Testimony*?

This is where we need to consider some of the subscripts mentioned above.

If Adam knew the events of Creation Days 1–6, they must have been revealed to him by God, as **Adam was not made until Day 6**, and so he could have known them only if God had told him. This view is reinforced by the words,

'These are the generations of [NIV: 'This is the account of'] the heavens and of the earth when they were created ...'

in *Genesis 2:4a*. The details of Day 7, the rest day, are included before this in *Genesis 2:2–3*, thereby completing (as we might expect) the record of a full seven-day week, before this subscript or closing signature appears.

Then follow the events of *Genesis 2:4b–5:1a*. This section tells us about Adam, his wife Eve, and their sons, and reads very much like a personal account of what Adam knew, saw, and experienced concerning the

Garden of Eden, and the creation of Eve (chapter 2), their rebellion against God (chapter 3), and the deeds of their descendants (chapter 4 to 5:1), albeit written in the third person. This section ends with the words,

'This *is* the book of the generations of Adam.'

Is it feasible that Adam could have written *Genesis 1:1–2:4a* as the result of his pre-Fall conversation with God, and *Genesis 2:4b–5:1* as the record of his own experiences? There is no problem concerning his ability to have done so.

Adam was created a mature man, endowed with all the DNA, knowledge and skill he needed to perform all the tasks assigned him by God. No cave-man he! Adam knew enough horticulture 'to dress and to keep' the Garden of Eden (*Genesis 2:15*), and ample intelligence to recognize and name the distinct kinds of animals (*Genesis 2:19*). He (and Eve) could converse with God without ever having learned an alphabet, and there is no reason to suppose that he was not fully skilled in writing also.

Are There Any Contradictions?

What about the supposed contradictions between the order of events in Genesis chapter 2 and the order given in chapter 1?

There are *none!*

If, with the NIV, we read

'Now the LORD God *had* planted a garden in the east …' (*Genesis 2:8*) and,

'Now the LORD God *had* formed out of the ground all the beasts of the field …'

(*Genesis 2:19* with emphasis added), it is clearly seen that chapter 2 states that the plants and animals were formed before Adam. When Adam named the animals (*Genesis 2:20*), they obviously were already in existence. There is no contradictory significance in the order of animals listed in Genesis 2:20; it is probably the order in which Adam met the animals, while the order of their creation is given in *Genesis 1:20–25.*

Dr "Henry Morris" comments:

'It was only the animals in closest proximity and most likely as theoretical candidates for companionship to man that were actually brought to him. These included the birds of the air, the cattle (verse 20—probably the domesticated animals), and the beasts of the field, which were evidently the smaller wild animals that would live near human habitations. Those not included were the fish of the sea, the creeping things, and the beasts of the earth mentioned in Genesis 1:24, which presumably were those wild animals living at considerable distance from man and his cultivated fields.'

Concerning the *names* of geographical sites, we have no idea what the configuration of the land or the rivers was before the Flood, because the pre-Flood world was completely destroyed. The land areas and rivers named before the Flood do not correspond to similarly named features after the Flood.

The *purpose* of *Genesis 2:18–25* is not to give another account of creation but to show that there was no *kinship* whatsoever between Adam and the animals. None was *like* him, and so none could provide fellowship or companionship *for* him.

Why not? Because Adam had not evolved from them, but was **'a living soul'** whom God had created 'in His own image' (*Genesis 2:7 and 1:27*). This means (among other things) that God created Adam to be a person whom He could address, and who could respond to and interact with Himself. Here, as in many other places, the plain statements of the Bible confront and contradict the notion of human evolution.

There is therefore enough evidence for us to conclude that Adam most probably was the author of *Genesis 2:4b–5:1*, and that this is his record of his own experiences with respect to events in the Garden of Eden, the creation of Eve, the Fall, and in the lives of Cain, Abel, and Seth.

The next section is from 5:1b to 6:9a, and deals with the line from Adam to Noah, ending with, 'These *are* the generations [or origins] of Noah.'

The next section is from 6:9b to 10:1a, and deals mainly with the Ark and the Flood, ending with, 'Now these *are* the generations of the sons of Noah, Shem, Ham, and Japheth.'

The wording of this subscript suggests that this portion was written by one of Noah's sons, probably Shem, as Moses was descended from Shem. These chapters read very much like an eye-witness account because of the intimacy of detail which they contain. Consider *Genesis 8:6–12* and note how this contains that ring of authenticity which is characteristic of an eye-witness account. It may even have been Shem's diary!

Genesis 8:6–12:

6 At the end of forty days Noah opened the window of the ark that he had made:

7 And sent forth a raven. It went to and fro until the waters were dried up from the earth.

8 Then he sent forth a dove from him, to see if the waters had subsided from the face of the ground.

9 But the dove found no place to set her foot, and she returned to him to the ark, for the waters were still on the face of the whole earth. So, he put out his hand and took her and brought her into the ark with him.

10 He waited another seven days, and again he sent forth the dove out of the ark.

11 And the dove came back to him in the evening, and behold, in her mouth was a freshly plucked olive leaf. So, Noah knew that the waters had subsided from the earth.

12 Then he waited another seven days and sent forth the dove, and she did not return to him anymore. (ESV).

Such meticulous details are the stuff of authentic eye-witness testimony. They have the ring of truth.

There is thus a substantial body of evidence that these portions of Genesis delineated by subscripts were written by the persons named therein, for the purpose of making and passing on a permanent record.

So then, were these first 11 chapters written as a *record of authentic historical facts*?

Answer: Yes, for several reasons.

I. Internal Evidence of the Book of Genesis

1. There is the internal evidence of the book of Genesis itself. As already mentioned, chapters 12–50 have always been regarded by the Jewish people as being the record of their own true history, and the style of writing contained in chapters 1–11 is not strikingly different from that in chapters 12–50
2. Hebrew scholars of standing have always regarded this to be the case. Thus, Professor "James Barr", Regius Professor of Hebrew at the University of Oxford, has written:

   'Probably, so far as I know, there is no professor of Hebrew or Old Testament at any world-class university who does not believe that the writer(s) of *Genesis 1–11* intended to convey to their readers the ideas that:

   a. creation took place in a series of six days which were the same as the days of 24 hours we now experience
   b. the figures contained in the Genesis genealogies provided by simple addition a chronology from the beginning of the world up to later stages in the biblical story
   c. **Noah's flood was understood to be world-wide and extinguish all human and animal life except for those in the ark,** Figure 30.23. Or, to put it negatively, the apologetic arguments which suppose the "days" of creation to be long eras of time, the figures of years not to be chronological, and the

flood to be a merely local Mesopotamian flood, are not taken seriously by any such professors, as far as I know.'

Fig.30.23: Noah's Ark

3. One of the main themes of Genesis is the Sovereignty of God. This is seen in God's actions in respect of four outstanding events in Genesis 1–11 (Creation, the Fall, the Flood, and the Babel dispersion), and His relationship to four outstanding people in Genesis 12–50 (Abraham, Isaac, Jacob, and Joseph). There is thus a unifying theme to the whole of the book of Genesis, which falls to the ground if any part is mythical and not true history; on the other hand, each portion reinforces the historical authenticity of the other.

II. Evidence from the Rest of the Bible

1. The principal people mentioned in Genesis chapters 1–11 are referred to as real—historical, not mythical—people in the rest of the Bible, often many times. For example, *Adam, Eve, Cain, Abel,* and *Noah* are referred to in 15 other books of the Bible.
2. The Lord Jesus Christ referred to the Creation of Adam and Eve as a real historical event, by quoting Genesis 1:27 and 2:24 in His teaching about divorce (Matthew 19:3–6; Mark 10:2–9), and by referring to Noah as a real historical person and the Flood as a real historical event, in His teaching about the 'coming of the Son of man' (Matthew 24:37–39; Luke 17:26–27).
3. Unless the first 11 chapters of Genesis are authentic historical events, the rest of the Bible is incomplete and incomprehensible as to its full meaning. The theme of the Bible is Redemption, and may be outlined thus:
   i. God's redeeming purpose is revealed in Genesis 1–11,
   ii. God's redeeming purpose progresses from Genesis 12 to Jude 25, and
   iii. God's redeeming purpose is consummated in Revelation 1–22.

   But **why does mankind need to be redeemed**? What is it that he needs to be redeemed from? The answer is given in Genesis 1–11, namely, from the ruin brought about by sin. Unless we know that the entrance of sin to the human race was a true historical fact, God's purpose in providing a substitutionary atonement is a mystery.

   Conversely, the historical truth of Genesis 1–11 shows that all mankind has come under the righteous anger of God and needs salvation from the penalty, power, and presence of sin.
4. Unless the events of the first chapters of Genesis are true history, the Apostle Paul's explanation of the Gospel in Romans chapter 5 and of the resurrection in 1 Corinthians chapter 15 has no meaning.

Paul writes: 'For as by one man's [Adam's] disobedience many were made sinners, so by the obedience of one [Jesus] shall many be made righteous' (Romans 5:19). And, 'For since by man *came* death, by man *came* also the resurrection of the dead. For as in Adam all die, even so in Christ shall all be made alive ... And so it is written, The first man Adam was made a living soul; the last Adam *was made* a quickening spirit'(1 Corinthians 15:21–22; 45). The historical truth of the record concerning the first Adam is a guarantee that what God says in His Word about the last Adam [Jesus] is also true. Likewise, the historical, literal truth of the record concerning Jesus is a guarantee that what God says about the first Adam is also historically and literally true.

Fig.30.24: A creation tablet found at Ebla in Syria and dating to the third millennium BC Curtsy Photo Dr Clifford Wilson

A creation tablet found at Ebla in Syria and dating to the third millennium BC, Figure 30.24. It ascribes the great works of creation to one great being, 'Lugal', literally 'the Great One.' It shows that both the creation story and the art of writing were well known to man up to 1,000 years before the time of Moses. It further shows that the liberal idea that the early chapters of Genesis were first put into writing hundreds of years after the time of Solomon is clearly fallacious.

We return to the question which, **Should Genesis be taken literally?**

Answer: If we apply the normal principles of biblical exegesis (ignoring pressure to make the text conform to the evolutionary prejudices of our age), it is overwhelmingly obvious that Genesis was meant to be taken in a straightforward, obvious sense as an authentic, literal, historical record of what actually happened.

Genesis: History ... or Mystery?

Philosopher "George Santayana" said, 'Those who cannot learn from history are doomed to repeat it.' But what would happen if there were no history to learn from?

The Bible records some 4,000 years of history. Even most non-Christian historians, archaeologists, etc. would agree that there is approximately

- 2,000 years of history between Jesus and the Patriarch Abraham. A simple addition of the genealogies from
- Abraham to Adam provides us with another approximately 2,000 years.

So fully half of the recorded time in the Bible occurs in Genesis chapters 1–11! Clearly this must be an important part of our history, Figure 30.25.

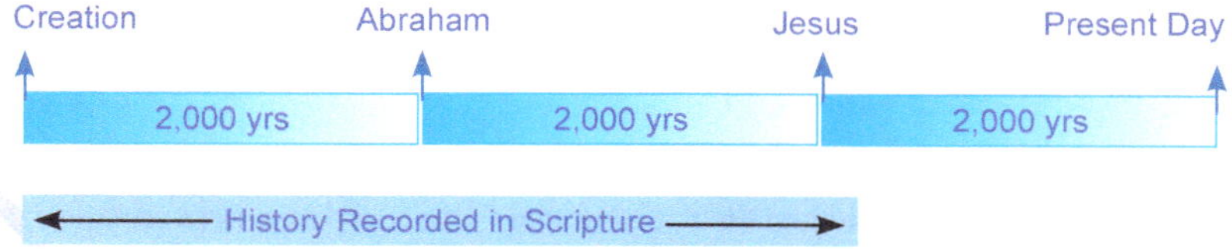

Fig.30.25: *Recorded History in Scripture*

CAN WE TRUST GENESIS 1–11 AS WRITTEN?

Many Bible scholars/commentators would declare that Genesis 1–11 is just a metaphor, allegory, poetry, mythology, or at the least that it cannot be understood as history as plainly written, even though they profess the Bible as the word of God. It's notable that none of these compromises was thought of before the rise of uniformitarian geology—And we see this in the excuses that are made for denying that God's Word means what it says:

'... God's upper level activity of issuing creative fiats from his heavenly throne is pictured as transpiring in a week of earthly days ... is a literary figure, an earthly, lower register time metaphor for an upper register, heavenly reality.'

'The study of paleontology has rendered it virtually impossible for a serious scientist to make a case for a **six-day creation** about six thousand years ago ... the first chapters of Genesis should use the kind of language technically known as 'mythological' to explain the origins of the universe.'

Many Christians believe the days described in Genesis were perhaps 'millions of years' of undetermined time rather than 24-hour days. They also claim that there are gaps in the genealogies, although once again, no one saw them before the rise of modern uniformitarian geology, precisely because they are not in the *text*.

Many also deny the global Flood, saying what is described in the Bible can be interpreted as merely a local event somewhere in Mesopotamia. But is Genesis really that hard to understand? Apparently not ...

'It is of course admitted ... it would be most natural to understand the word [day] in its ordinary sense; but if that sense brings the Mosaic account into conflict with facts, and another sense avoids such conflict, then it is obligatory on us to adopt that other.'

'... the most straightforward understanding of the Genesis record ... is that God created heaven and earth in six solar days, that man was created in the sixth day, that death and chaos entered the world after the Fall of Adam and Eve, that all of the fossils were the result of the catastrophic universal deluge which spared only Noah's family and the animals therewith.'

So, the reinterpretation of Genesis is the result of Christians accepting the evolutionary dogma taught in public schools and through the media and trying to fit it into Scripture.

But an abandonment of Genesis 1–11 as 'real' is an admission that approximately 50% of the time recorded in the Bible isn't real history!

How then can Christians that say half of the Bible's history is just a metaphor (or cannot be taken as plainly written) really expect intelligent people to accept the 'other half' as 'real'—events that are critical to salvation, atonement, etc.?

"Thomas Huxley" (Darwin's bulldog) wasn't impressed with this attitude, and he pointed out the inconsistency in books and lectures over 100 years ago.

'I confess I soon lose my way when I try to follow those who walk delicately among "types" and allegories. A certain passion for clearness forces me to ask, bluntly, whether the writer means to say that Jesus did not believe the stories in question, or that he did? When Jesus spoke, as of a matter of fact, that "the Flood came and destroyed them all," did he believe that the Deluge really took, place, or not?'

'If Adam may be held to be no more real a personage than Prometheus, and if the story of the Fall is merely an instructive "type," ... what value has Paul's dialectic.'

Huxley was intellectually consistent and made valid points, which many skeptics have repeated since.

'It becomes clear now that the whole justification of Jesus' life and death is predicated on the existence of Adam and the forbidden fruit he and Eve ate. Without the original sin, who needs to be redeemed? Without Adam's fall into a life of constant sin terminated by death, what purpose is there to Christianity? None.'

And Darwin's modern pit bull terrier, leading misotheist "Clinton R (Richard) Dawkins," says:

'Oh but of course the story of Adam and Eve was only ever symbolic, wasn't it? Symbolic?! Jesus had himself tortured and executed for a symbolic sin by a non-existent individual. Nobody not brought up in the faith could reach any verdict other than barking mad!'

So, what are the conclusions that many have made?

'The biblical story of the perfect and finished creation from which human beings fell into sin is pre-Darwinian mythology and post-Darwinian nonsense. The virgin birth, understood as literal biology, makes Christ's divinity, as traditionally understood, impossible. The view of the cross as the sacrifice for the sins of the world is a barbarian idea based on primitive concepts of God and must be dismissed. Resurrection is an action of God. Jesus was raised into the meaning of God. It therefore cannot be a physical resuscitation occurring inside human history.'

Are these quotes the ranting of an ardent atheist? Well, yes, but they are from a 'Christian' Bishop, "John Shelby Spong." Heretical as they are, these conclusions are the inevitable end result of applying the idea of millions of years and evolution to Scripture.

The Bible gets modified, while the belief in 'evolution' remains.

Most Christians in the western world would admit that despite many Christian influences (churches, schools, radio stations, book stores etc.) the culture is becoming less 'Christian' all the time. As God's people we must make a united stand to affect those around us for Christ, but that stand can only succeed if it is based on the authority of the Word of God.

Fig.30.26: John Calvin – Curtsy - Image Wikipedia.org

A non-believer growing up, stated *"I was not impressed by what I saw as the hypocrisy of Christians wanting me to believe one area of the Bible as plainly written while squirming and re-interpreting other parts. I was an atheist but I wasn't stupid. Atheists know that without Genesis as history, Christianity will fall."* No wonder the opponents of Christianity have applied so much effort to discredit this area of Scripture.

'Christianity is—must be! totally committed to the special creation as described in Genesis, and Christianity must fight with its full might, fair or foul against the theory of evolution.'

Calvin said: Genesis means what it says

Some professing evangelical Christians accuse creationists of taking a naïve literalistic view of Genesis, and claim that creationism is a 20th century aberration. Nothing could be further from the truth. A straightforward view of Genesis was the view of Moses (Exodus 20:8–11), the Apostle Paul (Romans 5:12; 1 Corinthians 15:21–22,45; 1 Tim. 2:13–14) and the Apostle Peter (2 Peter 3:3–7), and the Lord Jesus Christ Himself (Matthew 19:3–6; Mark 10:6–9; Luke 17:26–27).

It was also the view of the vast majority of the Church Fathers, including the faithful defender of the Trinity, "Basil the Great."

And the great leaders of the 16th Century Protestant Reformation, in returning to biblical authority, also accepted a straightforward view of Genesis. This includes the Father of the Reformation, "Martin Luther."

One of the most influential of the Reformers was the French lawyer and theologian "John Calvin" (1509–1564), Figure 30.26. He became leader of Geneva (Switzerland), which became a refuge for 6,000 Protestants. Calvin founded the University of Geneva in 1559, which attracted many foreign scholars, and still does today. His monumental *Institutes of the Christian Religion* (1559) proclaimed the grace of God and salvation in Jesus Christ. He was also a skilled commentator on books of the Bible, including Genesis. His teachings influenced many confessions, catechisms, preachers, leaders of modern Christian revivals, and were brought to America by the Pilgrim Fathers.

It's very interesting that on every point on which Creation Ministry International disagrees with much of modern Christendom, Calvin took our side. For example, Calvin believed that:

- **The Earth is 'Young'**:

  'They will not refrain from guffaws when they are informed that but little more than five thousand years have passed since the creation of the universe.'

- **God Created in Six Consecutive Normal Days**:

  'Here the error of those is manifestly refuted, who maintain that the world was made in a moment. For it is too violent a cavil to contend that Moses distributes the work which God perfected at once into six-days, for the mere purpose of conveying instruction. Let us rather conclude that God himself took the space of six days, for the purpose of accommodating his works to the capacity of men.'

  'I have said above that six days were employed in the formation of the world; not that God, to whom one moment is as a thousand years, had need of this succession of time, but that he might engage us in the contemplation of his works.'

- **The Day-Night Cycle was Instituted from Day 1** — before the sun was created [commenting on 'let there be light' (Genesis 1:3)]:

  'Therefore, the Lord, by the very order of the creation, bears witness that ***he holds in his hand the light, which he is able to impart to us without the sun and the moon***. Further, it is certain, from the context, that the light came to existence as to be interchanged with the darkness … there is, however, no doubt that the order of their succession was alternate …'

- **The Sun, Moon and Stars were Created on Day 4** — after the earth — and took over the role as light dispensers to the earth [commenting on 'let there be lights …' (Gen. 1:14)]

  'God had before created the sun, moon and the stars, but he now institutes a new order in nature, that the sun should be the dispenser of diurnal light, and the moon and the stars should shine by night. And he assigns them to this office, to teach us that all creatures are subject to his will, and execute what he enjoins upon them. For Moses relates nothing else than that God ordained certain instruments to diffuse through the earth, by reciprocal changes, that celestial light which had been previously created. The only difference is this, that the light was before dispersed, but now proceeds from lucid bodies; which, in serving this purpose, obey the commands of God.'

- **The Creation was Originally 'Very Good,' Lacking Any Evil** [commenting on Genesis 1:31]:

  'On each of the days, simple approbation was given. But now, after the workmanship of the world was complete in all its parts, and had received, if I may so speak, the last finishing touch, he pronounces it perfectly good; that we may know that there is in the symmetry of God's works the highest perfection, to which nothing can be added.'

- **Suffering on the Earth is the Result of Sin** [commenting on Gen. 3:19]:

  'Therefore, we may know, that whatever unwholesome things may be produced, are not natural fruits of the earth, but are corruptions which originate from sin.'

- **Physical Death is the Result of Sin:**

  'And therefore, some understand what was before said. "Thou shalt die," in a spiritual sense; thinking that, even if Adam had not sinned, his body must still have been separated from his soul. But since the declaration of Paul is clear, that "all die in Adam, as they shall rise again in Christ" (1 Cor. xv. 22), this wound was inflicted by sin. ...Truly the first man would have passed to a better life, had he remained upright; but there would have been no separation of the soul from the body, no corruption, no kind of destruction, and, in short, no violent change.'

- **God Created Adam and Eve Directly** [commenting on Gen. 5]:

  '... [Moses] distinguishes between our first parents and the rest of mankind, because God had brought them into life by a singular method, whereas others had sprung from previous stock, and had been born of parents.'

- **The Flood was Global** [just a small part of an extensive discussion on the real, historical nature of the Flood and Ark]:

  '*And the flood was forty days, &c.* Moses copiously insists on this fact, in order to show that the whole world was immersed in the waters.'

It is thus clear that if we accept the authority of Scripture alone, we must believe that Genesis should be taken at its plain meaning. Christians who deny this are imposing outside ideas onto Scripture. This is shown by the frank admission by the 'progressive creationist' "Pattle Pun:"

'It is apparent that the most straightforward understanding of Genesis, *without regard to the hermeneutical considerations suggested by science,* is that God created the heavens and the earth in six solar days, that man was created on the sixth day, and that death and chaos entered the world after the fall of Adam and Eve, and that all fossils [*sic* — creationists would say 'most'] were the result of the catastrophic deluge that spared only Noah's family and the animals therewith.'

Sadly, one hotbed of anti-creationist, theistic evolutionary/long age ideas even includes a college named after Calvin — "Calvin College" in Grand Rapids, Michigan, USA. Some of their staff have even invoked Calvin in support, although, as we have seen, Calvin opposed all such compromises.

Today the church needs a new Reformation to return to the authority of the Bible, the written Word of God, rather than trusting the fallible conjectures of unbelieving scientists.

The age of the Earth, according to naturalists and old-Earth advocates, is 4.5 billion years. Young-Earth creationists contend that the Earth is on the order of thousands, not billions, of years old. Is there evidence to support the young-Earth creationists' premise?

First, as we have shown elsewhere, the biblical narrative implies that the Universe was created with an immediate **appearance** of age in many ways. Adam and Eve were not mere zygotes, but walking, talking, working, and procreating individuals. The trees of the Garden were bearing fruit so that Adam and Eve could eat from them, light from distant stars was viewable on Earth, and daughter elements were possibly in the various rocks. That said, while certain attributes of the Earth would appear old, the biblical model suggests that other features of the Universe would highlight its youth. Here are numerous such examples:

THE WORD OF THE CREATOR

I. Bible's Teaching

If the Bible is the inspired Word of God, then whatever it teaches can be known to be true—including what it teaches about the age of the Earth. The evidence indicates that the Bible is in fact God's Word. Simple addition of the genealogies in Genesis 5 reveals that from Creation to the Flood was 1,656 years, give or take a few years. The genealogies of Genesis 11, which do not use precisely the same terminology as that of Genesis 5, account for roughly 400 to 5,000 years, ending with the birth of Abram. From Abram to Christ is roughly 2,000 years, and from Christ to present day is roughly 2,000 years. Therefore, the age of the Earth is 6,000-10,000 years.

Paleontology/Archaeology

II. Polystrate Fossils

Perhaps the most widely used argument for a millions-of-years-old Earth historically has been the rock layers of the geologic column. It would take millions of years for the thousands of meters of material beneath us to accumulate and lithify—or so the argument goes. Is that true?

A polystrate fossil is a single fossil that spans more than one geologic stratum. Many polystrate tree trunk fossils have been discovered, as well as a baleen whale, swamp plants called calamites, and catfish. Polystrate fossils prove that both the rock layers of the geologic column and the surfaces between them do not require millions of years of slow and gradual accumulation and lithification. After all, how could a tree escape its inevitable decay while sticking out of the ground for millions of years with its root's dead and lithified, while it waited to be slowly covered with sediment? Polystrate fossils provide evidence that the rock strata have formed rapidly—fast enough to preserve organic materials before their decay.

III. DNA in "Ancient" Bacteria Support a Young Earth

In 2000, a bacterium was discovered that is thought to be from the Permian Period of Earth's history—250 million years ago. The problem is that, according to geomicrobiologist of the University of Bristol John Parkes, "[a]ll the laws of chemistry tell you that complex molecules in the spores should have degraded to very simple compounds such as carbon dioxide" in that amount of time, and yet the bacterium's DNA was still intact. Further, the "Lazarus" bacterium actually revived in spite of its supposed great age. Not only was the bacterium revived, but analysis of its DNA indicated that the bacterium is similar to modern bacteria—it had not evolved in "250 million years." Critics verified that the DNA of the bacterium does in fact match that of modern bacteria, but respond that "unless it can be shown that [the bacterium] evolves 5 to 10 times more slowly than other bacteria," the researchers' claims should be rejected. So according to critics, the evidence does not match the "theoretical expectations for ancient DNA" predicted by the evolutionary model. Therefore, the bacterium cannot be ancient regardless of the evidence. Another plausible option: the bacterium is not 250 million years old.

IV. Human Population Statistics

Evolutionists argue that humans (i.e., the genus *homo*) have been on the Earth for roughly two to three million years. Using statistics, one can arrive at an estimate for how many people would be predicted to be on the Earth at different points in history. For example, accounting for factors such as war, disease, and famine, and assuming humans have been on the planet for only one million, rather than two to three million, years, we find that there should be $10^{2,000}$ people on the planet today. There are, however, not even 10^{10} people on the Earth. In fact, if three-feet-tall humans with narrow shoulders were squeezed into the Universe like sardines, only 10^{82} people could fit into the entire Universe. It would take $10^{1,918}$ (minus one) other Universes like ours to house that many humans.

It might be tempting to argue that the Earth could only sustain roughly 50 billion people, resource-wise, and therefore, all humans above that number would die off. If that were the case, however, there should be evidence that the Earth's resource capacity had been met many times in the past in the form of billions upon billions of hominid fossils. Hominid fossils, however, are acknowledged to be "hard to come by." In fact, "meager evidence" exists to attempt to substantiate the origin of the entire genus *homo*. Even after over a century of searching for *homo* fossils, one evolutionary scientist admitted several years ago, "The fossils that decorate our family tree are so scarce that there are still more scientists than specimens.

The remarkable fact is that all the physical evidence we have for human evolution can still be placed, with room to spare, inside a single coffin." Is belief in an old Earth reasonable or irrational?

Ironically, if our calculations are adjusted based on the predictions of the biblical model, roughly 4,350 years ago a Flood ensued that wiped out man from the face of the Earth. If the planet then began to be repopulated by six people (namely the sons of Noah and their wives), statistics show that there should be roughly 6.7 to 8.1 billion people on the planet today. As of today, the U.S. Census Bureau documents that the world's population is 7.5 billion people.

V. Carbon-14 in "Ancient" Fossils and Materials

At current rates, it takes 5,730 years for half of the radioactive element carbon-14 (C-14), from an organic sample like a bone or piece of wood, to break down into its daughter element, nitrogen-14. With such a "short" half-life, after 57,300 years (10 half-lives), less than 0.1% of the original C-14 atoms are left in any specimen. Current technology does not allow scientists to detect C-14 in specimens thought to be older than 60-100 thousand years in age—all of the measurable carbon-14 is gone. If C-14 is detected in any uncontaminated specimen, therefore, the specimen cannot be older than 100,000 years (assuming, as evolutionists do, a constant nuclear decay rate of C-14 into nitrogen-14—an assumption which would not hold in the biblical Flood scenario). The discovery of C-14 in fossils that are believed to be 10's to 100's of millions of years old is, predictably, shocking to those who accept the conventional dating scheme and its underlying techniques.

No matter how much care is taken to ensure that the specimens have not been contaminated, the fossils still reveal the presence of C-14. Fossilized wood from the Cenozoic era (up to 65 million years old, conventionally), fossilized wood, dinosaur fossils, and ammonite shells from the Mesozoic era (66-252 million years old, conventionally), and fossilized wood, reptiles, and sponges from the Paleozoic era (252-541 million years old) have been discovered with C-14 present. Similarly, coal from the Paleozoic era (thought to be 40-320 million years old), and even diamonds thought to be billions of years old, have yielded C-14 upon examination.

It is notable that regardless of where the specimens are found in the geologic column, the C-14 ages all fall within the range of 10-60 thousand years old (again, assuming a constant nuclear decay rate). While one might predict that deeper in the strata would correspond to an older age, the depth in the strata does not appear to correlate to the measured age of the specimen, supporting the creationist contention that the entire fossil record and geologic column from the Paleozoic up into the Cenozoic era likely formed during the single year of the biblical Flood. The geologic column and fossil record are not a record of life through time, but of death during the Flood a few millennia ago.

VI. Soft Tissue/Blood Vessels in Dinosaur Fossils

The last uncontested dinosaur fossil is found in the Cretaceous period of the geologic column, below the K-Pg boundary that marks a mysterious extinction event that wiped out some 70% of the planet's species. The dinosaur era (i.e., the Mesozoic) extends from roughly 252 million years ago to the K-Pg boundary, roughly 65-66 million years ago according to the evolutionary timescale. Obviously, no flesh could conceivably survive 100,000 years without decay, much less one million years, much less 65 million years, much less 200 million years. As of 2005, however, many dinosaur fossils have been "**cracked open**" and studied, only to find **collagen and blood vessels with red blood cells intact, original proteins, and soft, stretchy, flexible tissue. The list has grown to include *T-rex*, *hadrosaur*, *mosasaur*, *triceratops*, *thescelosaurus*, *psittacosaurus*, *archaeopteryx*, and *seismosaur* fossils.** While certain sterile conditions could conceivably preserve organic remains for hundreds or thousands of years, the fossils being studied were not discovered in sterile, laboratory environments, but rather harsh environments like the mid-western U.S., with large temperature differentials, erratic weather, and climate conditions that accelerate decay. No reasonable explanation has been offered, and yet the evidence has continued to mount.[25] The most plausible explanation is that the geologic strata that host the dinosaurs do not date to 66+ million years ago, but rather, to a few thousand years ago.

VII. Human/Dinosaur Co-existence

According to the evolutionary, old-Earth timeline, dinosaurs went extinct some 65 to 66 million years ago. Modern-day mammals and many other living organisms did not yet exist, since they are not found in the strata that house the dinosaurs. Humans (i.e., the genus *homo*) only arrived on the scene two to three million years ago according to that paradigm. No human, therefore, ever saw a dinosaur. If, however, evidence was discovered that proves humans and dinosaurs in fact co-existed in the recent past, then the evolutionary timeline telescopes down millions of years and the geologic strata in which the dinosaurs are found are shown to represent a time period in the not-too-distant past. Sure enough, physical, historical, and biblical evidences are available to substantiate the co-existence of humans and dinosaurs in the recent past.

Geology

VIII. Tightly Folded Rock Strata

If the rock layers of the geologic column represent millions of years of slow accumulation, lithification, and erosion, one would expect the layers beneath the surface layer to be "brittle," as rock layers are today. Plate movement would, therefore, result in the fracturing of those rock layers, rather than bending them—rocks do not bend, but rather, break. In several places on the Earth, however, rock layers have been discovered that are bent and folded at radical angles without fracturing (e.g., the Tapeats Sandstone and Muav Limestone of the Grand Canyon). These thick layers of sediment that eventually lithified—representing millions of years of time, conventionally—must have been laid down rapidly and had not yet had enough time to lithify before being bent by the rapid plate movement predicted to occur during the Flood.

IX. Rapid "Slow" Processes

Any old-Earth/evolutionary dating technique relies on the uniformitarian assumption: whatever processes we witness occurring today must be used to explain the past. If petrifying a tree, forming oil, carving a canyon, transforming the parent isotopes of a radioactive rock into their daughter elements, or moving a continent several miles would take millions of years at the lithification, transformation, erosion, decay, and "drift" rates we see today, then the Earth must be at least millions of years old. If, however, each of these processes are shown to occur rapidly under catastrophic conditions (as predicted by the young-Earth biblical model), then those processes cannot be used to prove an old Earth.

Sure enough, as creationists have predicted would be the case, each of these processes has been empirically verified as occurring rapidly under catastrophic conditions like those of the biblical Flood model. Petrification has been found to be able to occur in mere months to a few years under catastrophic conditions. Oil has been shown to form in hundreds to thousands of years. The rapid carving of canyons has been verified to occur under catastrophic conditions as well. Studies have verified that the nuclear decay rates of radioactive materials can be accelerated under catastrophic conditions, and evidence for the rapid movement in the past of the plates upon which the continents reside has been verified as well. If each of the chronometers that are said to prove "old" ages of the Earth is contradicted by the evidence, then where is the evidence of an old Earth?

X. Amount of Salt in the Sea

Ocean water is salty. Each year, hundreds of millions of tons of sodium are added to the oceans and only about 27% of it is removed by other processes, leaving an annual accumulation of 336 million tons of sodium. Starting with a zero sodium content in the sea and using the old-Earth assumption of uniformitarianism, the current concentration of sodium in the ocean would be reached in only 42 million years. According to the U.S. National Oceanic and Atmospheric Administration, however, the ocean is 3.8 billion years old.

The response to this fact, as must be the case in other examples in this list, would obviously be that accumulation and/or dissemination rates must have been different in the past. The average salt accumulation, however, would have to be over 90 times slower than present rates in order to accommodate the alleged age of the ocean. This conjecture simply does not hold up under scrutiny and, even if it did, it would merely prove the creationist contention that **uniformitarianism is not a reliable assumption**.

Present processes are not the key to understanding the past and, therefore, no old-age dating technique can be trusted, since they all rely on uniformitarianism. Since the Flood happened, catastrophism, not uniformitarianism, is a more reasonable assumption in interpreting physical evidence. Intimately tied to catastrophism are rapid processes and, therefore, young ages.

XI. Amount of Sediment on the Sea Floor

As water and wind scour the continents each year, 20 billion tons of material is estimated to be deposited in the oceans. As the tectonic plates of the Earth move, subduction occurs, with one plate slowly diving under another towards the mantle. One billion tons of material is estimated to be removed from the sea floor each year from that process, leaving 19 billion tons of sediment accumulating each year on the ocean floor. On average, the sediment thickness on the ocean floor is 1,500 feet. Based on the current rate of sediment deposition, however (i.e., assuming uniformitarianism once again), the sediment on the ocean floor would accumulate in only a small fraction of the alleged 3.8-billion-year age of the ocean (i.e., 0.5% or 19 million years). The average annual sediment accumulation would have to be 197 times smaller to match an ocean age of 3.8 billion years. The amount of sediment on the sea floor simply does not support a billions-of-years-old ocean, but fits well with a young Earth when the accelerated erosion rates during and immediately after the Flood are accounted for.

XII. Lack of Erosion Evidence Between Strata

When making a multi-layer cake, the adjoining surface between layers is smooth. If you made your cake outside over several weeks, waiting several days between new layers and leaving the cake open to the elements in the meantime, the surface of each layer would exhibit the indicators of time—decay, loss of cake from scavengers, erosion from rain water, etc. Similarly, if geologic strata are formed over millions of years, the surface between adjoining layers would not be smooth, but would exhibit proof of time passing in the form of, for instance, erosional and depositional surfaces.

However, the layers, by in large (e.g., at the Grand Canyon), display smooth contact surfaces—indicating rapid deposition without enough time for erosion. Those surfaces which show evidence of erosion match the type of erosion that would be predicted if the lower surface had not yet lithified when a rapid erosion event occurred above the surface, prior to further rapid deposition. Bottom line: the Grand Canyon exhibits evidence of a young Earth.

XIII. Helium in Zircon Crystals

Zircon crystals are considered to be some of the oldest minerals on Earth—thought to be billions of years old. They are very hard and resistant to deterioration, and are also able to preserve their contents well, making them safer from contamination. Within zircon crystals, a portion of the zirconium atoms are replaced by uranium while the crystals grow. As radioactive uranium-238 decays into its daughter element, lead-206, alpha particles are released that combine with nearby electrons. Helium is subsequently formed, which can then be detected in zircon crystals. While zircon crystals are able to preserve their contents well, helium is known to behave as a "slippery" material. Helium atoms are small and are in constant motion as gas particles. They are, therefore, hard to contain, and they diffuse quickly. Upon examination of zircon crystals that are thought to be 1.5 billion years old, however, scientists have discovered the presence of unusually high concentrations of helium.

If the crystals were billions of years old, the helium should have been diffused from the crystals and released into the atmosphere, since high concentrations of helium can only be sustained, theoretically, for a few thousand years without significant diffusion. The presence of high concentrations of helium illustrates the fact that at some point(s) in the relatively recent past, the nuclear decay rate of uranium-238 was accelerated, producing larger amounts of helium that have not yet had time to diffuse. If radioactive decay rates were accelerated at some point in the past (e.g., during the Flood), then radioactive materials will appear deceptively old, while actually being relatively young.

XIV. "Orphan" Radio-halos

As a radioactive atom of uranium decays into polonium within a solid crystalline material, alpha particles are released and "halos" form, marking the different stages of nuclear decay. Parentless radio-halos, however, are found in many granitic rocks, implying accelerated nuclear decay in the past and a young age for the Earth.

XV. Clastic Dikes

In sedimentary rock strata, open fractures often exist, and in some cases, other sedimentary material is injected into those cracks at a later time, filling them with a different type of sedimentary rock. These are called clastic dikes. The Ute Pass fault, west of Colorado Springs, for example, exhibits over 200 sandstone dikes, some of which are miles in length. The dikes are comprised of Cambrian Sawatch sandstone (allegedly 500 million years old) that injected rock from the Cretaceous period (allegedly 65-66 million years old).

Is it reasonable to presume that 500-million-year-old sediment remained unlithified for over 400 million years while Ordovician, Silurian, Devonian, Carboniferous, Permian, Triassic, Jurassic, and Cretaceous strata were laid down on top of it before intruding into the Cretaceous strata? Or is it more reasonable to infer that the layers of the geologic column from the Cambrian to the Cretaceous were laid down rapidly on top of one another during a global, aqueous catastrophe before they had lithified? Then, during the rapid uplift of the Rocky Mountains later in the Flood, the Cambrian Sawatch material was injected through the overlying layers forming the clastic dikes of the Ute Pass. Bottom line: the geologic column was formed rapidly—the Earth is young.

Astronomy/Astrophysics/Geophysics

XVI. Faint Young Sun Paradox

As the hydrogen within the Sun fuses into helium, the Sun gradually increases in temperature. Calculations show that (at current rates) 3.5 billion years ago, the Sun would have been 25% dimmer and would have heated the Earth less, dropping Earth's temperature some 31°F. Earth would have been below freezing! According to contemporary thinking, however, Earth, initially molten, was **hotter**, not colder, prior to 3.5 billion years ago, and was gradually cooling, not heating up. Not only is there no evidence that Earth was ever frozen, but if it had been frozen 3.5 billion years ago and beyond, according to evolutionists, life could not emerge 3.5-4 billion years ago since it relies on **liquid** water.

XVII. Rapid Decay Rate of Earth's Magnetic Field

Scientists have been measuring the strength of the Earth's magnetic field with precision since 1835. The magnetic field is decaying at an exponential rate with a half-life of roughly 1,100 years.[47] By implication, when we follow the exponential function back in history, doubling the Earth's magnetic field intensity every 1,100 years, we reach a point 30,000 years ago when the Earth's magnetic field strength would have been comparable to that of a neutron star, creating immense heat that would have prohibited life from existing and possibly even compromised the internal structure of the Earth. The Earth cannot be millions of years old.

XVIII. Lunar Recession Rate

The Moon is presently moving away from the Earth at a rate of approximately 4 cm per year. The recession rate is not linear. As the Moon moves further from the Earth, it recedes slower. Based on the equation that describes the Moon's recession rate, scientists can calculate where the Moon would have been compared to the Earth at different times in history. For example, 6,000 years ago, the Moon would have been 750 feet closer to the Earth than it is today—resulting in little effect on the Earth. If, however, the Moon has the contemporary age of 4.5 billion years old, there is a significant problem, because 1.55 billion years ago the Moon would have been touching the Earth. It would be physically impossible, therefore, for the Moon to be older than 1.55 billion years old based on the known recession rate of the Moon. In response, those who wish to maintain the contemporary belief in deep time must argue that present recession rates did not hold in the past. In so doing, however, they abandon uniformitarian thinking (i.e., "the present is the key to the past") which undergirds every deep time dating technique. They are, therefore, once again admitting that every evolutionary dating technique is suspect and does not prove an old Earth.

XIX. Atmospheric Helium Content

Helium is gradually accumulating in the Earth's atmosphere as radioactive isotopes beneath the Earth's surface decay, emitting alpha particles that attract electrons and form helium. The amount of helium in the atmosphere has been measured, the rate at which helium is introduced in the atmosphere has been measured, and the theoretical rate of helium release to space has been calculated as well. Using the typical old-Earth assumption of uniformity over time, it is easy to calculate an upper limit on the age of the atmosphere. The atmosphere can be no older than two **million** years—as opposed to the alleged age of 4.5 **billion** years.

XX. Spiral Galaxies

Earth is located in the Milky Way Galaxy—a spiral galaxy. According to the Big Bang model, galaxies began forming within a billion years after the Big Bang, making many of them over 12 billion years old. Of all of the galaxies that scientists have observed, some 77 percent of them are spiral galaxies.

The oldest spiral galaxy is thought to be roughly 11 billion years old. If you have ever sprinkled cinnamon on a hot, foamy drink and then stirred the drink with a straw or stick, you will notice the formation of the characteristic spiral galaxy shape. You may also notice that the portion of the spiral that is closer to the center rotates faster than the portion of the spiral that is close to the edge of the cup. That "differential rotation" causes the arms of the spiral to begin blurring closer to the center of the spiral over time.

After a few rotations, the center of the spiral is no longer recognizable. Similarly, spiral galaxies are spinning slowly. If spiral galaxies are as old as is claimed by secular cosmologists, after a few hundred million years the arms of the spirals should no longer be recognizable—and yet many of them are. *Space.com* admits: "The exact mechanism for the formation of the spiral arms continues to puzzle scientists. If they were permanent features of the galaxy, they would soon wind up tightly and disappear in less than a billion years." Apparently, the observational evidence does not harmonize with the deep time proposition of the Big Bang model.

XXI. Comet Contradiction

The solar system is comprised of hundreds of thousands of objects that are orbiting the Sun. Over 3,000 of those objects are comets. Comets are balls of ice and dirt moving through space in elliptical orbits around the Sun. They are believed to be "leftovers from the material that initially formed the solar system about 4.6 billion years ago." As comets in their orbit move close to the Sun, solar winds and radiation from the Sun "blow" material from the comet, creating the characteristic tail we observe.

Since material is removed from a comet with each cycle around the Sun, obviously the comet will eventually disintegrate—completely sublimating. The typical lifespan of a comet is 10,000 years. How, then, can the solar system be 4.6 billion years old if thousands of comets—thought to have formed when the solar system formed—are still orbiting the Sun? Scientists speculate the existence of a source for new comets that lies outside of the solar system, but no observational evidence has substantiated that claim. The biblical model, of course, provides a plausible explanation that harmonizes with the evidence: the solar system is less than 10,000 years old.

This list is but a small sample of the available evidences for a young Earth. Keep in mind that the assumption of uniformitarianism undergirds many of the arguments in this list. Uniformitarianism is a fundamental assumption of **evolutionary dating techniques**, not of the biblical Creation model. The creationist would argue that uniformitarianism is extremely unreliable due to the effects of catastrophic phenomena (especially that of the Flood; cf. 2 Peter 3:3-6). The biblical Creation perspective advocates, instead, catastrophism as the reliable way to interpret scientific evidence from the past.

By illustrating that uniformitarianism simultaneously proves a young Earth (according to the examples above) and disproves a young Earth (according to standard evolutionary dating techniques), the unreliability of uniformitarianism is substantiated. The old-Earth advocate is forced to abandon uniformitarianism, or be guilty of holding to it blindly without evidence of its reliability.

In abandoning uniformitarianism, however, the person who believes in an old Earth has now yielded his primary evidence for an old Earth and must embrace the contention that one simply cannot know the age of the Earth using science.

We would concur with that conclusion, but remind the old-Earth advocate that while science cannot provide the age of the Earth, there is another source of information that can provide its age. Recall the first point in our list: since we can know that ***the Bible is from God***, what it says is true. It gives us enough information to know the relative age of the Earth, on the order of thousands, not billions, of years.

There is never a reason to doubt the Bible. True science will always support it. If the Bible indicates that the Earth is young, then a fair, thorough assessment of the evidence will substantiate that truth, even if assessing the evidence requires time and effort. That evidence is readily available.

THE TRUTH IS REVEALED

The Creation movement has increasingly caused many to face up to the powerful, Biblical arguments for such things as:

- All living things were created (about 6,000 years ago) in six literal Earth-rotation days.
- There was no death, bloodshed or suffering before Adam's Fall.
- Noah's Flood covered the whole globe, and would have laid down a vast number of fossils.

However, many of the Christians who now accept the above points are still overawed by certain arguments from astronomy for billions of years. This seems to have compelled a number of writers to come up with novel 'interpretations' of the Bible to try to harmonize it with the idea that there were 'billions of years' before the creation of living things during the six days of Creation Week.

We are not talking here about the classical 'gap' (or ruin-reconstruction) theory, which has long been 'on the ropes.' Rather, we are sadly addressing recent publications by Christian writers trying to find room in the Bible for vast ages, who say that **the sun, moon and stars were all made long before Day 1**.

Nonetheless, what does the Bible actually say? God's historical record in Genesis 1:1–5 reads:

'In the beginning, God created the heavens and the earth. The earth was without form and void, and darkness was over the face of the deep. And the Spirit of God was hovering over the face of the waters. And God said, "***Let there be light***," and there was light. And God saw that the light was good. And God separated the light from the darkness. God called the light Day, and the darkness he called Night. And there was evening and there was morning, the first day.'

THE BEGINNING

The first thing God tells us in the Bible is that there *was* a beginning. Not a beginning to God, but a beginning to ***time*** and to the ***Earth*** and to the ***space-time*** environment in which we live. These words assure us that, as linguist "Charles Taylor" says, 'The universe was no accident, though many evolutionists think so, and some Eastern religions suggest so, with a near-eternal universe and gods emerging from it.'

The first Hebrew word in Genesis 1:1 is *bereshith*; it occurs without the article and so is a proper noun, meaning 'absolute beginning.' Why is this important?

Answer: Because the construction does *not* allow it to be translated 'In the beginning of God's creating' or 'When God began creating,' as some theistic long-agers would prefer.

What does 'God created the heaven(s) and the earth' mean?

The phrase 'heaven(s) and earth' in Genesis 1:1 is an example of a Hebrew figure of speech called a *merism*, in which two opposites are combined into an all-encompassing single concept. Throughout the Bible (e.g., Genesis 14:19, 22; 2 Kings 19:15; Psalm 121:2) this means the totality of creation, not just the Earth and its atmosphere, or our solar system alone. It is used because Hebrew has no word for 'the universe' and can at best say 'the all.'

One of the words in this Hebrew figure of speech is the plural noun *shamayim*, which signifies the 'upper regions' and may be rendered 'heaven' or 'heavens,' depending on the context. The essential meaning is everything in creation apart from the Earth. The word translated 'the earth' is *erets*, and here refers to the planet on which we now live.

The opening sentence of the Bible ('*In the beginning God created the heaven and the earth.*') is thus a summary statement (the details follow) that God made everything in the universe. The rest of Genesis 1 gives the details of how this happened over a period of six days.

Is This Exegesis Justified?

Answer: Yes, a summary statement can be either at the end of the list of happenings it summarizes, or at the commencement. Inspired by God, Moses put it first. By analogy we might say, "At the beginning of this year I built a garden shed. On the first day I laid the foundations. On the second day I erected the walls. On the third day I put the roof on. On the fourth day I installed the lights. On the fifth day I added some fish in a tank and some birds in a cage. On the sixth day I added some rabbits in a hutch. On the seventh day I rested."

Some long-agers have claimed that this summary statement in Genesis 1:1 means that the sun, moon and stars were created over a vast time period, before there were any days on Earth. Is this valid?

- First, note that the Hebrew text does not allow there to be any gap in Genesis 1 between verses 1 and 2, i.e., before Day 1.
- Second, observe that the Hebrew makes no mention of the greater light (sun), the lesser light (moon) and stars until Day 4. There is therefore no mandate for anyone to add these items to the text of Genesis 1:1.

When was 'the Beginning'?

The term 'heaven(s) and earth' is used by Moses in Exodus 20:11:

'For in six days the Lord made the heavens and the earth, the sea, and all that is in them.'

The Bible here unequivocally states that everything in the universe was created within a time period of six days (the same phrase '**six days**' is used in an earlier part of the passage to refer to our working days), **and thus nothing was created before these six days.** Since Adam was created on Day 6, and we have the genealogies from Adam to Christ, this verse totally precludes 'billions of years.'

A recent defender of the view that God created the stars and planet Earth long before the six days of Genesis 1 argues that 'the heavens', in Exodus 20:11, means the atmosphere. But this overlooks the context—that 'heavens' is not used in isolation but is combined with 'earth.' As said above, this combination is a term for the entire creation. Also, the 'heavens' of Exodus 20:11 is the same Hebrew term that he elsewhere tries to make mean the sun and the stars when it occurs in Genesis 1:1—since Exodus 20:8–11 is clearly referring to Genesis 1, 'heaven(s) and earth' must mean the same thing in both passages.

Genesis 1:1 tells us that the totality of creation was 'in the beginning.' Exodus 20:11 tells us that the totality of creation took six days. Genesis 1:14–19 tells us that the sun, moon and stars were created on the fourth of these six days.

Those who promote the long-age view often claim that there is a radical difference between the meaning of the Hebrew words ***bara*** ('create out of nothing' in Genesis 1:1) and *asah* ('make' in Exodus 20:11 and Genesis 1:16). They say that therefore, the latter refers to an *appearing* of the sun on Day 4, when dark clouds surrounding the Earth dissipated. However:

- *Asah* and *bara* are often used interchangeably, as in Genesis 2:4:

  'These are the generations of the heavens and of the earth when they were created [*bara*] in the day that the Lord God made [*asah*] the earth and the heavens.'
- *Asah* means 'make', not 'appear', throughout Genesis 1.
- In Genesis 1:9, when 'appear' is meant, a different word is used, i.e., *ra'ah*.

THE FORMATION OF LIGHT

Genesis 1:3 reads

'And God said, Let there be light. And there was light.'

This verse alone should be sufficient to undermine the various 'billions of years' scenarios. If the sun, moon and stars were all created (and so shining) prior to Genesis 1:2, why was it necessary for God to say *let there be light*, as recorded in verse 3?

Some long-agers invoke Job 38:4,9:

'When I laid the foundation of the earth … I made clouds the garment and thick darkness its swaddling band'

to say that the sun shone on an opaque Earth until God commanded light to penetrate the blanket of darkness. However, this is using a Hebrew poetic expression (in Job), and putting a twist on it to justify changes to the plain meaning of the Genesis 1 text regarding both the creation of light on Day 1 and of the sun, moon and stars on Day 4.

Long-agers overlook the words of Jesus in Mark 10:6, 'But from the beginning of the creation God made them male and female.' The Lord Himself obviously did not envisage any 'billions of years' prior to the creation of Adam and Eve.

The whole concept of the need to allow for 'billions of years' shows 'wrong-way-round thinking.' It is the outcome of Christians' using humanistic evolutionary scientific opinions to determine the meaning of the Bible, rather than vice versa.

LIGHT BEFORE THE SUN?

Can It Be True?

Genesis 1:3 reads: 'And God said, *Let there be light*. And there was light.' This light source must have been ***independent*** of the sun, which was not made until Day 4.

Sceptics and long-agers like to ask how there could have been light before the sun. The Bible provides at least four other examples of events involving non-solar light and God:

1. Exodus 14:19–20: when the Israelites were fleeing from Egypt a 'pillar of cloud', which brought darkness to one side and 'gave light by night' to the other, stood between them and the Egyptian forces.
2. Luke 2:9: when an angel announced the birth of Jesus to the shepherds at night, 'the ***glory of the Lord*** shone around them'.
3. Matthew 17:2: during the transfiguration of Jesus, 'His ***face shone*** as the sun, and His ***clothing*** was ***white as the light***.'
4. Revelation 21:3: 'the city [the New Jerusalem] had no need of the sun, nor of the moon, that they might shine in it, for the ***glory of God illuminated*** it, and its ***lamp is the Lamb*** [i.e. the Lord Jesus Christ, cf. John 1:29]'.

We therefore conclude that the first light that illuminated the Earth was an act of God quite in keeping with His several other acts, recorded in the Bible, involving ***light without the sun***.

Day 1 is described as involving an evening and a morning, so we conclude that the Earth was now rotating. Also, that the light was coming from one direction in relation to the Earth, thereby giving it a night/day cycle. Presumably on Day 4, when God created the sun, this first light source ceased.

The description of day and night before the existence of the sun gives a stamp of authenticity to the Genesis account. There was no way that a secular Jewish writer would have proposed a night/day cycle without the sun.

The ancients worshipped the sun as the source of light, warmth and life. Moses was brought up in all the wisdom of the Egyptians (Acts 7:22), who revered the sun god Re (or Ra). Nevertheless, Moses rejected this pagan notion of the deity of the sun and, inspired by God, wrote that God said: let there be light before He made the sun.

Light Spectrum

1. Visible light is a small segment of the electromagnetic spectrum, which includes X-rays, ultraviolet radiation, visible light, infrared radiation, microwaves and radio waves. It is therefore probable that when God said, 'Let there be light', the whole electromagnetic spectrum came into existence.
2. Moses not only defined the term 'day' (Hebrew *yôm*) the first time he used it by the words '***evening and morning***' and a number, he defined it similarly *every* time he used it (Genesis 1:5, 8, 13, 19, 23, 31). It cannot therefore, here, refer to an age or a succession of ages.

Do I have to believe in a literal creation to be a Christian?

Do I need to believe that God created the world in six normal days, that there was then a real rebellion against God by an actual Adam and Eve, followed by the entry of death, suffering and carnivory into a once-perfect world, and that the first chapters of Genesis are a literal, historical account of all this? So, does one have to believe this to be a Christian?

BECOMING A CHRISTIAN

In the New Testament, a Christian is seen as someone who does two things:

1. Believes that Jesus Christ (who was fully God and fully man), through His death on the Cross and His Resurrection, has paid the penalty for our sin. Cf. 'If you confess with your mouth the Lord Jesus, and believe in your heart that God raised Him from the dead, you shall be saved' (Romans 10:9).
2. Obeys the command to *repent* (literally, changes one's mind—about God and sin), that is, to acknowledge that one has lived and acted in rebellion against God, and to ask God's forgiveness for this. Cf. 'God ... commands all men everywhere to repent' (Acts 17:31b).

Is it then enough just to believe in Jesus and repent? Well, these two things imply that there exists a holy God against whom we have rebelled and so incurred the penalty for doing so.

But why does our disregard for God and our failure to keep His laws merit any penalty at all, let alone the death penalty?

Answer: When we read the first three chapters of Genesis, we see that our first parents, Adam and Eve, were created with a holy, godly nature and they lived in fellowship with God. However, they chose to rebel against Him and so became corrupt in themselves, hostile to God, and guilty before Him. Their rebellion was an affront to the holiness of God who had created them, and it earned them the death penalty (Genesis 3:17–19), about which they had been warned (Genesis 2:17, 3:3). This corruption, hostility and guilt involved the whole human race (Romans 5:12–19), and we have inherited their death penalty also (Romans 6:23), which we deserve.

It is true that one can go through the steps of becoming a Christian without accepting or even knowing the Genesis account of Creation and the Fall. However, such a minimal belief system misses out on the full measure of what God has provided as the basis for our coming into a right relationship with Him. This includes the totality of His Word to us, incorporating the logical foundation which Genesis supplies to the whole doctrine of Salvation.

This leads to a shallow faith that has little root in the Word of God and so has little foundation to resist the attacks and ridicule of skeptics, atheists, liberal religious leaders, fellow students, or work-mates, etc. In the parable of the Sower, it was *because they had no root* that the seeds which fell on stony ground withered away when they sprang up and were scorched by the sun (Matthew 13:5–6, 20–21).

Today Genesis is under attack as never before—not only by skeptics who ridicule it, but also by 'Christian' teachers who, in their books and sermons, blatantly misrepresent what the text says, to try and make it conform to the conclusions of some of modern science, with its antitheist presuppositions. So then, how can one tell what any part of God's Word, and in particular Genesis, actually means?

Answer: The key to understanding the meaning of any book of the Bible is to ask, 'What was the intention of the author?' When we do this with Genesis (e.g., by comparing the style of the first chapters with that of the rest of Genesis), it is very evident that Moses' purpose, under God, was to write an authentic, historical and factual account, beginning with the creation of the universe and Earth, and then narrating the history of mankind from the creation of Adam to the death of Joseph (Genesis 50:26).

LIVING AS A CHRISTIAN

Let us now examine some of the problems faced by a Christian when he or she does not accept what Genesis says about a literal Creation.

1. There is a slippery slope into unbelief that accompanies disbelieving *any* part of the Word of God. If some part of the Bible is not true because it does not mean what it says, how do we know that other parts, such as the Virginal Conception of Jesus or the forgiveness of sin, *are* true? In the 1940s and '50s, American evangelist "Charles Templeton's" preaching helped lead thousands of people to profess faith in Christ, but then he began to compromise with long-age evolutionary concepts. As a result, he had no answer to the spurious claims of evolutionist scientists and their atheistic theories about the origin of the Earth, life, etc., and no answer to the problems of suffering and evil in the world.

   In his autobiography, *Farewell to God*, Templeton tells how this compromise led him into total apostasy. He came to deny the accounts of Creation, Noah's Flood, and the supernatural origin of the ***Ten Commandments***.

He also ended up denying the Virginal Conception of Jesus, miracles, the Resurrection, the Ascension, Salvation, the efficacy of prayer, and the existence of the Trinity. He concluded:

'I believe that there is no supreme being with human attributes—no God in the biblical sense—but that all life is the result of timeless evolutionary forces, having reached its present transient state over millions of years.'

It is important to know that some of those who profess faith at mass evangelism rallies are like Templeton, and make an emotional decision without a proper foundation. Thus, it is not surprising that many professing converts 'fall away.' like Templeton did.

Evangelism

2. As a Christian, how do you propose to share the Gospel with those who hold the worldview that there is no God because matter formed itself in a 'big bang' billions of years ago, and life began spontaneously from chemicals in some primordial pond where the conditions were 'just right'? How are you going to convince them that there is a Creator God to whom we must give account, without first showing that their worldview is incorrect? How are you going to do this, if you say that Genesis allows for everything to begin with a 'big bang'?

   How will such people be convinced of the relevance to them of the death of a man on a cross 2,000 years ago, and of their need to repent, if they are not first given good reasons to abandon their atheistic worldview and to replace it with one based on the existence and claims of the Creator God of Genesis? An essential part of the Gospel is that there is a coming Day of Judgment. Genesis shows us not only that God has the authority to judge us, but also that He has the power to enforce our attendance!

   In fact, chapters 6–8 of Genesis show that God has already judged mankind with the global Flood, and Jesus (Luke 17:26–27) and the Apostle Peter (2 Peter 3:5–7) said that the coming judgment would be just as real. However, if Noah's Flood was just mythical or local, why should we believe in judgment to come? Thus, those who teach that Genesis describes a 'big bang' and a local Flood undermine the urgency of God's coming judgment.

3. How do you propose to share the Gospel with those whose worldview says that there is no such thing as sin because there is no absolute difference between right and wrong? How are you going to convince them that they are lost and need a Savior? Without the Genesis account of the rebellion against God by our first parents, and the Curse that followed, there is no basis for the origin of sin, or the fact that sin needs to be atoned for, or that Christ did this by His death on the Cross and His subsequent Resurrection (1 Corinthians 15:21–22).

4. If you believe only what you choose to believe in God's Word, how will you wield the Sword of the Spirit (Ephesians 6:17), when it is blunted or broken by non-acceptance of its total truth?

5. How are you going to share the Gospel with those who see the pain and suffering of people and animals, and say that if God exists, He cannot be a God of love as the Bible says (1 John 3:8)? Without the Genesis account that the world God originally made was 'very good' (Genesis 1:31), and the Genesis account of the Curse that fell on the Earth as a result of Adam's sin (resulting in death and suffering), so that the present world is not the way God made things, there is no answer. Long-age compromise views, even if anti-evolution, undermine the Gospel at this very point.

In fact, God made man 'in His own image' (Genesis 1:27), so that He and man could have communion together. When man turned away from God in disobedience, God was grieved, so much so that in His love He gave His Son to be our Redeemer and win us back to Himself (John 3:16).

Do you have to believe in a literal Creation to be a Christian? The short answer is 'No.'

EXOPLANET KEPLER

The "Scorching Lava World"

The Apostle Peter warned of a time when scoffers would come, deliberately ignoring the fact that it was *by **God's Word*** that the heavens and earth came into existence (2 Peter 3:3–5).

Today, in fact, there are many in the public sphere who assert no Creator was necessary. And rather than accept the Bible's account of creation ***in six days some 6,000 years ago***, they instead claim that the universe formed by naturalistic processes (i.e., by itself) over billions of years.

However, scientists today repeatedly encounter evidence that thwarts theories of evolutionary origins and long-age timelines. An example is the ***exoplanet Kepler-78b***, about 20% larger than Earth and weighing twice as much, recently discovered in the constellation Cygnus. Its discovery is so confronting to planetary formation theories that a media release by the Harvard Smithsonian *Center for Astrophysics* (*CfA*) proclaimed, "Kepler-78b is a planet that shouldn't exist."

The evolutionary astronomers are primarily fretting about the closeness of "this scorching lava world" to its star, circling it in only eight and a half hours—***one of the tightest known orbits in the universe.*** This makes it "an abomination" to current planet formation theories, because Kepler-78b's present orbit would have been *inside* its own star early in the mooted evolutionary origins of this planetary system, when the star is said to have been much larger. That is of course impossible, as CfA astronomer "Dimitar Sasselov" explained:

It couldn't have formed in place because you can't form a planet inside a star. It couldn't have formed further out and migrated inward, because it would have migrated all the way into the star. This planet is an *enigma.*

Fellow CfA astronomer "David Latham" echoed, "This planet is a complete mystery. We don't know how it formed or how it got to where it is today."

Kepler-78b is only one such 'complete mystery' of *many* 'exoplanet enigmas' discovered of late. It underlines once again the failure of all attempts to explain the formation of planets naturalistically. Surely there's a message there, for anyone willing to receive:

"The heavens declare the glory of God, and the sky above proclaims his handiwork." (Psalm 19:1)

OLDEST-EVER' FOSSILIZED TREE —

Falsely - Millions-of-Years

However, these mites, Figure 30.27, are so 'mighty' in amber that they could really remain intact for the claimed 230 million years? We most definitely think *not.*

Fig.30.27: These Photomicrographs are of the Two New Species of Ancient Gall Mites in 230-million-year-old amber droplets from northeastern Italy, taken at 1000 x magnification. The gall mites were named (left) Triasacarus fedelei and (right) Ampezzoa triassica. Curtsy - University of Göttingen/A. Schmidt

Amber (or fossilized tree resin) has been known to entomb many things including ants, 'Gladiator' insects, crustaceans, water beetles, barnacles, oysters, clams and water striders.

Evolutionists sometimes express surprise at how amber can preserve its contents and remain intact for millions of years. This is especially true of a recent find.

Amber droplets excavated from outcrops high in the Alps of northeastern Italy rocked evolutionary scientists by revealing a pair of gall mites (arthropods) supposedly *100 million years older* than any other arthropod encased in amber ever found (the oldest prior arthropod found is dated at 130 million years).

If these tiny creatures evolved from some common arthropod ancestor, then one would surely reasonably expect that the fossil record should show a variety of transitional forms from that arthropod ancestor to today's gall mites. However, when scientists recently described some of the earliest gall mites from these fossils, they found the contrary. The ancient gall mites look just like modern ones. Study lead author "David Grimaldi" said, "they're dead ringers for (modern) gall mites."

So, for the duration of the claimed 230 million years, there have been no significant changes in these organisms. This is the classic '***living fossil***' syndrome, highlighting the problem of 'evolutionary stasis.'

Does this mean that these evolutionary scientists are going to dismiss the 'millions of years' paradigm?

Not at all. Grimaldi stated, *"Amber is an extremely valuable tool for paleontologists because it preserves specimens with microscopic fidelity, allowing uniquely accurate estimates of the amount of evolutionary change over millions of years."* Or in this case, ***no evolutionary change***. "And that's surprising because the world has changed a lot from when these bugs were alive."

It may be surprising to those who imagine millions of years of evolution. However, this is not surprising at all to biblical creationists who expect ***mites to produce mites***.

Scripture tells us that God created everything in six ordinary days, roughly 6,000 years ago. God told these creatures to reproduce "after their kind" (10 times in Genesis 1).

However, sin entered the world through Adam's actions and as a result the whole of creation was cursed. After **about 1656 years**, God had had enough of mankind's sin and destroyed the earth with a ***global flood***. The fossils found inside amber are much better explained as dating from this global Flood around 4,500 years ago.

Uprooted trees, smashing against each other in the swirling currents and waves, would lose their bark and release copious quantities of tree resin. While still fluid, the resin would have enveloped organisms displaced from their usual habitat by the floodwaters. Also, heat is said to have been a likely factor in promoting resin flow from wood. Perhaps the Flood waters—heated in places by the 'fountains of the great deep' (Genesis 7:11)—provided ideal conditions for large quantities of liquid amber to ooze from mats of floating logs, enveloping the likes of these mites, and other flood debris, before hardening.

DAYS OF GENESIS 1 ARE LITERAL - WHILE SUN WASN'T CREATED UNTIL DAY 4?

We know today that all it takes to have a day-night cycle is a rotating Earth and light coming from one direction. The Bible tells us clearly that God said: *Let There Be light* on the first day, as well as the Earth. Thus, we can deduce that the Earth was already rotating in space relative to this provided light.

God can, of course, provide the ***Light of His Glory***, without a secondary source. We are told that in the new heavens and Earth there will be no need for sun or moon (Rev 21:23). In Genesis, God even defines a day and a night in terms of light or its absence.

'Progressive creationists' sometimes use the argument that the days are really long periods, although God could have used words for that if He had really meant that.

The creation of the sun after the Earth undermines progressive creationists' attempts to harmonize the Bible with billions of years. So, they must explain this teaching away. Some assert that what really happened on this fourth 'day' was that the sun and other heavenly bodies 'appeared' when a dense cloud layer dissipated after millions of years.

This is not only fanciful science, but bad exegesis of Hebrew. The word *'asah* means 'make' throughout Genesis 1, and is sometimes used interchangeably with 'create' (*bara'*), e.g., in Genesis 1:26–27.

It is pure desperation to apply a *different meaning* to the *same word* in the *same grammatical construction* in the *same passage*, just to fit in with atheistic evolutionary ideas like the 'big bang'.

If God had *meant* 'appeared', then He presumably would have *used* the Hebrew word for appear (*ra'ah*), as when the dry land 'appeared' as the waters gathered in one place on Day 3 (Genesis 1:9).

We have checked over 20 major translations, and all clearly teach that the sun, moon and stars were *made* on the fourth day.

The evidence that ordinary days are being referred to is so overwhelming that even liberal Hebrew scholars admit that the author can have had no other intent — particularly when the words 'evening' and 'morning' are used from the first day.

On the fourth day the present system was instituted as the Earth's temporary light-bearers were made, so the diffused light from the first day was no longer needed. This shows that once again, skeptics just repeat arguments long ago refuted by Bible believing scholars.

This would have been very significant to pagan world views which tended to worship the sun as the source of all life. God seems to be making it pointedly clear that the sun is secondary to His Creator-hood as the source of everything. He doesn't 'need' the sun in order to create life (in contrast to theistic evolutionary beliefs).

This unusual, counter-intuitive order of creation (light before sun) actually adds a hallmark of authenticity. If the Bible had been the product of later 'editors,' as many critics allege, they would surely have modified this to fit with their own understanding. It is only recently that the astronomical fact has been realized that a day-night cycle needs only *light plus rotation*. Having 'day' without the sun would have been generally inconceivable to the ancients.

APPENDIX

i. The use of the third person is no problem. Moses wrote the long account of his own life in Exodus to Deuteronomy in the third person, and many classical authors like Julius Caesar also wrote in the third person.

ii. 'It is significant to note that the elements which anthropologists identify as the attributes of the emergence of evolving men from the stone age into true civilization—urbanization, agriculture, animal domestication, and metallurgy—were all accomplished quickly by the early descendants of Adam and did not take hundreds of thousands of years.'

iii. Letter from Professor James Barr to David C.C. Watson of the UK, dated 23 April 1984. Copy held by the author. Note that Prof. Barr does not claim to believe that Genesis is historically true; he is just telling us what, in his opinion, the language was meant to convey.

iv. The Bible does not set out to prove the existence of God—in the beginning no-one doubted God. Instead, the Bible observes: 'The fool has said in his heart, There is no God!' (Psalm 14:1). Romans 1:20–21 says that the evidence for God can be plainly seen, but is deliberately rejected.

v. Genesis 1:2 begins with the Hebrew *waw* which can mean 'and', 'now', 'but', 'then', etc. Wherever *waw* precedes a noun (as in v.2 *waw* 'and' + *erets* 'the earth') it has the meaning of an explanation (called a *waw* disjunctive or *waw explicativum*, i.e. explanatory *waw*). It is ***not*** a sequence of events such as 'then the earth became' (which would require a *waw* consecutive, where *waw* precedes a verb). It compares with the old English expression 'to wit'; it could be translated by 'Now' or even with the use of parentheses as follows: 'In the beginning God created the heaven and the earth (the earth was without form and empty …).' Moses used the two *waw* constructions very deliberately in Genesis 1. Verse 2 has the only *waw* disjunctive. All 28 other verses beginning with 'And' have the *waw* consecutive.

vi. Acts 3:19; 2 Peter 3:9; etc. Not all 'salvation' texts in the Bible contain both instructions, but they are often used synonymously, as each implies the other. Thus, faith in Christ's death for us involves our acknowledging that He died for our sin, and repentance in the Biblical sense involves there being a means whereby we may be justly forgiven, namely that Jesus has paid the penalty for our sin.

vii. This shows the error of those who teach that sin is a feeling of low self-esteem. Not everybody has low self-esteem, but the Bible says that 'all have sinned and come short of the glory of God' (Romans 3:23).

viii. Many missionaries begin Bible translation with one of the Gospels and many tribal people have been converted as a result. However, many more genuine (i.e. persevering) conversions usually occur when missionaries begin their evangelism by teaching about Creation and the Fall from Genesis. See McIlwain, T., *Firm Foundations: Creation to Christ*, New Tribes Mission, Sanford, Florida, 1991.

ix. One of the main themes of Genesis is the sovereignty of God. This is seen in four notable events (Creation, the Fall, the Flood, and Babel), and in God's relationship to eight notable people (Adam, Abel, Enoch, Noah, Abraham, Isaac, Jacob and Joseph). There is thus a unifying theme to the whole of Genesis, which fails if any event is not true history; each event reinforces the historicity of all the other events.

x. Genesis 1–50 gives the foundational history of the Jewish people; other nations are covered only as far as the end of chapter 10.

xi. By means of data sent from the telescope on NASA's Kepler spacecraft launched in 2009, and later confirmed using measurements from the high-precision spectrometer HARPS-North at the Roque de los Muchachos Observatory on La Palma and the HIRES spectrograph at the Keck Observatory, Hawaii.

REFERENCES

1. *Annual Reviews of Astronomy and Astrophysics* 20:4Â5, 1982.
2. Behe, Michael, *Darwin's Black Box*, Free Press, New York, 1991.
3. Branden, Carl and John Tooze, *Introduction to Protein Structure*, Garland, New York, 1991.
4. Crick, Francis, *Life Itself*, Simon and Schuster, New York, 1981.
5. Einstein, Albert, *The World As I See It*, Philosophical Library, New York, 1949.
6. *Encyclopedia of Science and Technology*, McGraw Hill, New York, Vol. 3, 1971.
7. Gould, Stephen Jay, *Wonderful Life; The Burgess Shale and the Nature of History*, Norton, New York, 1989.
8. Hickman, Cleveland, Larry Roberts, and Allan Larson, *Integrated Principles of Zoology*, Wm. C. Brown, Dubuque, IA, 1997.
9. Hoyle, Fred, *The Intelligent Universe*, Holt, Rinehart and Winston, New York, 1983.
10. Irion, Robert, Ocean Scientists Find Life, Warmth in the Seas, *Science* 279:1302–1303, 1998.
11. Jastrow, Robert, *God and the Astronomers*, W.W. Norton Company, New York, 1978.

12. Kayzer, W, A Glorious Accident, *Understanding Our Place in the Cosmic Puzzle*, W.W. Freeman and Co., New York, 1997.com/store_redirect.php?.
13. Knight, Jonathan, Cold Start; Was Life Kick-Started in Frozen Seas Rather Than Boiling Vents? *New Scientist* 2142:10, July 11, 1998.
14. Komfield, E.C, The Evidence of God in an Expanding Universe, *Look*, January 16, 1962.
15. Levy, David, Four Simple Facts Behind the Miracle of Life, *Parade Magazine*, June 12, 1998, p 12.
16. Levy, Matthew and Stanley Miller, The Stability of the RNA Bases: Implications for the Origin of Life, *Proceedings of the National Academy of Sciences USA* 95: 7933–8, 1998.
17. Lewis, Ricki, Primordial Soup Researchers Gather at the Watering Hole, *Science* 227:1034–5, 1997.
18. Overman, Dean, *A Case Against Accident and Self-Organization*, Rowman & Littlefield Pub., New York, 1997.
19. Ruse, Michael, Answering the Creationists, *Free Inquiry* 18(2):28–32, 1998.
20. Spetner, Lee, *Not by Chance!* Judaica Press, Brooklyn, NY, 1997.
21. Trefil, James, *1001 Things Everyone Should Know about Science*, Doubleday, New York, 1992.
22. Yockey, Hubert, *Information Theory and Molecular Biology*, Cambridge University Press, Cambridge, 1992.
23. The seminal author on the colophon concepts was P.J. Wiseman, *Creation Revealed in Six Days,* Marshall, Morgan & Scott, London, 1948, pp. 45–53. For an excellent evaluation of this by a evangelical linguist see *The Oldest Science Book in the World,* by Dr Charles V. Taylor, Assembly Press, Queensland, 1984, pp. 21–23, 73, 121.
24. This discussion of Hebrew poetry was adapted from J. Sidlow Baxter, *Explore the Book,* Vol. **1**, pp. 13-16.
25. Aramaic paraphrases of the OT originating in the last few centuries BC, and committed to writing about AD 500. See F.F. Bruce, *The Books and the Parchments*, (Westwood: Fleming H. Revell Co., Rev. Ed. 1963), p. 133.
26. A.G. Fruchtenbaum, *Apologia* 2(3):54–58, 1993.
27. Adam and Eve knew how to sew fig-leaf 'aprons' for themselves (Genesis 3:7). Within a few generations, Adam's descendants founded a city (Genesis 4:17), were tent-makers, cattle farmers, musicians with the ability to make both stringed and wind instruments, and metallurgists with the ability to smelt the ores of copper, tin and iron and then to forge all kinds of bronze and iron tools (Genesis 4:20–24). Dr Henry M. Morris comments in *The Genesis Record* (Baker Book house, Grand Rapids, Michigan, 1976, pp. 146–147):
28. Adapted from J. Sidlow Baxter, *Explore the Book,* Vol. 1, pp. 27–29.
29. Meredith Kline, Space and Time in the Genesis Cosmogony, *Perspectives on Science and Christian Faith* 48:2–15, 1996.
30. Scripture Union, *Salt Magazine*, Milton Keynes, UK, p. 29, Jan/Mar 1998.
31. Charles Hodge, *Systematic Theology*, p. 571.
32. Pattle P.T. Pun, A Theology of Progressive Creationism, *Perspectives on Science and Christian Faith*, 39(1):14, March 1987.
33. Thomas H. Huxley , *Science And Hebrew Tradition Essays*, pp. 207, 208, 1897. Return to text.

34. G. Richard Bozarth, The Meaning of Evolution, *American Atheist*, p. 30, Feb. 1978.
35. Shelby Spong, Episcopal Bishop of Newark, *A Call for a New Reformation*, from home page for the Episcopal Diocese of Newark, 4/9/99.
36. Batten, D., Genesis means what it says: Basil (AD 329–379), *Creation* 16(4):23, 1994.
37. Citing Martin Luther, in Jaroslav Pelikan, editor, 'Luther's Works,' *Lectures on Genesis Chapters 1–5*, 1:3,6, Concordia, St. Louis, MO, USA, 1958.
38. Packer, J.I., John Calvin and Reformed Europe; in: *Great Leaders of the Christian Church*, Ed. Woodbridge, J.D., Moody Press, Chicago, IL, USA, pp. 206–215, 1988.
39. Calvin, *Institutes of the Christian Religion* 2:925, ed. John T. McNeill, Westminster Press, Philadelphia, PA, USA, 1960.
40. Calvin, J., *Genesis,* 1554; Banner of Truth, Edinburgh, UK, 1984, p. 78.
41. Pun, P.P.T., *Journal of the American Scientific Affiliation* 39:14, 1987; emphasis added.
42. Leupold, H.C., *Exposition of Genesis*, 1:41, Baker Book House, Michigan, 1942, who cites similar usage in Jeremiah 10:16; Isaiah 44:24; Psalm 103:19, 119:91; and Ecclesiastes 11:5.
43. E.g., Gray, G., *The Age of the Universe: What Are the Biblical Limits?* Morning Star Publications, Washington, pp. 18 ff, 2000.
44. For a fuller treatment: Grigg, R., From the beginning of the creation: Does Genesis have a gap? *Creation* 19(2):35–38, 1997; creation.com/gap.
45. Grigg, R., Why did God impose the death penalty for sin? *Creation* 15(1):32–34, 1992.
46. It is true that the Bible contains poetry (as in the Psalms), parables (as recorded in the Gospels), and metaphors (as when Jesus said, 'I am the bread of life'). However, Genesis is not written in any of these distinctive styles. Grigg, R., Should Genesis be taken literally? *Creation* 16(1):38–41, 1993.
47. Batten, D., Age of the earth—101 evidences for a young age of the earth and the universe, creation.com/age 4 June 2009.
48. Mystery world baffles astronomers, Harvard Smithsonian Center for Astrophysics media release no. 2013-25, cfa.harvard.edu, 30 October 2013.
49. O'Neill, I., Kepler-78b: Mystery exoplanet shouldn't even exist, news.discovery.com, 30 October 2013.
50. Planets baffle big bangers, *Creation* 36(3):9, 2014; creation.com/focus-363#planets.
51. Hartnett, J.G., Planetary system formation: exposing naturalistic storytelling; creation.com/planet-formation, 14 April 2016. Also, Psarris, S., Our created solar system.
52. The Amber Mystery, *Creation* 25(2):51–52, 2003; creation.com/gladiator-an-extinct-insect-is-found-alive#amber.
53. Fossil ant found alive! *Creation* 28(4): 56, 2006; creation.com/fossil-ant-found-alive.
54. Gladiator—an 'extinct' insect is found alive, *Creation* 25(2): 51–53, 2003; creation.com/gladiator-an-extinct-insect-is-found-alive.
55. Amber needed water (and lots of it), *Creation* 31(2): 20–22, 2009; creation.com/amber-needed-water.
56. 320-million-year-old amber has flowering plant chemistry, *Creation* 24(2): 16, 2010; creation.com/amber-with-flowering-plant-chemistry.

57. American Museum of Natural History, Oldest occurrence of arthropods preserved in amber: Fly, mite specimens are 100 million years older than previous amber inclusions, *ScienceDaily*, www.sciencedaily.com, 27 Aug. 2012.
58. Prehistoric bugs from time of dinosaurs found frozen in amber, foxnews.com, 27 August 2012.
59. Sarfati, J., Why Bible history matters (and the timing of the Fall, and Ark-building), *Creation* 33(4):18–21, 2011; creation.com/bible-history-fall-ark.

THE BEGINNING AND THE ENDING OF THE UNIVERSE

ETERNITY – INFINITY – GOD

Finite Minds

Here are some deep issues about God, eternity and infinity and the like which, while not logical paradoxes, are mind-stretching outside of our normal experience. They include: God's triune nature; what the spiritual realm might consist of; how He can be ***infinite***, Figure 31.1, and yet at the same time personal; and more.

Fig.31.1: *The Infinity*

PRESENCE OUTSIDE OF TIME

One could also mention the mysterious fact that God can be outside of time (as its Creator) and yet involved in activities which to us are always time-sequential. For example, John's Gospel repeatedly indicates that there was love between the three persons of the Godhead prior to creation. However, it is generally understood that, before this space-matter-time continuum was created, God was existing in a timeless eternity, mirroring an eternity yet future. It is vital to understand that God is infinite. Mathematical infinity is human perception, which means any undetermined value that could not be calculated through human calculation. God's infinity contains the entire visible and invisible entities, matters and non-matters including the Eternity. He is far above and the infinite of infinities!

The entire Universe and its contents is established by the principle of "God is Love," infinitely.

Yet our experience of communication is one word after another, i.e., a sequence in *time*. In fact, in our time-bounded cosmos, even wordless expressions of ***human*** love have a beginning and ending, both of which involve ***time***. So how can there be love in timelessness?

Similar mind barriers arise, too, when we start to think deeply about the concept of 'infinity,' even though this might appear at first glance to be much more straightforward. It includes the concepts that God is infinitely 'old' (eternal), infinitely knowledgeable (omniscient) and infinitely powerful (omnipotent).

It is both common and reasonable to resort to the notion that our finite minds simply can't be expected to have the 'processing power' to grapple with such things, any more than a goldfish can be expected to grapple with genetic algorithms, or a cat comprehend calculus. Skeptical materialists, of course, will generally paint that as a copout. To them, it just shows that the concepts themselves are irrational, hence unreal.

MODERN PHYSICS

Increasingly, however, mankind's God-ordained investigations are discovering things which show that even this present world is far more mysterious than we generally think. These include the way in which time actually flows at different speeds in different parts of a gravity field. Several creationist physicists have made use of this experimentally proven concept (without adjusting for it, GPS locations would drift by meters each day), Figure 31.2. Because of it, it becomes both feasible and understandable for light to have reached us from the most distant stars in a universe that is 6,000 years old (by Earth clocks).

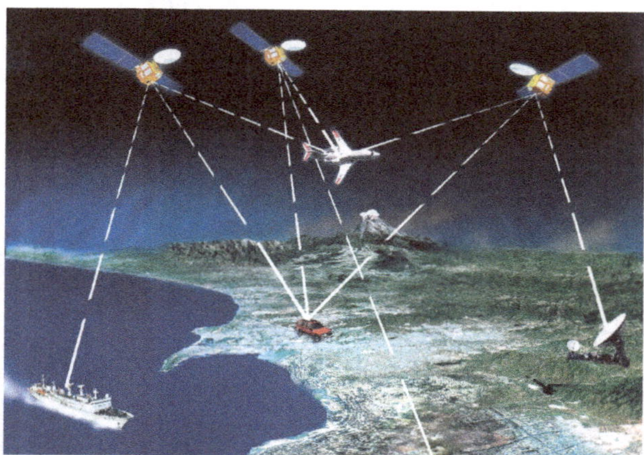

Fig.31.2: *GPS - Boeing-built Satellite into Orbit as part of the U.S. Air Force's Global Positioning System (GPS). Curtsy United Launch Alliance (ULA)*

QUANTUM MECHANICS

Then there is the field of physics known as quantum mechanics (QM), Figure 31.3. QM is immensely successful and useful, and at the same time stunningly mysterious and counter intuitive. Experiments seem to show, for example, that a single particle can go through two slits at once, provided there is no way it can be 'observed' to do so.

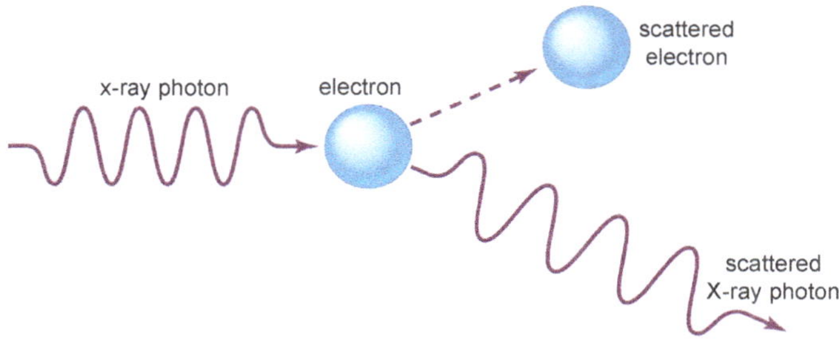

Fig.31.3: Atom - Wave-Particle Duality, Quantum Mechanics, Curtsy Britannica

When the cleverest physicists try to make sense of it, they start to come up with notions so bizarre that one could almost question their grip on reality. In all seriousness, many of them maintain that the best way to understand what happens in QM experiments is to invoke the idea of '***parallel universes***.' This means that when something happens in this universe, all other possibilities are covered in other universes. Say for instance that you were to roll a '6' on one of a pair of dice (rather than any of the other five numbers), then at that moment five alternative ('***parallel***') universes spring into existence (or have already existed among an infinity of such universes). That way, all of the other five numbers can come up, too—one in each universe.

That interpretation of QM is, perhaps surprisingly, fairly popular among physicists. It might be easy to dismiss it as being fairly ***obviously irrational***. But there is no doubt that the things this idea is trying to explain include some really weird stuff. Like the observation that a particle widely separated from its '***twin***' is *instantly* affected by something that happens to the other particle—regardless how far apart.

THE CONCEPT OF MATHEMATICS – THE UNIFIED THEORY

Despite the weirdness of the world such things reveal, optimists might hope that some future theory could unify all these phenomena. They would like to be able to reduce everything to a much simpler set of mathematical expressions. This quest and its link to deeper meaning was parodied around 1980 in the science/philosophy spoof series *The Hitchhiker's Guide to the Galaxy*. The universe's ultimate supercomputer was assigned the task of finding the meaning of "life, the universe and everything." After countless eons of number-crunching, the answer emerged as '42.'

Fig.31.4: Kurt Gödel - Curtsy Wikipedia.org

But it is precisely in the field of mathematics itself, where rigorous proofs *are* possible, that we find evidence of an 'irreducible complexity' to the world. This evidence stands in bold defiance of all of our efforts to make the world understandable to our minds. The mathematician "Kurt Gödel's" Incompleteness Theorem, Figure 31.4, famously proved that it was impossible for mathematics (or *any other system*) to formally prove or understand itself within its own rules.

It's as if Gödel was formalizing the idea expressed in Ecclesiastes 3:11b:

" ... he has put eternity into man's heart, yet so that he cannot find out what God has done from the beginning to the end."

This is often thought to refer to our hunger to understand things, especially the deep timeless truths of the way the world is and operates. It is God, the preacher says, who has placed this quest for knowledge into the hearts of humanity. But He has done things in such a way that no person is able to find out everything about His works from beginning to end. Some things are going to remain forever out of reach.

MATHEMATICS AND INFINITY

At least, we might have thought, mathematics has a straightforward way of handling ***infinity***. Start counting 1,2,3,4 ... and so on, and it's not hard to see that the number you could reach is infinitely large, i.e., the number of possible numbers is 'infinity.' There's even a mathematical symbol for it, ∞. Therefore, this seems to offer hope that we can at least begin to comprehend the infinite characteristics of God.

Fig.31.5: George Cantor - Curtsy Wikipedia.org

Nonetheless, here, too, it's more complicated than that. It turns out that mathematics reveals not just the reality of infinity, but even deeper (or greater, or higher if you like) infinities—even though that hardly seems possible at first glance. How can it do that? In all those numbers you can count, there are 'sets' of numbers. For example, all the even numbers, or all numbers divisible by 8, and so on. The number of such possible sets can be shown to also be 'infinity.' But mathematician "George Cantor," Figure 31.5, formally proved that this 'infinity' is actually *greater* than the first 'infinity' mentioned above. And that's not the end of it; by looking at this infinite number of sets, there is a higher infinity again above that. In other words, there is a ladder, or infinite hierarchy, of infinities. And above and beyond that still, mathematicians are catching glimpses of something so high above this '***infinity of infinities***' that they themselves are saying that it will probably never be approachable by the normal laws of mathematics.

Contemplating all this should make even the most hardened materialist pause and draw breath. Even given our limitations, our finite minds can, it seems, use the tools of mathematics to confirm that there are things that, though real, are out of our conceptual reach. In this way, we are allowed at least a very tiny glimpse into some of the eternal and unsearchable depths of God and the world He made. But no more than a glimpse; it's as if we have discovered a sign saying, 'There are wondrous things beyond this point, but for you it's this far, and no farther for now.'

As the Apostle Paul put it, Figure 31.6, talking about our present existence, before we are enlightened in that new creation: "For now we see through a glass, darkly; but then face to face: now I know in part; but then shall I know even as also I am known" (1 Corinthians 13:12). How spectacular that we can, right now, through Christ, truly *know* the ***infinite-personal Creator God***. We can know the One who knows all about all of these things—and all about each one of us—already.

Fig.31.6: The Apostle Paul – His Writing Impacted the Church

THE UNIVERSE'S BEGINNING

Once you let go of the Bible as history, all Christian doctrine begins to disintegrate. The major factor in the waning of the Christian faith is its continuing insistence on a supernatural God—the Almighty, the lawgiver and judge. There are alternatives to 'fundamentalism' to consider (i.e., believing in the God of the Bible); 'Traditional beliefs about God cannot be sustained in the light of the latest scientific and critical thinking.' And of course, the 'latest scientific thinking' rests on the foundation of discounting the time scale of Genesis creation.

Fig.31.7: Reality of The Universe's Creation Vs Big Bang Hypothesis of the Universe's Beginning and Death

However, the Bible gives us a *measured* time scale, an eye-witness record, of history, a foundation far more secure than any modern scientific *estimate*. Figure 31.7, A measurement wins over an estimate any day!

THE EARTH'S BEGINNING

A child's birth gives a measured time scale for any age. It is an eye-witnessed statement that a child was born on such date. Similarly, other eye-witnesses may have maintained a record of the earth having circled the sun 61 times since then.

Fig.31.8: The Earth Creation Vs Big Bang Hypothesis – Curtsy CMI

Likewise, the Bible gives us a 'birth account' for the universe—an eye-witness statement that ***God created*** it in six ordinary-length days in the time of Adam, Figure 31.8. The family histories and patriarchal ages in Genesis continue this record. God then confirmed it to Moses and wrote it down with His own finger in stone in the Sabbath Commandment (Exodus 20:11; 31:18; 32:16).

Jesus then confirmed the authenticity of the Old Testament scriptures—in detail and in its entirety—by correctly predicting his own death and resurrection on their foundation. That is, death entered the world only as the penalty for Adam's sin at the Fall, and Jesus, the last Adam (1 Corinthians 15 and Romans 5:6–21) took our place and paid that penalty for us, thus restoring us to eternal life.

Once the penalty was paid, death no longer had any hold on Him and He rose from the dead. Jesus' Resurrection authenticates Genesis as real history. The Resurrection of Jesus, an attested fact of history (Acts 17:31), is thus our guarantee of the *measured* biblical time scale for the universe (Luke 24:27,44).

SCIENTISTS ASSUME AND CONJECTURE

No scientist has any alternative or better 'birth account' for the earth or the universe. All scientific estimates of earth and universe age require a whole lot of assumptions. The key assumption is uniformitarianism, which is atheism disguised as science, because it assumes no miraculous interventions in history.

Christians have no reason to accept, and every reason to reject, atheistic assumptions about the universe.

Jesus' Resurrection validates the accuracy of the Bible, especially its history of Creation and Fall. No Bible scholar since Jesus has risen from the dead to validate any alternative point of view. Many may attempt to this valuable gift that God has given us in the Bible.

NUMBERS OF ATHEIST MIRACLES/MAGICS

Atheists often promote themselves as intelligent, logical, scientific, rational, etc. They even had a proposal to call themselves 'brights'!

The aggressive 'new Atheists,' such as "Richard Dawkins," "Sam Harris" and company, like to portray those of us who believe in a supernatural Creator as irrational, unscientific, unintelligent, ignorant, or even 'needing help' (Dawkins). The entertainment industry often reinforces these perceptions by portraying 'religious' people (Christians particularly, and especially church leaders) as buffoons or hillbillies (almost never as a university professor, for example).

Fig.31.9: Man of Faith and Science, Sir Isaac Newton. Curtsy - wikipedia.org

Reality runs against these perceptions. "Isaac Newton," Figure 31.9, the greatest scientific mind of all time, was a Christian believer, as were other founders of modern science. Surveys have consistently shown that people with a strong adherence to the Bible's authority are the least likely to be superstitious, in contrast to the average *de facto* Atheist. Indeed, one Atheist expressed his chagrin that "some of the most intelligent and well-informed people" he knew were Christians.

There is much more to say. Atheists believe that *everything* came about by purely material processes—the universe, life, mind, and morality. However, do they have a rational, logical basis for this belief?

ILLUSION NEEDS AN ILLUSIONIST

Magic Needs Magician

They actually believe in miracles without any reasonable cause for the miracles. That is, they believe in ***magic***, or the occurrence of things without a sufficient cause. What we commonly call '*magic*' is actually ***illusion***. For example, a rabbit does not just appear from an empty hat; there has to be a logical physical explanation; a sufficient cause. Illusion needs an illusionist.

Stuff does not happen without something to cause it to happen. Even young children understand this law of causation. Magic, where things 'just happen,' is the stuff of fairytales—there is no such thing.

Here are numbers examples of materialists believing in magic (and there are more), or miraculous events without any sufficient explanation or cause for those events.

I. Origin of the Universe

Materialists (Atheists) once tried to believe that the universe was *eternal*, to erase the question of where it came from. The famous British Atheist "Bertrand Russell," for example, took this position. However, this is not tenable. The progress of scientific knowledge about *thermodynamics*, for example, means that virtually everyone has been forced to acknowledge that the universe had a beginning, somewhere, sometime—the big bang idea acknowledges this (ideas like the multiverse only put the beginning more remotely, but do not get rid of the pesky problem).

The big bang attempts to explain the beginning of the universe. However, what did it begin from and what caused it to begin? Ultimately, it could not have come from a matter/energy source, the same sort of stuff as our universe, because that matter/energy should also be subject to the same physical laws, and therefore decay, and it would have had a beginning too, just further back in time, Figure 31.10.

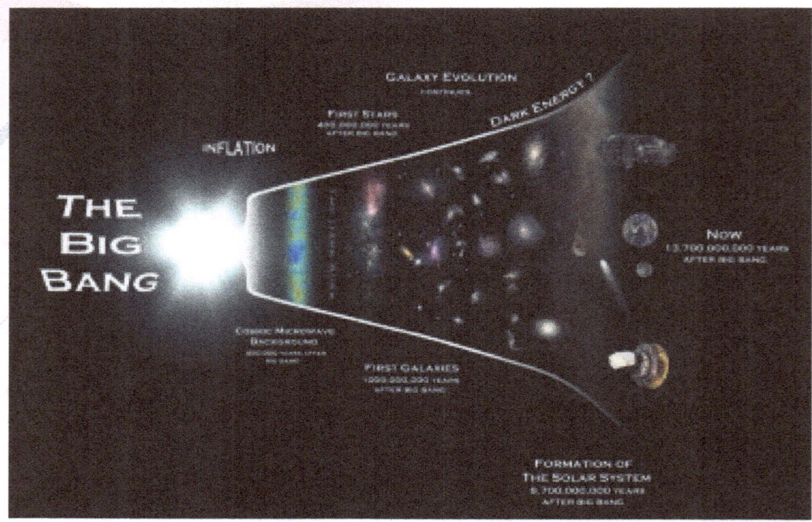

Fig.31.10: *Quantum Fluctuations – Universe Came out of Nothing? Curtsy Neil Shenvi*

So, it had to come from? Nothing! Nothing became everything with no cause whatsoever. *Magic*!

"The universe burst into something from absolutely nothing—zero, nada. And as it got bigger, it became filled with even more stuff that came from absolutely nowhere. How is that possible? Ask "Alan Guth." His theory of inflation helps explain everything."

So, proclaimed the front cover of *Discover* magazine, April 2002.

Physicist "Lawrence Krauss," one of the loud 'new Atheists', has tried to explain how everything came from nothing; he even wrote a book about it. However, his 'nothing' is a 'quantum vacuum', which is not actually nothing. Indeed, a matter/energy quantum something has exactly the same problem as eternal universes; it cannot have persisted for eternity in the past, so all their theorizing only applies after the universe (something) exists. Back to square one!

Materialists have no explanation for the origin of the universe, beyond 'it happened because we are here!' *Magic*: just like the rabbit out of the hat, but with the universe, a rather humungous 'rabbit'! 'Stuff happens!'

There are other aspects of the big bang, the 'mainstream' model of the universe's origin, that are also miraculous. The 'standard model' has a period of very rapid expansion called 'inflation' (which "Alan Guth," mentioned above, invented). There is no known cause for the initiation of this supposed expansion, no known cause for it to stop and no physical mechanism for the extremely rapid expansion (many orders of magnitude faster than the speed of light). However, these three associated miracles *must have happened* or the big bang does not work because of the 'horizon problem'. More *magic*!

"In the beginning God created the heavens and the earth" (Genesis 1:1). This is not magic, because God, who is eternal and omnipotent, is a ***sufficient cause*** for the universe. And He can exist eternally (and therefore has no beginning) because He is a non-material entity (God is spirit, as the Bible says in many places), Figure 3.11.

Fig.31.11: *In The Beginning God Created The Heavens and The Earth Gen 1:1*

II. Origin of Stars

According to the big bang, the 'only game in town' to explain the origin of stars, there had to have been two phases of star formation, Figure 31.12.

Fig.31.12: *Is Big Bang theory false or real? Did the origin of universe come from the biggest Explosion ever? Curtsy bixabay*

- **Phase I** involved the formation of hydrogen/helium stars (which are called Population III stars). Here is the first problem: how do you get gases formed in a rapidly expanding primordial universe to coalesce together to form a critical mass so that there is sufficient gravitational attraction to attract more gas to grow a star? Gases don't tend to come together; they disperse, especially where there is a huge amount of energy (heat). Cosmologists invented 'Dark Matter', which is invisible undetectable 'stuff' that just happens to generate a lot of gravitational attraction just where it is needed. More magic! Nonetheless, Dark Matter/ Dark Energy is a phenomenon that keeps the Universe balanced and held together. The author strongly believes that Dark Matter/Dark Energy is the manifestation of the

Fig.31.13: *The Bible tells us that God made the stars on the fourth day of Creation Week - Curtsy iStockphoto*

Lord Jesus Christ, within the God-Headship of the Trinity; (Col. 1:17).

However, we have countless stars—like the sun—that are not just hydrogen and helium, but contain the heavier elements.

- **Phase II** supposedly comes in here. Exploding stars (supernovas) from phase 1 produced sufficient pressure to force hydrogen and helium together to make new stars that made all the heavier elements (which astronomers call 'metals'), including the elements of which we are made. These stars are called Population I and II stars.

Now here is another problem: how do exploding stars, with matter flying at great speed in all directions, cause stars made of all those new elements to form? There has to be a coming together of the elements, not a flying apart. Pieces hitting one another would bounce off rather than coalesce. Most hypotheses involve multiple supernovas from phase I in close proximity, such that sufficient material collided together to form enough of a proto-star with sufficient gravity to overcome the tendency to fly apart and attract more matter and so grow a normal star.

However, supernovas are not common events, especially multiple ones at the same time in close proximity. Thus, this scenario requires a huge number of very improbable events to account for the vast numbers of the heavier stars.

This is more *magic*; *miracles* without a *miracle worker*.

God made the sun and the stars on the fourth day of Creation Week. Again, this is not magic or superstition, because God is able to do such things, Figure 31.13.

III. Origin of Life

Astrobiologist Professor "Paul Davies" said,

"*How did stupid atoms spontaneously write their own software ... ? Nobody knows ... there is no known law of physics able to create information from nothing.*"

Not only must the DNA code be explained (how can a coded information storage system come about without intelligent design?), but the incredible machinery that reads the information and creates the components of life from that information has to be explained as well, Figure 31.14.

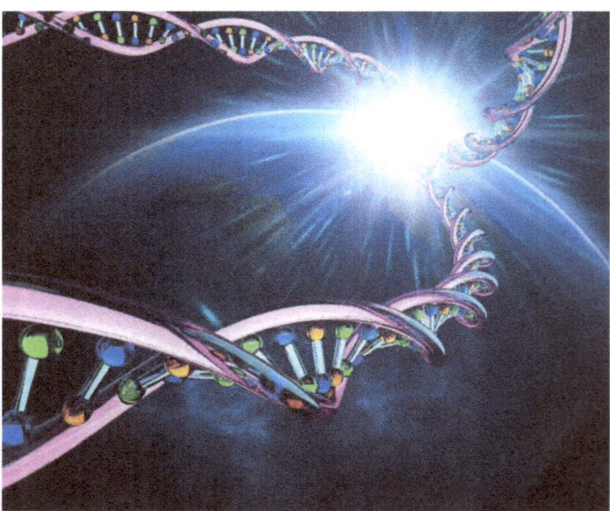

Fig.31.14: The Origin of Life – Curtsy Getty

Former hard-nosed English Atheist philosopher "Antony Flew," Figure 31.15, abandoned Atheism/materialism because of the growing evidence for such design in living things. He said,

"It now seems to me that the findings of more than fifty years of DNA research have provided materials for a new and enormously powerful argument to design."

This research, "has shown, by the almost unbelievable complexity of the arrangements which are needed to produce (life), that intelligence must have been involved."

That is, only an incredibly intelligent *designer* could account for the information systems in living things.

Fig.31.15: The Notorious Atheist Who Changed his Mind, Antony Flew. Curtsy Photo from researchintelligentdesign.org

Well-known American Atheist philosopher, "Thomas Nagel" said,

"What is lacking, to my knowledge, is a credible argument that the story [of cosmic evolution] has a nonnegligible probability of being true. There are two questions. First, given what is known about the chemical basis of biology and genetics, what is the likelihood that self-reproducing life forms should have come into existence spontaneously on the early earth, solely through the operation of the laws of physics and chemistry?"

The scientific knowledge of life grows daily, and as it does the prospects of a naturalistic (materialistic/atheistic) explanation for its origin recede into the distance. The origin of life is another miracle. 'Stuff happens'? More *magic*.

The origin of life demands a *super*-intelligent cause, such as the Creator-God revealed in the Bible.

IV. Origin of the Diversity of Life

The origin of life is only the beginning of the problem for the materialist. Along with other atheistic biologists, "Richard Dawkins" has spent his life trying to deny that living things exhibit supernatural design. In the book that 'put him on the map,' he wrote,

"*Biology is the study of complicated things that give the appearance of having been designed for a purpose.*"

The diversity of life is a huge problem. How did a microbe change itself into every living thing on earth, ranging from earwigs to elephants, from mites to mango trees? For almost a hundred years, mutations and natural selection, the mechanisms of 'neo-Darwinism,' or 'the modern synthesis,' have been said to explain this diversity of life. However, with our modern knowledge of living things, this has proved useless as an explanation.

In July 2008, 16 high profile evolutionists met, by invitation, in Altenburg, Austria. They had come because they realized that mutations and natural selection did not explain the diversity of life, and they had come together to discuss this crisis in evolutionary biology. The only consensus was that there is a major problem, a crisis.

"Thomas Nagel" (continuing from the earlier quote) put it this way:

"The second question is about the sources of variation in the evolutionary process that was set in motion once life began: In the available geological time since the first life forms appeared on earth, what is the likelihood that, as a result of physical accident, a sequence of viable genetic mutations should have occurred that was sufficient to permit natural selection to produce the organisms that actually exist?"

Think of the supposed origin of humans from a chimp-like ape in six million evolutionary years. Modern comparison of the genomes shows such large differences (of at least 20%) that this is just not feasible, even with highly unrealistic assumptions in favors of evolution.

Actually, it was not even feasible when the difference was incorrectly trumpeted to be about 1%.

Materialists have no sufficient explanation (cause) for the diversity of life. There is a mind-boggling plethora of miracles here, not just one. *Every basic type of life form is a miracle.*

Genesis 1 tells us that God, the all-powerful, all-knowing Creator, made the various kinds of life to reproduce "after their kind." Here is a sufficient cause, but even the description of the nature of living things to reproduce according to each kind has been confirmed with every witnessed reproductive event (billions of humans alone), and also in the fossil record where the transitional forms are missing and 'living fossils' testify to consistent reproduction 'after their kind' in thousands of species.

V. Origin of Mind and Morality

The origin of ***mind and morality*** from ***energy and atoms*** has long been a problem for the materialist. It is a major theme of philosopher Thomas Nagel's book, ***Mind and Cosmos***, already referred to.

A fig tree produces figs, not apples. That seems obvious. Likewise, physics and chemistry produce physical and chemical outcomes. However, ***mind and morality*** are ***not just matters of physics and chemistry***. Sure, creatures that are physical and chemical have mind and morality, but how did such non-material things arise from the material?

This is a serious problem for materialism, and the Atheist Nagel candidly admits it, to the extreme annoyance of his atheistic colleagues.

The famous (and reluctant) convert from Atheism to Christianity, "C.S. Lewis," put it well when he wrote,

"If the solar system was brought about by an accidental collision, then the appearance of organic life on this planet was also an accident, and the whole evolution of Man was an accident too. If so, then all our present thoughts are mere accidents—the accidental by-product of the movement of atoms. And this holds for the thoughts of the materialists and astronomers as well as for anyone else's. But if their thoughts—i.e., of materialism and astronomy—are merely accidental by-products, why should we believe them to be true? I see no reason for believing that one accident should be able to give me a correct account of all the other accidents. It's like expecting that the accidental shape taken by the splash when you upset a milk-jug should give you a correct account of how the jug was made and why it was upset."

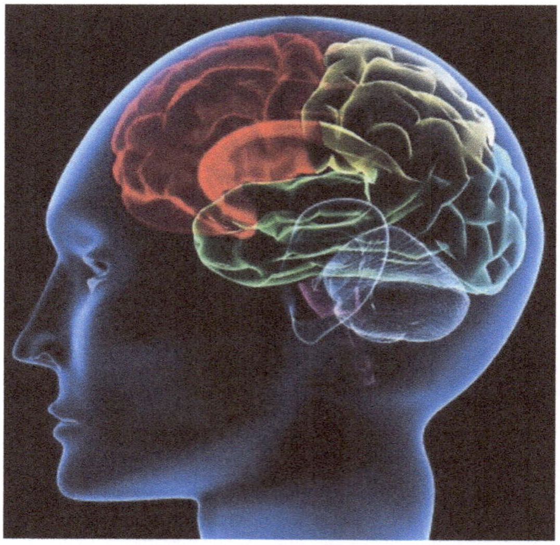

Fig.31.16: Mind and Morality – Curtsy (YAKOBCHUK VASYL)

The Atheist has no sufficient cause to explain the existence of mind and morality, Figure 31.16. Magic happens!

Why do apparently intelligent people resort to believing in magic—uncaused events—at so many points? By not believing in God, they have put themselves into an irrational philosophical corner. Romans 1:21 in the Bible says that when people deny that the Creator-God exists, they end up with 'futile thinking.'

Richard Lewontin admitted that (leaving God out of the picture), *"We take the side of science in spite of the patent absurdity of some of its constructs ... "* (he confuses 'science' with materialism).

WHERE DO WE GO FROM HERE?

God made man "*in His image*," a creature with a mind and morality (Genesis 1:27). As such, we are able to think about God and 'know' Him, Figure 31.17. That is the very reason for our existence. Isaiah 1:18 records God speaking to the people of Israel who had turned away from Him,

"*Come now, let us reason together, says the Lord: though your sins are like scarlet, they shall be as white as snow; though they are red like crimson, they shall become like wool.*"

Fig.31.17: Man Made in God's Image

THE ULTIMATE FOLLY

Trying to live life as if God does not exist is the ultimate rebellion (sin), and the ultimate folly. The good news is that *God is in the forgiving business* for those who will admit their error and seek Him:

"Seek the Lord while he may be found; call upon him while he is near; let the wicked forsake his way, and the unrighteous man his thoughts; let him return to the Lord, that he may have compassion on him, and to our God, for he will abundantly pardon." (Isaiah 55:6–7).

Jesus Christ came into the world to make possible our forgiveness! This is not magic.

How Does it End?

Fig.31.18: How Does End – Curtsy - iStockphoto

The End-Time

Many people enquire about end times. In most cases they want to find out if the particular view about when Christ is going to return, whether there will be a rapture, and the sequence of events, Figure 31.18.

Of course, one's view of end times is important—otherwise the Scripture wouldn't spend so much time telling us about it.

Many who are passionate about eschatology and who have formed a definitive view, often use it as a type of litmus test for orthodoxy. The mere fact that sincere, Bible-believing Christians can hold such a range of views should indicate how challenging it is to interpret these passages definitively—after all, with much of it we are dealing with the future and it hasn't happened yet!

However, we creationists mandate is to focus on providing the *foundation* for a biblical view of end times, which goes to the heart of what we hope for in eternity. For that we need to understand the Bible's big picture of Creation, Fall, Redemption, and Restoration. Get this big picture wrong and one will not be able to have any sound biblical view of end times. As we explain this, it will be important for you to look at the biblical references we cite, because these will be common elements that all Bible-based end times views share—and these elements are related to this big picture view.

THE EARTH WILL PASS AWAY

Before the Fall described in Genesis 3, the earth was a perfect paradise with no death, disease, thorns, or carnivory. Adam's sin brought death on not just his descendants (all of us), but also on the whole world. Today's world is scarred by the global judgment of the Flood. Romans 8:22 tells us:

"For we know that the *whole creation* has been groaning together in the pains of childbirth until now."

Fig.31.19: *Heaven and Earth will Pass Away – Curtsy Google Image*

Peter says that the world will be judged again in the future—this time with fire, not water (2 Peter 3:10–14). Just like at the time of the global Flood, the entire earth will be destroyed. But this destruction will pave the way for the final restoration of the earth, Figure 31.19. This has to happen, because a sin-cursed earth that bears the marks of judgment is not a suitable home for resurrected, perfected people.

To some this seems overly dramatic. But just think, as long as a fossil remains in the ground it is a reminder of death and sin, and also Satan's legacy on the earth, so it has to be erased completely. Just as God raises our dead-destroyed bodies and restores them to be perfect bodies that will never sicken or die, God will completely restore the destroyed earth.

NEW HEAVENS AND NEW EARTH

Even the phrase 'new heavens and earth' indicates the nature of the restoration we look forward to. The Bible uses the phrase 'heavens and earth' to talk about the entire physical universe in a comprehensive way (Genesis 1:1). So, when the Bible talks about a *new* heavens and earth (2 Peter 3:13, Revelation 21–22), even the term indicates that it is *like* the first heavens and earth—otherwise there would need to be a different term for it, Figure 31.20.

Fig.31.20: New Heaven and New Earth

So, 'new heavens and earth' means we should look to what God originally intended for the physical creation to understand what things will be like—what Eden was like before Adam sinned. Eden was a picture of the perfect paradise God created for people to live in. It was a place especially suited for humans to live comfortably and engage in easy, pleasant work (Genesis 2:15).

In the well-watered garden with plentiful food, all their needs were provided for. Best of all, they enjoyed direct, *intimate fellowship with God*. There was no sin, no death, and no barrier between men and God. This has to happen again, and it is most certainly something we all look forward to. Accordingly, the 'new heavens and earth' will be a *restoration* back to how things were before the Fall.

PHYSICAL REGENERATION

When people think of heaven and our resurrected state, many imagine an ethereal or ghost-like spiritual realm where we'll all sit on clouds and play harps, Figure 31.21. However, that fails to appreciate God's purpose in creating the earth to be a perfect home for humanity, and that fails to appreciate God's plan to restore the original lost paradise.

Fig.31.21: The Resurrected Redeemed – Eternally with Jesus

Jesus' mission was not a 'Plan B.' God is not going to scrap this world for an ethereal, spiritual state that we can't imagine, He's going to *restore* it to a state even better than the initial creation, because there will never be the possibility of another Fall.

There are many good aspects of creation that we can appreciate and enjoy even in this fallen world. And many of the things we love about this earth are things we're still going to enjoy on the new, restored earth.

We will have spiritual and physical bodies (we actually have those now), sensory experiences, and be able to appreciate the beauty of God's creation as He always intended. That's why the Resurrection of Christ is vitally important—He was raised with a glorified physical body—the same sort of body that we will receive (1 Corinthians 15).

RESTORATION OF GOD'S PLAN

The restored earth must again be physical because if God does not restore the physical world and at least what was lost, then Satan would 'win' in a certain sense, Figure 31.22, because he would have foiled God's original purpose in creating the earth. Instead, God will undo everything Satan did, and He will make creation consistent with His consistent plan. This is why the resurrection of Christ was so important. It gives us an inkling of what God will do when Christ returns. Only the Creator of the universe has the power over death.

Fig.31.22: Restoration of God's Plan

The end of Revelation gives us a glimpse of what is to come. A world without sin, and where God is praised because of His mercy and grace, and Jesus is glorified as the Savior of the nations. Those who have trusted in Christ are resurrected in glorious bodies that will never sicken or die, to live righteous lives in intimate fellowship with our Creator.

COMPROMISE IN GENESIS CAUSES PROBLEMS FOR END TIMES

It is absolutely clear that Revelation 21–22 presents the New Heavens and Earth as the physical restoration of Edenic perfection. If God's original design included millions of years of death and suffering so that evolution could progress, and God called that 'very good,' then the idea of a 'restoration' is not really something to look forward to. Is He again going to use a process of millions of years of death and suffering? Revelation 21:1 says:

"*Then I saw a new heaven and a new earth, for the first heaven and the first earth had passed away,* "

How can a millions-of-years view be biblically consistent when God promised that there will be no more death or pain in the new heavens and earth? Even long agers and theistic evolutionists believe in this promise, but how can they be consistent if they think God used a process of death and suffering over millions of years to originally create?

GOD'S PLAN WITHOUT SIN AND DEATH!

All these elements are common to all Christian views about eschatology. Scripture's teaching about the future New Heavens and Earth only makes sense in light of the Bible's big story of Creation, Fall, Redemption, and Restoration, and most importantly starting with a literal, good creation. If God created a perfect world with no sin or death, it makes sense that the final goal would be its restoration to its pre-sin state, Figure 31.23.

Fig.31.23: God's Plan to Remove Sin and Death Eternally

RESTORING THE PHYSICAL CREATION

In one form or another form, eschatology is core to creationists in one sense—creationists absolutely assert that the goal of Jesus' work was to restore not just believers, but the entire physical creation. And this will happen when Jesus appears again. No matter what various people believe about the timing, or how to interpret various prophecies, getting the biblical foundations right means we all should look forward to the day when we stand on the restored earth with our resurrection bodies, in the presence of our awesome Creator.

FUTURE OF TIME, SPACE, AND MATTER

Certain "progressive creationists" have been accusing Biblical Christians of believing that the universe never had a beginning—that time, space, and matter have always existed in some form.

That is not only wrong—it is false! It can't be merely a misinterpretation of something the bible has clearly stated. It can hardly be anything but a deliberate misrepresentation. It has been stressed repeatedly—the basic truth that Genesis 1:1 is the divine record of the absolute beginning of time, space, and matter.

" *In the beginning God created the heaven and the earth.*"

Time itself, according to this most profound (yet simple) of all declarations, had a *beginning*. Space (i.e., "the heaven") and matter ("the earth") began simultaneously with time. Before that beginning, there was *nothing*—that is, nothing except God!

At this point, human reason must defer completely to faith in God. The skeptical philosopher may feign intellectual curiosity in asking the believer: "But who made God?"

A naïve child can ask the same question, of course. We cannot comprehend with our minds the concept of an eternal, omnipotent, omniscient, God, but there is certainly no better answer to the problem, and we *can* comprehend it intuitively with our hearts.

Such a God is an adequate uncaused First Cause to explain the existence of the universe; but nothing else is. By the universal scientific principle of cause-and-effect, the existence of personalities and moral values in the universe requires a Personal, Moral Being as their Cause. ***That's God***!

Not only did God create the universe of time, space, and matter, He did it instantaneously, by His own omnipotence and spoken will.

"By the word of the Lord were the heavens made; and all the host of them by the breath of His mouth. . . . For He spake, and it was done; He commanded, and it stood fast" (Psalm 33:6,9).

Also, God's statement to the prophet Isaiah.

"Mine hand also hath laid the foundation of the earth, and my right hand hath spanned the heavens: when I call unto them, they stand up together" (Isaiah 48:13).

There was no gradual evolution of the stars and galaxies, nor of the solar system and its planets. They all *stood up together* when God spoke them into being. God did extend His many creative works over a ***six-day*** period. But each time He spoke, the commanded action followed immediately. For example,

"God said, Let the earth bring forth grass, the herb yielding seed, and the fruit tree yielding fruit . . . and it was so" (Genesis 1:11).

The reason His work was spread over ***six days*** was to serve as a pattern for man's work week later on. This definitive truth was actually placed in His Ten Commandments, written in stone directly by God Himself.

"Six days shalt thou labor, and do all thy work: . . . . For in six days the Lord made heaven and earth, the sea, and all that in them is" (Exodus 20:9,11).

All Ten Commandments, including this, were on

"tables of stone, written with the finger of God" (Exodus 31:18).

It is dangerously presumptive for anyone—including *Christian progressive creationists*—to claim that these plain words really mean that God was *creating everything during billions* of years and therefore, we should work six days out of seven! That would be an obvious non sequitur.

But this is just what many such Christians are at least implying—especially those educated at *Harvard Divinity School*, Union Theological Seminary, or many other such American and European temples to the unknown God. They may speak knowingly about parallel cosmogonies on the Sumerian tablets and so forth, but the fact is that the Genesis record was given by divine inspiration, most likely through the antediluvian patriarchs and then eventually compiled and edited by Moses into its present form. It thus precedes all these Near-Eastern cosmogonic myths, which are at best mere corruptions of the straightforward, original, true account in Genesis.

The latter was indirectly confirmed by the Lord Jesus Christ, who, as Creator, was there when it happened! Especially Matthew 19:4-6; Mark 10:6-9; Luke 17:26-27; also, II Peter 3:4-6.

Not only were all things created by divine fiat, essentially instantaneously, they were also pronounced by God, after it was all completed, to be ***"very good"*** (Genesis 1:31). Therefore, there can be no fossil remnants of that creation period which speak of suffering and death.

This is surely the greatest heresy of progressive creationism.

To accept the geological ages with their multi-billion-year testimony (all the fossils, which are used to identify the various ages, speak eloquently of suffering and death) is in effect accusing God of wanton cruelty.

It even compromises the gospel, which requires the substitutionary suffering and death of Christ as the awful wages of sin.

I am not charging individual progressive creationists with heresy, of course. One can perhaps sincerely (though unjustifiably) interpret the creation *days* to be *ages*. But the geological ages with their billions of fossils are a problem!

To accuse God of causing all this long before sin entered the world—that's something that seems very close to heresy, if not blasphemy.

"By man came death" (I Corinthians 15:21), the Bible says.

In any case, we trust the above repetition of many previous expositions will make it very clear that we believe (along with most other Biblical literalists) that the universe has *not* existed forever. It had a wonderful beginning, created instantaneously by the omnipotent God several thousand years ago.

There are some "evolutionists" who do believe in an infinitely old universe, holding usually to some form of the steady state theory, and we have referred to them occasionally, not as believers in creation but as opponents of the big-bang theory. The latter concept is held by most evolutionists and, unfortunately, accepted also by many progressive creationists, who think that Genesis 1:1 refers to the assumed big bang. It does not, of course.

QUANTUM FLUCTUATION

Only a very small minority of cosmologists and astrophysicists believe in God at all. Most physicists and cosmologists now believe that the primeval space-time universe evolved via a quantum fluctuation out of nothing, Figure 31.24. As far as matter is concerned, however, they are still at a loss as to what to believe about that.

Where did matter come from? . . . The best theories of the origin of the universe still fail to explain how it managed not to turn up empty.

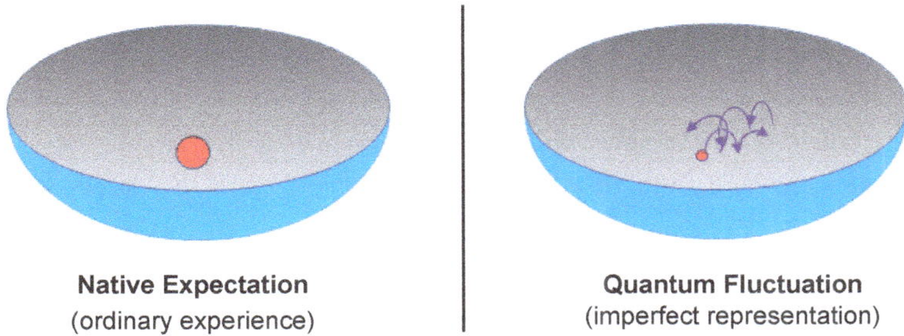

Fig.31.24: *Quantum Fluctuation*

As far as the origin of life is concerned, they haven't a clue there either.

Nobody knows how a mixture of lifeless chemicals spontaneously organized themselves into the first living cell.

The hope that life may have evolved somewhere else than on Earth has spawned a comprehensive, but futile, search by radio telescopes.

None of these searches has detected a bona-fide signal from an E.T. intelligence.

Evolutionists seem pathetically eager to find some way of accounting for the universe and its life forms without resorting to God and creation. But they must inevitably fail, and some at least sense that they will fail.

In our attempts to understand the nature of the universe, theorists must often admit to reaching a possible dead end—a question that we may never satisfactorily answer.

It is obvious that Genesis 1:1 is a satisfactory answer, but our atheistic physicists and astronomers keep trying since they feel they must find an answer that does not involve God. We can understand *their* motivation, but it is really distressing that some *creationists*—no matter how *progressive* they think themselves to be, still think that God would use the big bang and the geological ages to *create*, refusing to take His account of creation literally.

ETERNAL UNIVERSE

Well, the universe has not existed from eternity past, but it *will* exist eternally in the future. There was a beginning, but there will be no end, Figure 31.25.

"Praise ye Him, sun and moon: praise Him, all ye stars of light. . . . for He commanded, and they were created. He hath also established them for ever and ever: He hath made a decree which shall not pass" (Psalm 148:3,5-6).

Even the ***earth*** and its atmospheric heavens will continue forever, once they have been ***made new again***.

". . . The new heavens and the new earth, which I will make, shall remain before me, saith the Lord. . . ." (Isaiah 66:22).

The present *"Heaven and earth shall pass away,"* of course (Matthew 24:35), for all the age-long effects of sin must be purged away (such as the fossils, for example), but then God will

Fig.31.25: While The Universe is Eternal, it Still Needs a Cause – Curtsy Thinking to Believe

"make all things new" again (Revelation 21:5).

Peter says that we can then

"look for new heavens and a new earth, wherein dwelleth righteousness" (II Peter 3:13).

And there, in the holy city on the new earth,

"His servants shall serve Him: And they shall see His face; . . . and they shall reign for ever and ever" (Revelation 22:3-5).

ATOM & DESTRUCTION OF PRESENT EARTH

Preparing for New Earth and New Heaven

Atom is a wonder of creation. It is highly organized material in our universe. The Universe and all its contents including our human bodies are made up of atoms. Every physical object around is made up of atoms. The Greek philosophers Democritus and Leucippus speculated about the existence of atoms. But for millennia, atoms remained mysterious and enigmatic.

However, in the modern age, scientists have proven the existence of atoms. Today using "STM" – Scanning Tunneling Microscopes – you can amazingly see the individual atoms. At present we understand the structure of atoms. Also, we have recognized the power of atoms. New elements were discovered. Periodic table was constructed. New materials were created – from chocolates and ice creams to super planes, supercomputers and spacecraft.

If we look at a chemical reaction, we should see the stunning precision of atoms as they interact with each other. Atomic clocks were built to measure time with amazing accuracy. Global wireless networks like 5G became possible because of the revolutions in atomic science and thermodynamics. Atomic energy now supplies electricity to millions of homes. Nuclear submarines depend exclusively on the power from atoms. Computer chips and transistors are built on atoms. Quantum entanglement and quantum computers are being developed based on atomic functions.

Fig.31.26: Atomic Destruction of The Earth – Curtsy Wikipedia

The atomic technology also revolutionized healthcare. MRI scan are readily available for any patient. The MRI machine uses powerful magnets to produce a strong magnetic field that forces protons in the body to align with that field. It takes vital images of our organs using this atomic technology. Based on the atomic reactions inside living cells, new medications are being created and new vaccines are being discovered. On the other hand, atoms also frighten us, Figure 31.26. Hiroshima, Nagasaki, Yuko-Hama, Chernobyl, Fukushima, Three Mile Island – those names may remind us of the destructive power of atoms.

Also, the Second World War, nuclear weapons were dropped over Japanese cities – Hiroshima and Nagasaki. Thousands of people were incinerated into gases within a few minutes. Millions more were affected by the radioactive debris in their neighborhoods.

Dangerously, a few days ago, China tested a hypersonic nuclear missile that can travel at five times the speed of sound. These nuclear missiles can enter into outer space and then can attack any destination on the earth. They are thousands of times more powerful than the ones dropped on Japan.

Currently, we are living in a new cold war between the United States and China. Both countries will sharpen their nuclear weapons in the coming years. Whoever can employ the atomic weapons faster will control the future of the world.

God gifted mankind with mind-blowing capacity to understand the structure of atoms and put its energy for their uses. But man has failed in doing God's will on this earth, and God must interact in His perfect timing.

First, let us examine the fascinating journey of great scientists in their understanding of atoms and their structure. We should start with Sir "Isaac Newton." He realized the power of mathematics in understanding our world. He invented calculus. From a mathematical viewpoint, Newton predicted the existence of atoms. He became the professor of Mathematics at Cambridge University, England. He wanted to thoroughly examine the relationship between physics and mathematics. His greatest work was titled the *Mathematical Principles of Natural Philosophy*. In this masterpiece, he laid out the mathematical foundation of the natural world. Then there was "Joseph Priestly" (1733 – 1804). Like Newton, Priestly believed that God created our observed world and by examining it and studying it rationally we may understand the nature of reality.

He was a minister and a scientist. Who said science and religion are enemies? While studying the Bible on one hand, Priestly discovered oxygen. His scientific laborious experiments led him to the discovery Oxygen gas. His discovery of oxygen unleashed a chemical revolution in science which facilitated the discovery of atoms.

Then there was "John Dalton" (1766 – 1844), an English chemist. Born into a Christian Quaker family, he was also a dissenter like Priestley. He noticed some mathematical regularities in how chemical compounds are formed. He formulated the law of multiple proportions based on his observations. Dalton stated discovered that 100 grams of tin combined with 13.5 grams of oxygen formed tin oxide. Also, 100 grams of tin combined with 27 grams of oxygen formed still tin dioxide. 13.5/27, that is the ratio of one to two. Oxygen as it bonds with tin is following a fixed ratio. There must be discrete units of oxygen to make this possible.

Nature is using simple ratios of counting numbers. Dalton thought, maybe elements are made up of countable bits. He formulated his atomic theory of matter based on this thinking. Remember how it all started: observing mathematical regularities in chemical reactions.

Then came "Robert Brown} (1773 – 1858). He was a Scottish botanist. He was spending much of his time around his microscope, studying plants, flowers and their tiny particles. In 1827, he found something interesting while looking through his microscope. Minute particles from the grains of pollen of a plant were making jittery motion in water. It is called Brownian motion. It was observed also in gases like motes in sunlight. For the next 78 years, nobody was able to explain this curious, seemingly erratic motion of tiny particles suspended in a fluid or in a gas. It fell on Einstein to explain the Brownian motion.

Before we discuss about Einstein, let us look at another great scientist named "James Clerk Maxwell" (1831 – 1879). His mathematics not only took us in the direction of the reality of atoms but also helped us know what is inside the atom. Maxwell was born in Edinburgh on the 13th of June 1831. His mother encouraged him to "look up through to God's Nature." She advised him to memorize the Bible. From his childhood, he knew the Bible by heart. At eight years of age, Maxwell could repeat the whole of Psalm 119th.

Maxwell was hands-on operation physicist. Examining some of the books on the history of science, you would realize the scientific revolution Maxwell brought in our world. In the 1860s, Maxwell had a series of papers designed to illustrate that matter consisted of what he called molecules. His equations led to the discovery of radio waves. They predicted the speed of light.

Historian Basil Mahon explains the importance of James Clerk Maxwell in his book, *The "Man Who Changed Everything."*

The influence of James Clerk Maxwell runs all through our daily lives. His electromagnetic waves bring us radio and television and provide the radar that makes safe air travel possible. Color television works on the three-color principle that he demonstrated. Pilots fly aircraft by control systems which derive from his work. Many of our bridges and other structures were designed using his reciprocal diagrams and photo elastic techniques.

Even more significant is his influence on the whole development of physical science. He started a revolution in the way physicists look at the world. It was he who began to think that the objects and forces that we see and feel may be merely our limited perception of an underlying reality which is inaccessible to our senses but may be described mathematically.

Maxwell was the first to use field equations to represent physical processes; they are now the standard form used by physicists to model what goes on in the vastness of space and inside atoms. He was also the first to use statistical methods to describe processes involving many particles, another technique which is now standard. He predicted, correctly, that light was a wholly electromagnetic phenomenon and that its speed was simply the ratio between the electromagnetic and electrostatic units of charge. His equations of the electromagnetic field were the chief inspiration for Einstein's special theory of relativity and, along with his kinetic theory of gases, played a part in Planck's discovery of the quantum of energy."

So, one can see the incredible influence of Maxwell on our world. If the 18th century belongs to Newton and the 20th century belongs to Einstein, the 19th century must belong to Maxwell. Maxwell would unleash a world led by triumph of abstract mathematics. His mathematics not only took us closer to the reality of the atoms, but also made more delightful endeavors become possible. You can peer inside the atoms.

Then there was "Albert Einstein." Einstein admired Maxwell so much that he placed a photograph of Maxwell on his study wall. Robert Brown and James Maxwell – the fruits of their work had come to ripen in the hands of Einstein. In 1905 he published a paper in which he showed that Brownian motion was caused by the minute particles being hit by water molecules.

In the year: 1905; It is called *Einstein's Miraculous Year*, because that year he published five great papers that changed the face of physics. Explaining the Brownian motion, Einstein convinced many physicists of the existence of atoms. In 1913, French physicist "Jean Perrin" did experiments and proved the accuracy of Einstein's predictions. He won a Nobel Prize for proving the existence of atoms.

The 20th century witnessed more stunning developments in our journey to understand the atom: the discovery of electrons, of protons, of neutrons, even subatomic particles like quarks, leptons and muons. Scientists like "J.J. Thompson," "Ernest Rutherford," "Niels Bohr," "Paul Dirac," "Werner Heisenberg," and "Erwin Schrodinger" changed the face of atomic theory of matter.

Quantum theory successfully predicted the properties of atoms, subatomic particles and their forces. Recently, we can confidently discuss the inside of the protons and neutrons. In 1989, scientists at IBM took individual xenon atoms and arranged them to spell out the company's name.

Looking through an electronic microscope, now you can see the miraculous arrangement of the atom structure of an individual atom that evaded human curiosity for millennia.

Mathematics of atoms also paved the way to understand the universe at large. When you observe a galaxy, you will see millions of stars. Their gravitational interactions can be astonishing. A mathematical analysis of those interactions led to the prediction of dark matter/dark energy. Once again, we will see the amazing power of Maxwell's equations. We came to know more amazing facts about nature: Atoms make up only about 3.8% of the the entire cosmos.

This fascinating journey to comprehend the structure of atom reveals the beauty of God's creation even in the tiniest things in nature.

THE FUTURE OF ATOMS – THE FUTURE OF THE EARTH AND HEAVENS

Secular scientists, hypothetically state that ultimately the universe will die in coldness, as entropy increases under the second law of thermodynamics.

Nonetheless, the Bible predicted different scenario. It is likely an atomic explosion will hit the Earth causing much damage during the Tribulation Period, which will last 7 years. The tribulation period will take place as soon as The Holy Spirit (God) is removed from the Earth together with all the believers in Jesus Christ (God). The universe has never experienced such difficult time in all history, as stated in 2 Peter chapter 3.

"The heavens and the earth, which are now, by the same word are kept in store, reserved unto fire against the day of judgment and perdition of ungodly men .... But the day of the Lord will come as a thief in the night; in the which the heavens shall pass away with a great noise and the elements shall melt with fervent heat, the earth also and the works that are therein shall be burned up. Seeing then that all these things shall be dissolved, what manner of persons ought ye to be in all holy conversation and godliness" - 2 Peter 3

TIME, SPACE, AND ATOMS (MATTER)

God brought all three into existence as we read in Genesis chapter 1. God integrated all three into the fabric of our universe. And one day in the future, God will allow them to be destroyed.

Einstein was Jewish boy. He knew the Old Testament. Professor Steven Gimbel wrote an excellent book on Einstein titled Einstein's Jewish Science. He describes the influence of Judaism on Einstein. Jews always believed that time is not absolute. It is relative. They understood it from Psalm 90.

But the day of the Lord (The Tribulation Period) will come as a thief in the night 2 Peter 3:10. The Day of the Lord is the most dreadful judgement day in the history of the universe. It is the judgment day of God, where the heavens, means space. The space will come to an end with a great noise, *"and the elements shall melt with fervent heat, the earth also and the works that are therein shall be burned up"* 2 Peter 3:10

On that day, elements will melt. On God's Day of Judgement, that is how this Earth and Heavens will burn up preparing to the New Regenerated of Earth and Heavens, where sin is abolished and decay is ceased to exist.

THE NEW EARTH

There is a lot of confusion over the doctrine of heaven and the future new heavens and earth. Many Christians, while they look forward to being with Jesus after our physical death here on earth, do not have a real idea of what our eternal existence will 'look like' or entail. Often, they read about the new heavens and earth described in Revelation 21–22 and then imagine existing forever in some sort of ethereal realm, instead of eternal existence in a real, restored, physical universe.

However, the Bible has much to tell us about what we have to look forward to, and understanding our future state also helps us to understand what we lost when Adam fell. This also has serious implications for those who want to allegorize the Creation events in Genesis in order to add millions of years of evolutionary history.

In 1 Corinthians 15, Paul takes a bold stance on the Resurrection:

"Now if Christ is proclaimed as being raised from the dead, how can some of you say that there is no resurrection of the dead? But if there is no resurrection of the dead, then not even Christ has been raised. And if Christ has not been raised, then our preaching is in vain, and your faith is in vain. We are even found to be misrepresenting God, because we testified about God that he raised Christ, whom he did not raise if it is true that the dead are not raised. For if the dead are not raised, not even Christ has been raised. And if Christ has not been raised, your faith is futile and you are still in your sins. Then those also who have fallen asleep in Christ have perished. If in Christ we have hope in this life only, we are of all people most to be pitied" (15:12–19).

Paul is saying that we must believe in the physical resurrection of Christ to be saved—it's that important. This is because our future resurrection—what we look forward to—is the same sort of resurrection as Jesus.' So, the resurrection of the dead is a Gospel issue; you can't be a Christian and not believe that we (believers) will live forever with Christ—in real, physical, resurrected bodies. But where will those bodies live? Scripture's testimony is clear and unanimous that the new heavens and earth (hereafter NHE) will be a physical (yet also spiritual) realm.

As we shall see, a fully restored creation, as outlined in the book of Revelation and elsewhere in Scripture. is unequivocally linked to the events in the Garden of Eden and *is part of all mainstream eschatological views*. The doctrine of a NHE has been a standard core doctrine of the evangelical church and all of the mainstream (non-cultic) denominations throughout all of Christian history.

THE END OF THE PRESENT WORLD

The Bible teaches that this present earth, indeed all of creation; cf. (Rom 8:22) is cursed because of the presence of sin and will be destroyed. Peter writes:

> *"But the day of the Lord will come like a thief, and then the heavens will pass away with a roar, and the heavenly bodies [or 'elements'] will be burned up and dissolved, and the earth and the works that are done on it will be exposed. Since all these things are thus to be dissolved, what sort of people ought you to be in lives of holiness and godliness, waiting for and hastening the coming of the day of God, because of which the heavens will be set on fire and dissolved, and the heavenly bodies will melt away as they burn! But according to his promise we are waiting for new heavens and a new earth in which righteousness dwells. Therefore, beloved, since you are waiting for these, be diligent to be found by him without spot or blemish, and at peace"* (2 Peter 3:10–14).

NOAH'S FLOOD – GOD'S JUDGMENT

Peter is not using figures of speech. Indeed, just before this, he reminds his readers that God had previously judged the whole globe with a cataclysmic flood in history, Figure 31.27. He does not say, "It will be as if the heavens are being burnt up." In straightforward language, he is saying that God is actually going to burn up the universe and set up a "new heavens and earth in which righteousness dwells." And he uses this fact to tell his audience to live a life befitting citizen of the new heavens and earth.

Fig.31.27: People perished out of Noa's Ark

It is not a hyperbole to call them an "uncreation" of the heavens and earth—many judgments in the Bible are reversals of creation; for example, the Flood reversed creation to the time before the land was separated from the seas on Day 2, and Jeremiah 4:23 alludes to a future uncreation that reverses the universe back to the state described in Genesis 1:2.

Again, depending on one's eschatological stance, many of the various judgments in Revelation (e.g., one third of mankind killed [9:15] may be presented either with some symbolic meaning or a more literal meaning—what happens *at the end*. It is not our point to discuss the details, but to clearly present the big picture, and this is that the present Creation will suffer terrible, utterly destructive judgment because of sin. But the destruction is not the end of the story, because God will create a new heaven and a new earth "for the first heaven and the first earth had passed away" (Revelation 21:1) and believers will live in that new creation for eternity.

WHY WILL THIS WORLD BE DESTROYED?

To destroy a whole universe seems a rather drastic solution to the problem of a fallen world. However, Scripture is clear that the whole creation fell. Romans 8:19–23 says:

"For the creation waits with eager longing for the revealing of the sons of God. For the creation was subjected to futility, not willingly, but because of him who subjected it, in hope that the creation itself will be set free from its bondage to corruption and obtain the freedom of the glory of the children of God. For we know that the whole creation has been groaning together in the pains of childbirth until now. And not only the creation, but we ourselves, who have the first fruits of the Spirit, groan inwardly as we wait eagerly for adoption as sons, the redemption of our bodies" (emphases ours).

When Adam sinned, the earth was cursed because of him (Genesis 3:17–19), and the earth was further polluted by murder, violence, and immorality (6:11–12; Leviticus 18:24–28; Numbers 35:33; Psalm 106:38; Jeremiah 3:2, 9; 16:18). But it is not just the dirt on the surface of the earth; it is the entire universe, all of creation (Greek *ktisis*), that is cursed. Thus, all of creation is in need of restoration.

Because the creation has been affected by the Curse and further polluted by man's sin, it is not a suitable place for resurrected, perfect people to live. How could we live among fossils, graveyards and reminders of death (even if their inhabitants were vacated), and a Flood-scarred earth that bore testimony to God's great judgment of sin? How can we live forever next to a star that has a limited lifespan or in a universe with built-in, non-eternal features that will eventually die of 'heat death'? All of this is a reminder that death is an enemy that beset all of Creation, and a reminder that Christ has conquered death, thus, giving us something to look forward to:

"For as by a man came death, by a man has come also the resurrection of the dead. For as in Adam all die, so also in Christ shall all be made alive. ... The last enemy to be destroyed is death" (1 Cor. 15:21–22, 26).

Just as our bodies die, and return to dust, they will be raised as new bodies that nevertheless have continuity with our former selves, likewise the earth is fallen and will be destroyed, but it will be destroyed in order to be renewed. The restoration of the earth is directly analogous with the resurrection of the redeemed in Christ. Just as we have to die before we are resurrected, the earth must be destroyed before it is renewed. It is not an 'ultimate' or final destruction; it is a destruction that clears the way for its re-creation.

LOOKING FORWARD TO THE NEW WORLD

Genesis uses the phrase "heavens and earth" to encompass all of the physical creation (the universe). When the Bible (2 Peter 3:13, Revelation 21–22) uses the phrase "new heavens and earth", it is having a similarly all-encompassing meaning. It is an indication of the *continuity* of the new creation with the old. But the word 'new' has the connotation of "superior" or "improved", such that the old will be obsolete.

We are given several images of the new world. If we want to see what an unfallen physical creation looks like, the obvious place to start is Eden. Eden is a picture of God's ideal paradise on earth. It was a place especially suited for humans to live comfortably and engage in easy, pleasant work (*Genesis 2:15*) and for the purposes of appreciating their Creator. All Adam and Eve's needs were provided for, and they were in regular, direct fellowship with God. There was no sin, no death, and no barrier to mankind's relationship with God.

In the NHE (New Heavens and Earth) there is a return to a ***sinless*** state with no suffering or evil of any kind, and unlimited access to God (*Revelation 21:3–4*). All of this is possible because of Christ's sacrifice in paying for our sins. But it is even better than the original creation, because it is not a simple return to Eden. Rather, God will redeem the best parts of *culture* as well as the earth.

The best advance in the new heavens and earth will be that there will never be any possibility for sin or another Fall. "Since 'the wages of sin is death' (*Romans 6:23*), the promise of no more death is a promise a promise of ***no more sin***. Those who will never die can never sin, since death is a punishment for sin. Sin results in mourning, crying, and pain. If those will never occur again, then *sin* can never occur again."

Will the New Earth be Physical?

As we mentioned earlier, some people think our eternal destination is an ethereal place populated by disembodied spirits. But that makes the mistake of confusing two places: the place where believers who die await the resurrection (variously called Paradise or Heaven), and the place we will exist after the consummation of all things eschatological. On this Paul wrote:

"So. we are always of good courage. We know that while we are at home in the body, we are away from the Lord ... Yes, we are of good courage, and we would rather be away from the body and at home with the Lord" (*2 Cor 5:6, 8*).

The Bible is clear that, although believers who die are "at home with the Lord," they still await the resurrection of the dead, when our bodies will be transformed to be like Jesus.' Again, Paul says:

Fig.31.28: *"So is it with the resurrection of the dead. What is sown is perishable; what is raised is imperishable," (1 Cor 15:42). Curtsy Flickr/Mikel Ortega*

We will also be morally perfected at that time so we will never sin.

The NHE will be as physical as the current heavens and earth. Just as the place where perfect people will live cannot be fallen, it also cannot be ethereal and non-physical. We will need a physical, material world to live in then just as much as we do now. And the Bible's descriptions of this world include *re-created animals* and trees, cities, streets, rivers, and other *physical* things. Conversely, the new earth is never described in 'ethereal' or ghostly terms, Figure 31.28.

What about the 'Spiritual Body'?

In *1 Corinthians 15:44*, Paul says,

"It is sown a natural body; it is raised a spiritual body."

Some people take this to mean that when we are raised, we will be some sort of ethereal being. This interpretation misunderstands what Paul means when he calls our earthly bodies "natural" and our resurrection bodies "spiritual" (see *Christ as the Last Adam*). It doesn't refer to the 'stuff' the body is made of, but of what motivates us and drives our desires. E.g., Paul previously referred to a 'spiritual' person, using the same Greek word *pneumatikos*, and it was obviously a physical person (*1 Corinthians 2:15*).

Let's remember that even in our sinful physical bodies we are still spiritual beings. Even the Lord Jesus was referred to as a 'life-giving" or "quickening spirit". The point being made that one can be physical and spiritual at the same time. And in the same way, the NHE will be both a *physical* and *spiritual* place. The spirit is not the sum of our being but part of it. This is why God first made the body of Adam from the dust of the ground, then breathed on him and this man became *'a living creature'* or 'living soul' (Hebrew *nephesh chayyah, Genesis 2:7*).

Currently, even at our best we're sinful, even though we're forgiven sinners. In the resurrection, our desires will be perfectly aligned with God's will. Not only will we not be able to sin, we won't *want* to sin. It will be incomprehensible to us to sin.

Will We Experience Time?

If the NHE will be made of matter, then it will take up space. And we know that space (size) and time are connected and related to each other. It is a common belief that eternal life will be timeless, but that is not really correct. Time, as we understand it, began with the creation of the physical universe. And the timeframe used in the Bible is the earth's. When the earth rotates with a light source on it, it defines a day. So only God is outside of time, because He preceded what He created—*only that which has no beginning in time can ever be outside of time.* Every created being experiences and will always experience life as a continuous series of events, one moment after another. When we sing a hymn in heaven, we will sing one word after another in time and we will need to count the bars of the song over time. If we go from one place to another, it will take us time to travel. **It's uncertain how that time will be *measured*,** but it will certainly be *experienced* in some unknown form to encompass eternity. Therefore, we will be awarded a form of eternity, while our God the savior is the infinite and eternal who is also, the giver of eternity to us. The Eternal Life to God's children is a gift.

GOD'S TRIUMPH IN THE RESTORED CREATION

Author and former pastor "*Randy Alcorn*" writes:

"God has never given up on his original creation. Yet somehow, we've managed to overlook an entire biblical vocabulary that makes this point clear.

- ***Redeem.***
- ***Restore.***
- ***Recover.***
- ***Return.***
- ***Renew.***
- ***Resurrect.***

Each of these biblical words begins with the *re*-prefix, suggesting a return to an original condition that was ruined or lost. God always sees us in light of what He intended us to be, and He always seeks to ***restore us*** to that design. Likewise, He sees the earth in terms of what he intended it to be, and he seeks to ***restore it*** to its original design."

One important *theological* reason that the new earth has to be physical is simply: if God does not redeem or restore the physical world, then Satan wins, because he would have foiled God's original purpose in creating. We're told that God will undo everything Satan did, and He will make creation even better than before. However, our new spiritual body would be able to be raised (raptured) to the highest heaves, as Jesus did and will return with him immediately with His Heavens armies before the millennium.

By the end of Revelation, sin is gone. In addition, God gets praised because of His mercy and grace, and Jesus is glorified as the Savior of the nations. Humans are resurrected. We're not just sinless in the Resurrection; we're positively righteous and can never fall again. This is attributed to the removal of the origin of sin altogether. The earth *must* similarly be restored, or there is a huge gap in how we perceive God's redemptive work.

COMPROMISE ON GENESIS CREATES A NHE PROBLEM

When one reads of the restored creation and the New Heavens and Earth, particularly in chapters *21* and *22* of Revelation, it is absolutely clear that this is analogous to what God originally did in *Genesis 1*. They are inexorably intertwined. He is sovereign and finally has His way and we can only marvel at His plan—even though He knew of Satan's plans in advance.

Those who believe that God somehow used a process of **"millions of years of evolution"** have a huge inconsistency problem here. Presumably they have no problem with what is often called the ***'Blessed Hope'***—this future eternal paradise where all believers will live forever in a restored universe, recreated miraculously and instantaneously by God. But because it is analogous to *Genesis 1*, then how could God have used a process of death and suffering to create the original?

Simply, is God going to restore things back to millions of years of death and suffering?

Moreover, if in the NHE we can see stars billions of light years away, do we then think that God 'recreated' and stretched out space over billions of years again (according to an old-Earth view of Genesis). Of course, that makes no sense. One option would be to allegorize the concept of a NHE just as is done with *Genesis 1*, but one then has to wonder why bother being saved because one cannot be sure of the future state or have any real hope of what there is to look forward to.

In short, if God's original design involved millions of years of death and suffering, then what is wrong with this creation?

Why destroy it and create a new one?

Fig.31.29: New Heaven and New Earth – Curtsy Google Image

The NHE doctrine only makes sense within a creation/fall/restoration framework. If God created a perfect world with no sin or death, it makes sense that God will restore it to a perfect world with no sin or death.

However, every old-earth or evolutionary view puts death before Adam was created. This does not come from Scripture but from *deference to a philosophical assumption called **naturalism** (the belief that natural processes can explain everything that has ever happened in the history of the universe).* So to some extent, it makes death part of God's originally 'very good' creation (cf. *Genesis 1:31*).

The NHE tells us what God's intended creation looks like—but in the view of the compromisers, why didn't He simply create it that way to begin with?

The NHE is a powerful testimony to the original very good creation, Figure 31.29.

Paul tells us,

"'What no eye has seen, nor ear heard, nor the heart of man imagined, what God has prepared for those who love him'—these things has God revealed to us through the Spirit" (*1 Corinthians 2:9–10a*).

The Bible tells us exactly what we have to look forward to when we trust in Jesus—eternal life in a perfect resurrection body, in a physical restored body, in perfect sinless fellowship with God.

GOD THE AUTHOR OF CREATION ...

Death, Bloodshed, Disease, and Suffering – "Very Good"?

There are many who have rejected the Genesis record of Creation on what they believe are 'scientific grounds.' They believe that the current secular (naturalistic/humanistic) teachings of evolution and millions of years of earth history are accurate. Nonetheless, they have involved God in their particular versions of creation, with theistic evolution and progressive creation being some of the more popular opinions circulating among churches today.

This 'slippery slope' is dangerous because they are not really defending creation, but are actually bringing into question the very character and nature of God. By trusting man's fallible opinions over God's inerrant Word, they are holding God responsible for millions of years of death, suffering, bloodshed, disease, etc.

The obvious question is:

'Would an all-powerful and loving God actually use this cruel and extremely wasteful process of evolution to create the world?'

According to Genesis, God called the creation 'good,' and when He finished, God proclaimed His creation 'very good!' If He did create the world using this sadistic method over millions of years, that would mean that suffering and death are 'good' and consistent with the character and nature of God. It is ironic that many of these same people would insist that there will be no death, suffering, sickness, and disease in heaven, as that would be contrary to even their sense of what is good.

"Tennyson" described our sin-cursed world as: 'Nature red [i.e., bloody] in tooth and claw.' According to these 'creationists,' science has proven that it has been that way for millions of years. That's because they start outside the Bible with our present 'groaning' world (Romans 8:22) and extrapolate back, using that as the basis to reinterpret the Scripture, believing that 'the present is the key to the past.' They fail to realize that it was God who was there, and He has given us His account of creation in Genesis. Hence, divine revelation is the only trustworthy key to the past, present, and future.

Sadly, many influential Christians, and most Christian colleges and seminaries, endorse the idea of millions of years. They are confusing our young people, because this teaches that they can believe that death and bloodshed are not the consequence of the Fall of Man, which invalidates the reason why Jesus Christ—the last Adam (1 Corinthians 15:45)—came to die and shed His blood on the Cross. If fossils resulted over millions of years, they are not from any worldwide flood. Therefore, the same teachers almost always deny the global Flood of Noah.

This compromise is occurring because the Church has allowed so-called 'science' (i.e., evolutionary beliefs) to gain acceptance. As a result, many object to a literal Genesis—but in doing so they undermine the basis of all Christian doctrine. The doctrinal foundations for marriage, original sin, the curse and death, the virginal conception of Messiah, and redemption itself logically depend on a literal Genesis. What they are really doing is questioning the authority of God and His Word, while defending the theories of men.

Paul warned the Colossians to 'Beware lest any man spoil you through philosophy and vain deceit, after the tradition of men, after the rudiments of the world, and not after Christ' (Col. 2:8). Furthermore, the Ephesian elders were told that false teachers will come in from the outside and even arise within the church, speaking 'perverse things, to draw away disciples after them' (Acts 20:30).

God is not the 'author of confusion' (1 Cor. 14:33)! He has spoken as plainly as possible regarding the Creation, the Fall, the Flood, and the future restoration. The straightforward reading of the Genesis record clearly teaches that God created in six literal days. Even in Exodus 20:9–11, God based our seven-day week on the days of creation: 'for *in* six days the Lord made heaven and earth, the sea, and all that in them *is*, and rested the seventh day.'

Why then the confusion? We are finite fallen creatures in a sin-cursed world using our fallible minds tainted with false humanistic assumptions. This often leads to Christians trying to redefine what the infinite all-knowing Creator clearly communicated in His infallible Word.

We need to examine our understanding of Genesis to see if any secular assumptions are influencing our beliefs about God and His creation as set forth in His Word.

HAWKING'S DEATH AND DECAY

End of the Universe

Professor Hawking gives the public his atheistic thoughts on how life began, and how he thinks it and the universe might end, Figure 31.30.

Fig.31.30: *Hawking's End of the Universe – Death and Decay*

HAWKING'S ORIGIN OF LIFE

The Professor begins with four unproven and unsupported, and hence unscientific, statements:

Just by chance some molecules bumped into each other at random until one found that it could copy itself. Then began the slow process of evolution that led to all the extraordinary diversity of life on Earth. Life seems to be what matter does, given the right circumstances. I think that life is quite common throughout the universe.

In fact, no one has ever advanced a single instance of life coming into being from non-life—i.e., from anything other than previous life. Nor is there the slightest shred of evidence of any life having arisen outside of the earth.

- First, it just puts the problem back a step. and
- second, it would have been burned up in Earth's atmosphere.

And so, the *better* (and indeed more scientific, because more rational) explanation is that life originated on Earth because it was created here by the *Living* God.

Hawking concedes that "on the face of it, life does seem to be too unlikely to be just a coincidence," even in his evolutionary scenario, which he gives as:

The earth lies at exactly the right distance from the sun to allow liquid water to exist on its surface, and the sun just happens to be the right size to burn for billions of years, long enough for life to have evolved. The solar system is littered with all the elements needed for life ... [from] older stars that have burned out, [which] only existed because of a tiny unevenness in the early primordial gas, that was itself produced by a one-in-a-billion imbalance in the sea of particles that came from the big bang.

He then asks: "So is there a grand Designer who lined up all this good fortune?" And he answers: "Not necessarily."

Readers, especially if theistic evolutionists, should realize that, for Hawking, God has no place, *even in his evolutionary belief system.*

HAWKING'S MULTIPLE UNIVERSES

Hawking advances the following argument in support of his atheism. He asks:

What if there were other universes, ones not as lucky as ours? Each of these universes could have come from its own big bang, with different laws of physics and different conditions. In some, gravity might not exist, so there could be no life. In others, hydrogen might not fuse, so there would be no stars and again no life. And for any number of reasons, universes could have come and gone without producing anything at all.

Frankly, this is nonsense. The word 'universe' means '*the totality of matter, energy and space*'. This means there is one, and only one universe, and there cannot by definition be more than one. Furthermore, seeing there is only one universe that we can observe, this rules out appealing to any others, whether real or hypothetical. So why do he and his fellow atheists invoke others? Prof. Hartnett replies:

Since there is only one universe and it looks designed for human life (and always has looked that way), there is no room for a chance explanation. Only if there were many universes, of which this just happens by chance to be one that contains human observers, could one advance a chance explanation, and so they do. They invoke either a 'multiverse'—a set containing many different universes where the laws of physics and the starting conditions on each are different to ours, or 'multiple universes,' of which ours is just one universe of the many, with a chance combination resulting in human observers.

A much better explanation is that the universe looks as if it was designed for human life on Earth because God's Word tells us that it was so designed (Isaiah 45:18), and that He, Almighty God, was the Designer.

THE PROVIDENCE OF THE UNIVERSE AND US

Professor Hawking now begins a new theme, with the words: "Cosmology tells us what lies in store for both the universe and us." And he refers to "the enormous challenges our species will face" because "we are puny organisms compared to the mighty universe that made us" and "the Earth that gave us life will not always be the blue sanctuary it is today."

Not correct, Professor. We are not "puny organisms", but people whom God made in His image and likeness (Genesis 1:26–28; see also Made in the image of God). And neither the universe nor the Earth gave us life, but the Almighty Living God. As to our future, God says that "it is appointed for man to die once, and after that comes judgment" (Hebrews 9:27–28).

Hawking next lists some of the things he sees as hazards to humans on Earth. These are:

1. Asteroids—billions of them, he says.
2. Technology—i.e., destroying ourselves with nuclear weapons.
3. Gamma rays from Star WR104, if and when it becomes a supernova.

In support of the latter, he says:

450 million years ago over half of all living organisms were wiped out in a grand extinction. One explanation is that a gamma ray burst irradiated the planet so badly that Earth's ecosystem virtually collapsed.

Evolutionary/uniformitarian geologists call this alleged event the *Ordovician–Silurian extinction*. This is supposed to be the second worst mass extinction in uniformitarian history, exceedingly even the *Cretaceous–Paleogene (K–Pg) extinction* claimed as the demise of the dinosaurs 65.5 million years ago (formerly Cretaceous–Tertiary (K–T), but in 2004, the Tertiary was abolished as an official period and replaced with two periods: Paleogene and Neogene). The worst is claimed to be the *Permian–Triassic (P–Tr) extinction*, 251 million uniformitarian years ago.

There was a mass destruction of animal and human life 4,500 years ago during Noah's Flood.

However, without any corroborating evidence that this particular burst ever occurred, this hypothesis is worthless, scientifically speaking. In fact, there was *one* mass destruction of life (long-age geologists often claim *five* mass extinction events). Also, ***Hawking and the uniformitarian geologists are wrong about the timing by a factor of 100,000***. It wasn't 450 million years ago but ***4,500. At that time***, all land animals that breathed through their nostrils (Genesis 7:22) perished except those aboard Noah's Ark. This mass destruction (***but not extinction***) of land-animal life, and the extinction of much marine life, wasn't due to a gamma ray burst, but to a ***worldwide Flood***, which was sent by God because of the evil of mankind at that time (Genesis 6:5-8). Dinosaur kinds were created on Day 6 of Creation Week, so they ***did not become extinct 65 million years ago***. Those which were preserved on Noah's Ark eventually became extinct following the Flood. For the true history of dinosaurs, see:

GETTING AWAY FROM IT ALL

Mars

Hawking's solution for "keeping humankind alive in an aging universe" is to look for somewhere else for the human race to live. His first candidate is ***Mars***. He lists the problems, but thinks that in 500 years we will have overcome these difficulties, and mankind will be living there, Figure 31.31. He then says, "Look further into the future and ultimately our solar system will follow the same path as countless billions of solar systems before it, and cease to exist." Countless billions? Well, there is some evidence of extrasolar planets around some other stars, but to make such an unqualified statement about something which was unobserved (and would have been largely unobservable) seems to be going considerably beyond the evidence.

Fig.31.31: The planet Mars is not hospitable to life. Curtsy NASA/JPL-Caltech/ASU

PLANET GLIESE 581D

Then, to ensure the survival of the human race for billions of years, he tells us we need to go to other solar systems. His next suggestion is for a spaceship to carry passengers to Gliese 581D, Figure 31.32. This is a large rocky Earth-like planet, the nearest known, seven times bigger than Earth, and orbits a star at just the right distance to allow water to exist on its surface. (Of course, that doesn't mean that water necessarily exists there.

It's more than 20 light-years (120 trillion miles) from Earth, so the fastest object we have ever put into space (the tiny unmanned dish radio antenna called *Voyager 1*) would take over 350,000 years to get there.

Fig.31.32: *Planet Gliese 581D - Super-Earth Planets are as the Name Suggest Massive Solid Planets – Curtsy NASA*

His solution is "new technology on an enormous scale", to find a way for a large manned spaceship to travel 1,000 times faster than *Voyager 1*, so the trip would only take 73 years.

Hawking tells us that the main challenge for building such a spaceship is not technical but financial:

The cost of building such a spacecraft would be huge, but for the society that made it there would be little payback. They would never see it again. So, constructing such a machine will either be the greatest act of generosity in history, or it will have to be funded by the travelers themselves.

The second problem is moral. Assuming you could get there in 73 years:

At least one generation of humans would have to spend their entire lives in space—one couldn't say they were volunteers. The ethics of sending a human cargo on such a voyage would have to be carefully considered.

Actually, solving these two problems might be difficult, because in the evolutionary worldview

a. altruism is very light on, and
b. there is *no* ultimate basis for morality.

Hawking's solution to survive long journeys and inhospitable worlds is that within the next 1,000 years, he thinks that:

- Genetic engineering will give us longer life spans and greater intelligence.
- Modifying our genes will give us skills that protect us from radiation, and give us the ability to breathe poisonous gases, and to resist infection.
- We might even develop sophisticated artificial life forms using synthetic DNA.

Then, in a computer simulation, his imagined future monster spacecraft splits up into about 200 smaller spaceships, presumably each with its human cargo, all off to colonize 200 other planets in what he calls "a true diaspora of life."

After the above diversion into science fiction fantasy, Hawking returns to his main theme of whether this cosmic world will go on forever. He tells us: "At 13.7 billion years, our universe is still in its youth. The earliest date we cosmologists think it could end is 30 billion years from now."

Wrong! professor Hawking. Our universe is—6,000 years old. And it will end when God decides to end it.

HAWKING'S PANACEA—DARK ENERGY

As to whether this cosmic world will go on forever, Hawking says:

I think the solution lies back where we began—with the big bang. What caused the explosion or inflation of the universe in the first place? ... The key to it all is something called dark energy—a mysterious form of energy that pushes space itself apart even as gravity is making matter clump together. It seems as if dark energy supplied the 'kick' that inflated the universe, ***although we're not quite sure how***.

What is certain is that the fate of the universe depends on how this ***dark energy appears***. If the dark energy slowly weakens, then gravity could get the upper hand and in 20 billion years or so the universe would go into reverse and drive everything back to whence it came. In a strange reversal of the big bang, space itself would contract. This theory is known as the ***big crunch.*** In the end, if the theory is right, 1000 billion years from now ... ***the entire universe would exist as one tiny point***, much as it was at the instant of the big bang.

But I think it is more likely the dark energy will drive the expansion of the universe forever, and that ultimately everything will just keep spreading out until the universe was cold and dark. Everything would become so far apart that even gravity would be defeated. I think a ***big chill*** is what we've got in store, not a big crunch. So, will this be the end of us and life as we know it, or will we have figured out how to navigate to a new universe before death?

I think we will only know when we truly understand why the universe exists at all. Perhaps then, when we finally unravel the whole cosmic puzzle, we will become masters not just of our universe, but the universe next door.

Yes, this universe is indeed running down continually—stars burning out and dying, for instance—heading towards 'heat death.' This unrestrained decline is in accord with it having been "subjected to futility" and "bondage to decay" in response to Adam's sin (Romans 8). But verse 20 of that same chapter tells us that this subjection was "in hope." This is the glorious hope that long before such a cold, dark future is reached, God will intervene, through Jesus Christ, the sinless last Adam—and create a New Heavens and Earth which believers will inhabit for a deathless eternity.

We also bring to readers' notice the following response to the big bang concept by a portion of the secular scientific community in 2004.

SECULAR SCIENTISTS OBJECT!

In *An Open Letter to the Scientific Community*, originally published in *New Scientist* **182**(2448):20, 22 May 2004, and now available at Cosmology statement, 34 secular scientists signed the statement, part of which is quoted below. Since then, it has been co-signed by a further 218 Scientists and Engineers, 187 Independent Researchers, and 105 Other Signers. It reads as follows:

The big bang today relies on a growing number of hypothetical entities, things that we have never observed—inflation, dark matter and dark energy are the most prominent examples. Without them, there would be a fatal contradiction between the observations made by astronomers and the predictions of the big bang theory. In no other field of physics would this continual recourse to new hypothetical objects be accepted as a way of bridging the gap between theory and observation. It would, at the least, raise serious questions about the validity of the underlying theory.

But the big bang theory can't survive without these ***fudge factors***. Without the ***hypothetical inflation field, the big bang does not predict the smooth, isotropic cosmic background radiation that is observed***, because there would be no way for parts of the universe that are now more than a few degrees away in the sky to come to the same temperature and thus emit the same amount of microwave radiation.

Without some kind of dark matter, unlike any that we have observed on Earth despite 20 years of experiments, ***big bang theory makes contradictory predictions for the density of matter in the universe.*** Inflation requires a density 20 times larger than that implied by big bang nucleosynthesis, the theory's explanation of the origin of the light elements. And ***without dark*** energy, the theory predicts that the universe is only about 8 billion years old, which is billions of years younger than the age of many stars in our galaxy.

What is more, the big bang theory can boast of no quantitative predictions that have subsequently been validated by observation. The successes claimed by the theory's supporters consist of its ability to retrospectively fit observations with a steadily increasing array of adjustable parameters, just as the old Earth-centered cosmology of Ptolemy needed layer upon layer of epicycles.

The statement continues, lamenting the fact that virtually all cosmological funding is for projects within the big bang framework, and ends with the plea that "Allocating funding to investigations into the big bang's validity, and its alternatives, would allow the scientific process to determine our most accurate model of the history of the universe."

Readers will observe that the scientists who signed this statement are not objecting to any theistic beliefs of Stephen Hawking's (he has none), but to his unsound science! Theistic evolutionist big bangers take note!

EVOLUTIONISTS DECEIVE STUDENTS TO BELIEVE IN EVOLUTION

There have been many examples of evolutionary falsehoods used to indoctrinate students into evolution. The list includes

- Forged Haeckel embryo pictures, still used in many textbooks
- Staged photos of peppered moths which wouldn't even prove goo-to-you evolution anyway but merely the creationist-invented theory of natural selection.
- Misleading analogies that cars and airplanes evolved when of course they were *designed* (Intelligent Design leader Phillip E. Johnson calls this 'Berra's Blunder', and Ian Plimer committed this blunder too).
- Claiming that creationists believe that God must have created cave fish as blind.
- Insinuating that creationists deny natural selection and variation.
- Piltdown Man, an obvious forgery not exposed for 40 years, and the peccary tooth dubbed 'Nebraska man'
- Archaeoraptor , the Piltdown Bird.

TEACHING LIES

But at least one evolutionist is happy to use falsehood, as long as the end result is more students believing in evolution. An evolutionary True Believer and educator, one Bora Zivkovic, Online Community Manager at PLoS-ONE, proudly stated:

'it is OK to use some inaccuracies temporarily if they help you reach the students.'

And by 'inaccuracies', he didn't mean approximations or simplifications (e.g., $\pi \sim 3$ or ($^{22}/_{7}$) for quick calculations, or the octet rule taught to beginning chemistry students), but outright falsehoods — using *analogies that he knows are inaccurate*, and *ideas he states are false*.

For example, he discusses a common evolutionary propaganda tactic, NOMA (non-overlapping magisterial), invented by the late Marxist "Stephen Jay Gould." This pretends that science and religion are two non-intersecting categories of thought, so cannot prove or disprove each other. We have shown that this is a form of the fallacious fact-value distinction, and is philosophically bankrupt. Zivkovic *agrees* that it's false, but justifies its pretense all the same:

You cannot bludgeon kids with truth (or insult their religion, i.e., their parents and friends) and hope they will smile and believe you. *Yes, NOMA is wrong, but is a good first tool for gaining trust.* You have to bring them over to your side, gain their trust, and then hold their hands and help them step by step. And on that slow journey, which will be painful for many of them, it is "OK" to use some inaccuracies temporarily if they help you reach the students. (Emphasis added)'

I.e., so never mind such archaic concepts as truth: the important thing is that they *accept evolution!*

Zivkovic continues by praising an account of a Florida teacher and fanatical evolutionary activist, "David Campbell" in the *New York Times*. This teacher used an argument about the changing face of Mickey Mouse as an example of 'evolution.' Of course, this is just another form of Berra's Blunder, and *Zivkovic agrees that it's fallacious*. Yet he justifies teaching it:

'If a student, like "Natalie Wright" who I quoted above, goes on to study biology, then he or she will unlearn the inaccuracies in time. If most of the students do not, but those cutesy examples help them accept evolution, then it is OK if they keep some of those little inaccuracies for the rest of their lives.

It is perfectly fine if they keep thinking that Mickey Mouse evolved as long as they think evolution is fine and dandy overall. Without Mickey, they may have become *Creationist activists* instead. Without belief in NOMA they would have never accepted anything, and well, so be it.

Better NOMA-believers than Creationists, don't you think?'

Once again, better to have them believe overt falsehoods than deny the evolutionary religion.

Therefore, what is Zivkovic's motivation? In his own words:

'Education is a subversive activity that is implicitly in place in order to counter the prevailing culture. And the prevailing culture in the case of Campbell's school, and many other schools in the country, is a deeply conservative religious culture.'

Translation: ***educrats*** like him are rather proud of trying to undermine Christianity, and so much the better if it means opposing the worldview of the parents of the students he teaches. This should be a lesson for Christian parents, as Christian author and columnist "Cal Thomas? points out:

'The tragedy is that too many conservative Christian ... parents who want their children to have a different worldview—their own—willingly participate in the destruction of their children's minds by turning them over to a way of thinking that is antithetical to their beliefs. Parents who worship at conservative churches on Sunday willingly send their children to schools five days a week where what they are taught undermines what they learned in church and at home. They would never think of taking their kids to a church that teaches doctrines opposed to their beliefs, but they don't give a second thought to doing the same thing by sending them to government schools. It makes no sense.'

Worse, the Christians parents ***pay*** the misotheists to program their children in a value system diametrically opposed to their own! It's like Moses handing over shekels to the Canaanites to teach paganism to the Israelite children.

OTHER EVOLUTIONARY PROPAGANDISTS

Other evolutionary propagandists are also on record as setting greater store on evolutionary indoctrination than critical thinking and learning facts. E.g., the atheistic anti-creationist "Eugenie Scott," leader of the atheist–founded-and-operated and pretentiously named "National Center for Science Education," tacitly admitted that if students heard criticisms of evolution, they might end up not believing it!

'In my opinion, using creation and evolution as topics for critical-thinking exercises in primary and secondary schools is virtually guaranteed to confuse students about evolution and may lead them to reject one of the major themes in science.'

She also knows the indoctrinatory value of NOMA:

'... I would describe myself as a humanist or a nontheist. ... I have found that the most effective allies for evolution are people of the faith community. One clergyman with a backward collar is worth two biologists at a school board meeting any day!'

It should be pointed out that not all atheistic evolutionists agree with teaching NOMA, e.g., "William Provine," biology professor at Cornell:

'Let me summarize my views on what modern evolutionary biology tells us loud and clear ... There are no gods, no purposeful forces of any kind, no life after death. When I die, I am absolutely certain that I am going to be completely dead. That's just all—that's gonna be the end of me. There is no ultimate foundation for ethics, no ultimate meaning in life, and no free will for humans, either.'

'... belief in modern evolution makes atheists of people. One can have a religious view that is compatible with evolution only if the religious view is indistinguishable from atheism.'

"Richard Dawkins" and "P.Z. Myers" are other misotheists who despise NOMA. And they made this, as well as their hatred of Christianity, very clear in their interviews shown in the movie Expelled. Some evolutionists have criticized *Expelled* for showing this: but these evolutionists' problem is not with the opinions of these two, but that *they give the game away*. Such evolutionists would clearly prefer Zivkovic's NOMA approach, but would probably rather he was not openly proud of his deliberate deception.

THE FOUNDATION

Many Christians expect evolutionists to be honest and fair. Indeed, many are. But we should not be too surprised whenever someone who denies an absolute moral Lawgiver chooses to transgress moral/ethical bounds deliberately, and what's more, proclaims it as a worthy act. As may have said in Bomb-building vs. the biblical foundation, the claim is not that atheistic evolutionists cannot be moral, but that they have no objective basis for their morality.

While some creationists have been known to lie, this is *contrary to* their professed belief system, and not something they will openly defend or promote, as Zivkovic does. When evolutionists lie, it is *consistent with* theirs.

As the Russian writer "Fyodor Dostoyevsky" (1821–1881) puts in the mouth of the Grand Inquisitor in *The Brothers Karamazov*, 'Without God, everything is permissible; crime is inevitable.' So when Christians debate atheists, or send their kids to secular schools, they should heed the warning of the 18th century British statesman and philosopher "Edmund Burke": 'There is no safety for honest men but by believing all possible evil of evil men' [meant inclusively in those days].

WOULD A LOVING GOD SEND PEOPLE TO HELL?

"If God is supposed to be loving, how could He send people to Hell just because they didn't worship Him?"

It's implied that it's deeply unjust for God to judge sin at all, and even worse to do so by sending the sinners to a place of eternal punishment, Figure 31.33. However, these critics often profess they don't believe in the God they are accusing, and also deny any objective standard of right and wrong anyway. Sometimes believers also struggle with this question, wondering how God could condemn someone who never heard of Jesus and so never had a chance to believe in Him, for example.

Fig.31.33: *Would a Loving God Send People to Hell? Curtsy - 123rf.com/Sorapong Chaipanya*

There are a lot of charges brought up against the God of the Bible; perhaps one of the most common is, The question of Hell is not an appealing one even for people who affirm its existence. No one *likes* the idea of many people suffering judgment in the life to come. But the *'Good News'* of the Gospel requires that there be *'Bad News'*—salvation in Jesus Christ would not bring glory to God if there were actually nothing to be saved from. So, it's important to be able to give an answer, even in the case of a subject that no one particularly likes contemplating.

What is Sin?

Simply put, sin is anything that doesn't conform to a holy God's standard of perfection. A sin can be something we do that is wrong, or something that we don't do that we should do. God gets to set this standard because He is the Creator—and this standard is not arbitrary, but has its source in God's own nature. We often make judgments about some sins being worse than others, but all sin is an offense against God because He is holy.

In addition, humans have a sin nature. This is a 'bent' toward sin. So, while we might not sin at every possible opportunity, or to the greatest possible extent, everyone will sin, given the opportunity (Romans 3:23).

In fact, the struggle against the human propensity to do things that we don't want to (warring against the flesh), should be reminder that we are not perfect and are born sinners. That's exactly why we need a Savior

Who Goes to Hell?

When unbelieving critics talk about Hell, they sometimes speak like it will be full of innocent people (like themselves!). However, the Bible doesn't indicate that innocent people will spend a single moment in Hell. Rather, Hell is God's answer to the fundamental injustice of this life. There are many murderers, rapists, and other people who wreak havoc in the lives of others, who never experience judgment in this life. Everyone knows that it is wrong that these people never be brought to account for what they've done; something in the human heart *demands* justice. Hell is God's answer.

Without Hell, justice would never overtake the unrepentant tyrants responsible for murdering millions. Perpetrators of evil throughout the ages would get away with murder—and rape, and torture, and every evil.

Even if we may acknowledge Hell as a necessary and just punishment for evildoers, however, we rarely see *ourselves* as worthy of Hell. After all, we are not Hitler, Stalin, Pol Pot, Bundy, or Dahmer.

God responds, *"There is no one righteous, not even one. There is no one who understands, no one who seeks God. All have turned away, they have together become worthless; there is no one who does good, not even one"* (Romans 3:10–12).

For the majority of people who are not guilty on the scale of these obviously (even to us) depraved people, it's hard to understand that we deserve punishment, too. But most people have grievances against others—if someone stole from you, or hurt your children, or if you were a victim of something fundamentally unjust, you would want justice; your sense of what is right would demand that the person at fault pay a penalty for wronging you.

Every time we break God's law, that's an affront to God, and He demands justice, just as we do imperfectly on a smaller scale. If you've ever said in your heart, "That person should *pay* for what he did!" then you fundamentally agree with the idea of Hell, because the doctrine of Hell says *somebody* is going to pay for every sin, eventually.

SIN - REBELLION AGAINST OUR LOVING CREATOR

God didn't create people to go to Hell, and He didn't create people to sin. In fact, the place He made for people originally was perfect. The Garden of Eden had everything Adam and Eve could ever want. It was safe and pleasant, God lovingly provided everything they needed, and they enjoyed a perfect relationship with the Creator. God gave them some simple commands (have children, tend the Garden, and don't eat the fruit from the Tree of the Knowledge of Good and Evil), and their continuing perfect relationship with God only required them to obey. It was a position that we can only imagine today.

Even though Adam and Eve had everything they could ever possibly need, they disobeyed God and ate from the Tree that God had forbidden. Sin immediately broke the perfect fellowship that they had enjoyed with God. They realized that they were sinful, and they were ashamed and aware of their wrongdoing—as is shown by their initial attempts to cover themselves with fig leaves.

God is holy, meaning that He is completely separated from anything sinful. And as their Creator, He had the right to judge them when they disobeyed—in fact, His nature and His justice *demanded* that He respond when they rebelled against Him. He could have instituted the death penalty instantly, and He would have been perfectly just if He had done so. But God is also loving and merciful, so He did not put a premature end to the human race. Adam and Eve had spiritually died, meaning that their relationship to God was broken, but they would continue to physically live long enough to have children, who would inherit their propensity to sin.

THE PURPOSE OF THE TREE OF THE KNOWLEDGE OF GOOD AND EVIL

Some say that a lot of trouble could have been avoided if God had just left the Tree of the Knowledge of Good and Evil out of the Garden. But this misunderstands the vital function of the Tree. The other commands God gave Adam and Eve were fairly self-explanatory and had pleasant outcomes for them, but what was the purpose of the command not to eat from the Tree?

It may seem surprising, but God had a loving purpose in putting the Tree in the Garden.

God created human beings to be in a relationship with Him. But a true loving relationship has to be freely given or chosen—one could program a robot to think it loves its programmer, but that would be meaningless because the robot didn't have a choice. God wanted human beings to love Him freely, for who He is, not just for what He had given and provided for them. But that required the chance to not love Him, to rebel.

The function of the Tree was to give Adam a chance to obey or rebel, and Adam chose to eat the fruit and to rebel against God.

There were two pivotal times in history when God freely gave and made a way that mankind could choose to have a relationship with Him, the Creation and the Incarnation. This also highlights why the battle of Creation is so important.

The Fall from grace in the original Creation should help us understand our plight in this sin cursed world, and make it that much easier to recognize what God has done through Jesus. But if people reject the Bible's account of origins, they will not understand humankind's plight and the choices that God gave/gives us.

Even after Adam and Eve sinned, God still loved them and provided for them. He agreed with them that their new sinful state required them to cover themselves; but the fig leaves were inadequate. He killed animals instead and made clothes out of their skins for Adam and Eve.

This is the first place in Scripture where anything is killed—and for thousands of years, animals would continue to be killed in an attempt to cover over man's sin, and to delay God's wrath against humanity.

God was not willing to leave all of humanity to perish (cf. 2 Peter 3:9), so He promised that Eve would have a descendant who would defeat Satan (Genesis 3:15). This is called the 'protevangelion' because it's the first hint of the Gospel in Scripture. The rest of the Old Testament can be characterized as God dealing with sin in various ways by judging it or putting off judgment, and getting ready for the descendant of Eve who would deal with the sin problem once and for all.

Adam and Eve were driven out of the Garden into a world that was now largely hostile to them. And when Genesis tells us that Adam had a son (Seth) "in his image and likeness", it leaves no ambiguity about whether the sinful state was in fact passed on from father to son.

JESUS - THE LOVING GOD INCARNATE

Scripture affirms that Jesus is the fulfillment of God's promise to Eve. The only way that humanity could be saved is if there was a single person who was both fully God and fully man, who was related to every single person through Adam—because He could only redeem us if He was related to us, and so would be qualified to be our Kinsman-Redeemer (Isaiah 59:20).

In addition, this person would have to live a perfect human life—avoiding every sin and perfectly obeying every command of God's Law. And this is precisely who Jesus was and what He did.

It's vitally important to understand that this was the only way that humanity could be saved. We can't save ourselves even by our best efforts; no other god or religion or philosophy can save us. If Jesus had not gone to the cross for us, and if He had not been raised on the third day, we would be completely and totally without hope.

When a person repents of their sin and trusts in Christ, God accepts Jesus' sacrifice as payment for that person's sins (Isaiah 53:6), and credits Jesus' righteousness to that person (2 Corinthians 5:21). This brings the person into a right relationship with Him in a legal sense; they have an 'innocent' instead of a 'guilty' sentence (this is called *justification*).

Furthermore, the Holy Spirit indwells that person and starts the process of actually *making* them righteous (this is called *sanctification*, and continues until the process is finished after death, and must not be confused with justification). The believer also has a host of privileges as part of being an adopted child of God.

REBELLING AGAINST THEIR LOVING CREATOR

Therefore, we see that because of Adam's rebellion, all people are born with a sinful nature which is offensive to God. We're not blameless, because we not only *have* the sinful nature, but we cooperate with it and *enjoy* sin. So, we are culpable for the sinful things we do. We deserve to go to Hell, every single one of us, but God in His love provides a way out, so that anyone who repents will not be judged for their sin, but rather Jesus' sacrifice pays for it. And He did this without us doing anything to merit it: "For while we were still weak, at the right time Christ died for the ungodly" (Romans 5:6) and "God shows his love for us in that while we were still sinners, Christ died for us" (Romans 5:8).

But there are many who don't repent. There are some who never hear the Gospel; there are others who hear it and reject it for any number of reasons. There are those who even descend into conscious hatred of God; they recognize Him and hate Him, much like Satan and the fallen angels. And many of these same ones complain that God is not loving because they chose Hell of their own accord.

If Jesus' sacrifice is the only way to salvation, yet someone rejects Jesus, what is God supposed to do with that person? God cannot apply the salvation they've rejected (because remember God gave people the choice to reject Him in the Garden of Eden, and so He won't override that when they actually do).

When someone sins and rejects salvation, the only option left is to punish the rebellious creature.

HELL - REJECTING GOD

It can be hard for the person who loves God to comprehend that there are people who hate God as much as we love Him. That there are people who hate Him so much that if they saw Him finally, they would not embrace Him and turn from their rebellion, but they would shake their fist all the more and damn themselves for eternity.

Just as through the Spirit, the believer is finally sanctified after death, something happens to the unbeliever at death that makes him unable to ever repent. He has chosen to hate God and he will hate God for all eternity.

Jesus reminded us that some will not believe even if He rose from the dead. The unbeliever cannot inhabit Heaven, because he embodies everything that can never enter Heaven; and to be in the presence of God is not Heaven for him in any case, but the most exquisite torment. He has lost the ability to experience God as anything but terrifying.

For such a person, Hell is God giving him what he asked for all along—a place where His presence is not manifested as it is in this life. But this also means that there are none of the blessings and providence that even the unbeliever experiences in this life.

THOSE WHO HAVE NEVER HEARD OF GOD

Many people ask how God can condemn people to Hell who have never heard of Him or who never had the chance to repent. This often involves the perception, however, that people are in a 'neutral' state and either choose for or against Him when they hear the Gospel.

In reality, everyone is in a 'default' setting of rebellion against God and His law, and only the work of the Holy Spirit is able to change that. Accordingly, the people who have never heard are already rebelling against what they know about God; and they will be judged proportionately to the revelation of God and His law they've had from nature. Romans 1:18–28 points out that some truth about God is obvious (in the heart) from creation, so that people are 'without excuse'. Romans 2:14–16 says that people also have a conscience, and don't even live up to their own standards, let alone God's.

Of course, the importance of preaching the Gospel and doing missions work is highlighted by the 'problem' of those who have never heard. The answer is instead of questioning God's justice (despite "Shall not the God of all the earth do right" Genesis 18:25), we should be spreading the message until everyone has heard.

GOD BECAME OUR SAVIOR

So why would a loving God send someone to Hell?

It is because that person has chosen in such a way that God has no other choice. The existence and reality of eternal judgment for the person who does not repent is sobering, and no one really wants to contemplate it too deeply.

But the person who goes to Hell must reject Christ, who died so that anyone who repents can be saved. Therefore, God is not to be blamed when an unrepentant, rebellious creature chooses a destructive path that leads to Hell. In fact, we all deserve hell due to our sin nature that separates us from God, but thank God for Jesus.

The good news is that anyone reading this article is still alive, and so if you haven't repented of your sins and trusted in Jesus yet for salvation, there is still time to avoid the terrible fate that awaits those who rebel against the Creator, or to tell your unbelieving friends or family members about the Gospel.

If you consider yourself a good person that doesn't need salvation then just consider the following questions:

- Have you ever told a lie?
- Have you ever stolen something,
- committed adultery,
- blasphemed etc.?

If you are truly honest with yourself you will find that you have failed to reach God's standard for holiness and entrance into eternity with Him. And if you haven't accepted Jesus as your Savior then you will have to represent yourself in God's 'courtroom' when you die and answer for your sins. But for those who believe, "we have an advocate with the Father, Jesus Christ the righteous" (1 John 2:1).

RECONCILED WITH GOD

In the decades after the death and resurrection of Jesus, the apostles brought out the significance of His life, ministry, death, and Resurrection in their writings, which are collected in our New Testament. In Colossians 1:11–23, Paul writes:

May you be strengthened with all power, according to his glorious might, for all endurance and patience with joy, giving thanks to the Father, who has qualified you to share in the inheritance of the saints in light. He has delivered us from the domain of darkness and transferred us into the kingdom of His beloved Son, in whom we have redemption, the forgiveness of sins.

He is the image of the invisible God, the firstborn of all creation. For by him all things were created, in heaven and on earth, visible and invisible, whether thrones or dominions or rulers or authorities, all things were created through him and for him. And he is before all things, and in him all things hold together. And he is the head of the body, the church. He is the beginning, the firstborn from the dead, that in everything he might be preeminent. For in him all the fullness of God was pleased to dwell, and through him to reconcile to himself all things, whether on earth or in heaven, making peace by the blood of his Cross.

And you, who were once alienated and hostile in mind, doing evil deeds, he has now reconciled in his body of flesh by his death, in order to present you holy and blameless and above reproach before him, if indeed you continue in the faith stable and steadfast, not shifting from the hope of the Gospel that you heard, which has been proclaimed in all creation under heaven, and of which I, Paul, became a minister.

THE RELATIONSHIP BETWEEN THE FATHER AND THE SON

This passage shows how salvation is grounded in the relationship of love between the Father (YHWH) and the Son. Jesus is the "beloved Son," echoing God's declaration at Jesus' baptism (Matthew 3:17) and the Transfiguration (Matthew 17:5). The fullness of God was "pleased to dwell in Him," and it is the Father's will that the Son be preeminent in everything. In turn, the Son reconciles to Himself all things to present believers "before him"—the Father. So, the relationship of the Persons of the Trinity to each other is at the very center of the drama of salvation.

Jesus is called "the image of the invisible God." One aspect of being made in God's image is that we are God's representatives, and Jesus is the *ultimate* representative of God, and even "the exact imprint of his nature" (Hebrews 1:3). The Old Testament repeatedly states that it is ***impossible to see God and live***, so Jesus as the incarnate God is the ultimate revelation of the Father. Jesus told His disciples, "Whoever has seen me has seen the Father" (John 14:9, also John 1:18).

OUR SINFUL STATE—SEPARATED FROM GOD

Paul gives a dismal picture of life outside of Christ. We were trapped in "the domain of darkness", "alienated and hostile in mind, doing evil deeds". There is no indication in Paul's writing that the people in this state are unhappy about it; rather, in Romans 1 and other places he explains how sinful people are willing rebels—not only are we trapped in sin, we *enjoy* it apart from God's grace. Not only can we do nothing about it, we wouldn't want to—hence Paul said "you were dead in the trespasses and sins" (Ephesians 2:1).

Accordingly, God must take the initiative. *He* must be the one to transfer us from the domain of sin into the Kingdom of Christ. *He* must be the one to reconcile us; *He* must make peace, because we cannot and would not.

CHRIST THE MEDIATOR

The preeminence of Christ is a focus of this passage. He is "the firstborn of all creation". This does not mean that He is the first created being (this would be the heresy of Arianism). Rather, the word *prototokos* describes His position of preeminence and authority over creation, i.e., the inheritance rights of a first-born son in that culture. The Old Testament uses the Hebrew equivalent *bekôr* to describe David, the youngest one, but chosen by God. Hebrews 1:6 uses "firstborn" about Jesus in the same sense—and this "firstborn" is likewise worthy of angelic worship, so likewise could not be a created being.

This authority is grounded in His role in the work of Creation. Jesus is consistently described as the agent of creation; this is brought out most fully in John 1. Paul makes it clear that nothing, whether physical or spiritual, came into being without His involvement. This also indicates that the Son is uncreated.

Jesus is not only the mediator of creation; He is also the *reason* for creation—all things were created *for* Christ. Not only is Jesus the *reason* for creation, He is also the organizing principle for creation—all things **hold** together in Him (also taught in Hebrews 1:3).

All this makes Christ supremely suitable to be the one to bring about the reconciliation between God and man. Jesus is called "the beginning, the firstborn from the dead". As before, "firstborn" indicates authority. "Christ stands at the head of the new creation as the *firstborn from the dead*, the one who initiates the eschatological resurrection (1 Cor. 15:20). 'Beginning' here thus implies 'founder.'"

The means for our reconciliation is the "blood of his Cross". In Ephesians 2:16, Paul again links reconciliation to the Cross.

THE GREAT RECONCILIATION

When we hear the Gospel and believe, God does several things, which Paul lays out in this passage. God "delivers us from the kingdom of darkness and transferred us to the kingdom of his beloved son." He qualifies us "to share in the inheritance of the saints in light. He reconciles us "in order to present you holy and blameless and above reproach in him". Accordingly, we go from being hostile slaves of sin and darkness to heirs and subjects of Christ.

This reconciliation is comprehensive. This passage is tied together with the Greek word *pas*, which is translated in English with the words "all" and "everything." Some variation of the merism "in heaven and on earth" occurs several times in the passage to emphasize the scope of Christ's work.

God accomplished this reconciliation with two great 'imputations,' i.e., crediting something from one person to another's account. All the sins that all believers will ever commit were ***imputed*** to Christ at the crucifixion, as prophesied in Isaiah 53:6. This cancelled our sin debt and nailed it to the Cross (Colossians 2:14). Simultaneously Jesus' perfect human righteousness is imputed to these believers (2 Corinthians 5:21).

LIVING IN THE LIGHT OF CHRIST

This reconciliation results not only in a changed status, but a changed way of life. Paul's exhortation to the Colossians assumes that they have experienced this reality. God's "glorious might" is the basis for Paul calling the Colossians to endurance and joyful patience and thankfulness. Christians can be changed in this way because Christ has reconciled us through His death on the Cross—no matter what sort of sins they had committed before. Paul warns the Corinthians that "the unrighteous will not inherit the kingdom of God," and lists a number of sins. Then he says:

And such were some of you. But you were washed, you were sanctified, you were justified in the name of the Lord Jesus Christ and by the Spirit of our God (1 Corinthians 6:9–11).

And this is so Jesus can present us as holy and blameless and above reproach before the Father. These are basically three ways of saying that Christ took our sin on the Cross so He could present us as sinless before the Father.

But then Paul for the first time talks about something the Colossian believers, and we as believers need to do—there's an element of conditionality. Paul exhorts the Colossian believers to continue in the faith, stable and steadfast, not shifting from the hope of the Gospel. This *presupposes* that the Colossian believers have experienced the objective reconciliation and that this means they are able to continue. Yet they still must continue! And we know they must because Paul tells them to continue. And so must we! As Paul says in Ephesians 2:8–10:

For by grace, you have been saved through faith. And this is not your own doing; it is the gift of God, not a result of works, so that no one may boast. For we are his workmanship, ***created in Christ Jesus for good works***, which God prepared beforehand, that we should walk in them.

OUR CREATOR IS OUR MEDIATOR!

In this passage, we see that our salvation is tied up in the relationship between the Father and the Son, and the relationship between God and His creation. In particular, as the rest of the New Testament explains in detail, God the Son took on human nature in the Incarnation (John 1:14), and became fully man as well as remaining fully God. This enabled Him to be the mediator between God and Man (1 Timothy 2:5). After His glorious resurrection, He sits at the right hand of God the Father, to be our "advocate", who "is the propitiation for our sins, and not for ours only but also for the sins of the whole world" (1 John 2:1–2). In the future, He will come again and create a new heavens and earth, where believers will enjoy resurrection bodies and eternal fellowship with Him, in a final victory over the Fall.

YOU ARE ABSOLUTELY SAFE – IF YOU CHOOSE!

Prominent Atheists have propagated several hypotheses to the "Origin of the Universe." The leading unscientific unproven theory is the Big Bang. These atheists are steadily attempting to erase the creator of the Universe from His existence, hoping to inject their own unscientific ideologies. Additionally, well known evolutionists conjectured a sacred life evolved from a non-life natural materialistic substance, also hoping to erase the Creator of all life in the Universe.

Atheists and evolutionists alike are diligently attempting to CANCEL God from our evolved western culture. While their plans have well succeeded to delude the minds of multitudes of people and even nations, they have particularly aimed to pollute the innocent minds of school children.

However, their debased plans and futile diligence would not change the Wholesomeness of the presence of the Creator to continually Hold and Sustain the entire Universe in the Palm of His Holy Hands.

Both the Big Bangers together with the Evolutionists, supported by Atheistic Media, proponent Research Institutions as well as several Government Departments engaged in spreading their hypothetical demise of the Universe in total destruction through few means, leading to a total annihilation, death, decay and catastrophically ending. The universe of the big banger, e.g., will end up in big chill, big crunch or big rip. While, the evolutionary Universe believes it will expand forever and will eventually run out of hydrogen; e.g., the stars eventually burn out, and the Universe cools down to a vast frozen graveyard of dead stars. The evolutionists will continue their death ceremonies and decay sermons condemning the vulnerable and the helpless.

Albeit the present Universe is polluted by sin, death and decay, the Creator has promised restoring it to its pure condition prior to the advent of sin and decay. Since the dawn of creation sin against God's desire wreaked havoc on all creation. The Creator, whom never failed a promise has assured His believers to revitalize this universe to a New Heaven and a New Earth.

According to Bible statistics, God has made approximately 7,500 promises to us as we add up his words in both Old and New Testaments. Of these scriptures, God has never broken a promise. We can count on God's word always and forever.

With respect to the end of the universe, extensive studies have been conducted to provide guidance to those who put their trust in the Creator as well as those who may not have comprehended yet the value to acknowledge the creator of the Universe.

The author of this book hopes and prays for such people that somehow, one day they may acquire the knowledge and the wisdom to recognize that the Creator is living and awaiting each individual to have a true and meaningful personal fellowship with Him.

Among the 7,500 promises of the Creator, one promise was given to both Daniel, who was the prime minister of Babylon during the reign of King Nebuchadnezzar few centuries prior to the birth of the Lord Jesus Christ, and given to the apostle John 95 years after the birth of the Lord Jesus Christ. The assured prophecy stated that the Lord Jesus Christ who have executed all the act of the entire creation of the Universe will descend from Heaven to take the believers with to join His glory in Heaven. The believers who died will be resurrected first to join the believers who are still alive. Both believers will be raptured in a twinkle of an eye to transfigure to an identical form as the form of the Lord Jesus Christ prior to entering the holiness threshold of Heaven.

This rapture signals the beginning of traumatic events, called the tribulation period for 7 years, leading to the second coming of the Lord Jesus Christ descending with all His believers to His created Earth.

The Lord Has No Equal (Isaiah 40:12-31)

12 ***Who else has held the oceans in his hand? Who has measured off the heavens with his fingers? Who else knows the weight of the earth or has weighed the mountains and hills on a scale?***

13 ***Who is able to advise the Spirit of the Lord? Who knows enough to give him advice or teach him?*** 14 ***Has the Lord ever needed anyone's advice? Does he need instruction about what is good? Did someone teach him what is right or show him the path of justice?*** 15 ***No, for all the nations of the world are but a drop in the bucket. They are nothing more than dust on the scales.***

He picks up the whole earth as though it were a grain of sand. 16 All the wood in Lebanon's forests and all Lebanon's animals would not be enough to make a burnt offering worthy of our God. 17 ***The nations of the world are worth nothing to him. In his eyes they count for less than nothing—mere emptiness and froth.***

18 ***To whom can you compare God? What image can you find to resemble him?*** 19 Can he be compared to an idol formed in a mold, overlaid with gold, and decorated with silver chains? 20 Or if people are too poor for that, they might at least choose wood that won't decay and a skilled craftsman to carve an image that won't fall down!

21Haven't you heard? Don't you understand? Are you deaf to the words of God— ***the words he gave before the world began? Are you so ignorant?***

22 *God sits above the circle of the earth. The people below seem like grasshoppers to him! He spreads out the heavens like a curtain and makes his tent from them.*

23He judges the great people of the world and brings them all to nothing.

24They hardly get started, barely taking root, when he blows on them and they wither.

The wind carries them off like chaff. 25 "***To whom will you compare me? Who is my equal?" asks the Holy One.*** 26 ***Look up into the heavens. Who created all the stars?***

He brings them out like an army, one after another, calling each by its name.

Because of his great power and incomparable strength, not a single one is missing.

27 O Jacob, how can you say the Lord does not see your troubles? O Israel, how can you say God ignores your rights? 28 Have you never heard? Have you never understood?

The Lord is the everlasting God, the Creator of all the earth. He never grows weak or weary.

No one can measure the depths of his understanding. 29 ***He gives power to the weak and strength to the powerless.***

30 ***Even youths will become weak and tired, and young men will fall in exhaustion.*** 31 ***But those who trust in the Lord will find new strength. They will soar high on wings like eagles. They will run and not grow weary. They will walk and not faint.***

CANCEL GOD – THE DEMISE OF NATIONS

Prominent Atheists have propagated several hypotheses to the "Origin of the Universe." The leading unscientific-unproven theory is the Big Bang. These atheists are steadily attempting to erase the creator of the Universe from His existence, hoping to inject their own unscientific ideologies. Additionally, well known evolutionists conjectured a sacred life evolved from a non-life natural-materialistic substance, also hoping to erase the Creator of all life in the Universe.

Atheists and evolutionists alike are diligently attempting to CANCEL God from our evolved western culture. While their plans have well succeeded to delude the minds of multitudes of people and even nations, they have particularly aimed to pollute the innocent minds of school children.

However, their debased plans and futile diligence would not change the Wholesomeness of the presence of the Creator to continually Hold and Sustain the entire Universe in the Palm of His Holy Hands.

Both the Big Bangers together with the Evolutionists, supported by Atheistic Media, proponent Research Institutions as well as several Government Departments engaged in spreading their hypothetical demise of the Universe in total destruction through few means, leading to a total annihilation, death, decay and catastrophically ending. The universe of the big banger, e.g., will end up in big chill, big crunch or big rip. While, the evolutionary Universe believe it will expand forever and will eventually run out of hydrogen; e.g., the stars eventually burn out, and the Universe cools down to a vast frozen graveyard of dead stars. The evolutionists will continue their death ceremonies and decay sermons condemning the vulnerable and the helpless.

Albeit the present Universe is polluted by sin, death and decay, the Creator has promised restoring it to its pure condition prior to the advent of sin and decay. Since the dawn of creation sin against God's desire wreaked havoc on all creation. The Creator, who never failed a promise has assured His believers to revitalize this universe to a New Heaven and a New Earth.

According to Bible statistics, God has made approximately 7,500 promises to us as we add up his words in both Old and New Testaments. Of these scriptures, God has never broken a single promise. We can count on God's word always and forever.

With respect to the end of the universe, extensive studies have been conducted to provide guidance to those who put their trust in the Creator as well as those who may not have comprehended yet the value to acknowledge the creator of the Universe.

The author of this book hopes and prays for such people that somehow, one day they may acquire the knowledge and the wisdom to recognize that the Creator is living and awaiting each individual to have a true and meaningful personal fellowship with Him.

Among the 7,500 promises of the Creator, one promise was given to both Daniel, who was the prime minister of Babylon during the reign of King Nebuchadnezzar few centuries prior to the birth of the Lord Jesus Christ and given to the apostle John 95 years after the birth of the Lord Jesus Christ. The assured prophecy stated that the Lord Jesus Christ who have executed all the act of the entire creation of the Universe will descend from Heaven to take the believers with Him to join His glory in Heaven.

The believers who died will be resurrected first to join the believers who are still alive. Both believers will be raptured in a twinkle of an eye to transfigure to an identical form as the form of the Lord Jesus Christ prior to entering the holiness threshold of Heaven.

This rapture signals the beginning of traumatic events, called the tribulation period for 7 years, leading to the second coming of the Lord Jesus Christ descending with all His believers to His created Earth.

THE RESURRECTION OF THE DEAD

Immediately after the fall of Adam and Eve death and decay entered the universe as a consequence of sin and disobedience. Death infected all living entities of humans, animals, and plants, which caused their lives to be limited on earth. The consequence of sin and disobedience extended to planets and galaxies. Also, the earth as part of the Milky way galaxy is cursed. Additionally, it is clearly observed the death of planets, stars, and galaxies in the universe.

While death has touched all God's creation due to the fall of Adam, other deaths have entered the world of angelic beings the moment Lucifer, the highest created being, rebelled against his creator desiring to be as God lusting after power. One third of the created angelic beings turned into evil demonic beings following the established earthly kingdom of Lucifer, who was named Satan after expelled from the heavenly Kingdom of the creator God.

God is not in the business of destruction, instead He is in the justice and redemption enforcement. As the human race was created in the image of God, He does not destroy His victimized human beings, as He is the God of Love. However, He is a perfect justice and perfect love. In the fullness of time and according to His plan, even prior to creation He allowed the redemptive act for those whom He created and loved. All human beings that were victimized by the deception of Satan were freely given the chance to substitute their eternal death as a consequence of sin to eternal life as a consequence of the redemptive act of God's love, in particularly for those who accept this redemptive act of substitution, where Christ Himself took the death on His body instead of the death of all whom He loved, and those who accept such miraculous redemptive act. Also, God planned the resurrection of the dead to eternal life.

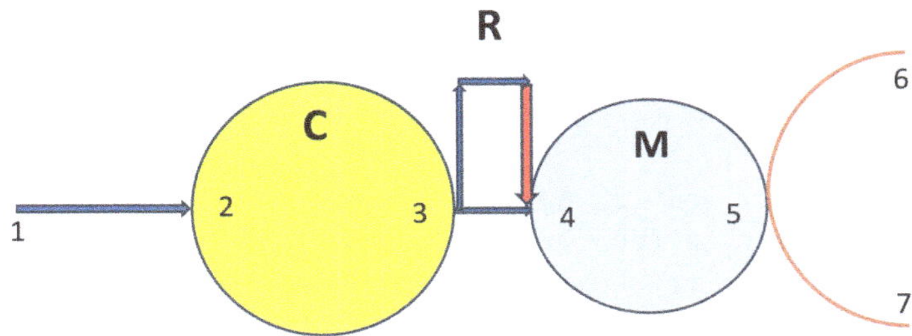

Fig.31.34: The Biblical Human History

There are several kinds of dead believers categorized as follows:

A. Resurrection of Old-Testament Believers

Refer to Figure 31.34, point 1 to point 2 represents the time of dead human-beings from Adam till the birth of the Lord Jesus. This era represents mainly the Israelite believers in YHWH and His Messiah. The Old Testament believers will not be resurrected at point 3, which is called the rapture. They will be resurrected at point 4, the moment where Chris second coming also takes place at this point of time. They are not resurrected at point 3. The main reason that Old Testament saints are not resurrected at point 3 because this point of time is dedicated to resurrect only the bride of Christ only, which is the church. The church believers are considered the New Testament saints as described below:

i. The object for the Resurrection of Church believers is to save them from the time of Tribulation and from the severe plagues falling upon earth dwellers. This is a promise from The Lord God. 1 Thessalonians 5:9-11 "For God did not appoint us to suffer wrath but to receive salvation through our Lord Jesus Christ. He died for us so that, whether we are awake or asleep, we may live together with him. Therefore encourage one another and build each other up, just as in fact you are doing."

ii. The Old Time Believers are already dead and secured in heaven with Christ. There, in heaven there is no Tribulation or Plages to be saved from.

iii. The believers in the Church-era are considered the "last" to follow the redeemer Jesus, while the Old Testament believers are considered the "first" believers in the Messiah the redeemer. Jesus stated in Matthew 20:16 "So those who are last now will be first then, and those who are first will be last." Accordingly, the Old Testament believers will be resurrected at the end of age. The end of age is marked by the second coming of the Lord Jesus Christ to earth to claim His kingdom on "Earth as it is in Heaven." This type of Old Testament resurrected believers will join the believers whom were resurrected at the end of the Church Era prior to the Tribulation.

B. Resurrection of Church Believers

Refer to Figure 31.34, the circle C, represents the time of Church believers Era. It is the time where Jesus was born as the Redeemer of all human beings marking the beginning of the Church era. Also, His death and resurrection represented His main object to redeem the mankind in all ages of history. The resurrected believers in this Church era consists of two types of resurrections.

i. The dead in Christ will be resurrected first and then

ii. the still alive believers will join them in the air and

both believers will join Jesus Christ to take them to heaven for 7 years prior to returning back to earth to begin the millennium era.

These two types of believers will be prepared to attend the "Wedding Supper of the Lamb in heaven." The Church is considered the bride, while Christ is considered the bride-groom. Revelation19:6-9 "6 Then I heard what seemed to be the voice of a great multitude, like the roar of many waters and like the sound of mighty peals of thunder, crying out,"

"Hallelujah! For the Lord our God - the Almighty reigns. 7 Let us rejoice and exult and give him the glory, for the marriage of the Lamb has come, and his Bride has made herself ready; 8 it was granted her to clothe herself - with fine linen, bright and pure — "for the fine linen is the righteous deeds of the saints. 9 And the angel said to me," "Write this: Blessed are those who are invited to the marriage supper of the Lamb." And he said to me, "These are the true words of God."

C. Resurrection of the Dead or Martyrs Believers During the Tribulation Era

Refer to Figure 31.34, point 3 to point 4 represents the Tribulation period, where God's judgment takes place, and where the Anti-Christ reveals himself. Point 4 represents the resurrection of all those who accepted the redemptive act of the Lord Jesus and denounced the Anti-Christ and refused to take his beastly-sign.

These types of believers, the tribulation believers – dead or martyred, together with the Old Testament believers are resurrected to join both the descending Christ with His Church believers, and the newly resurrected Old Testament believers. They will all Join the Descending Christ the King to establish His kingdom and to rule the earth.

Additionally, at point 4 Christ the King will chain Satan and all his living followers including the demonic spirits and powers before the Millenium begins.

However, Satan will be released for a short time at the end of the Millenium. Revelation 20:7-10"7 When the thousand years come to an end, Satan will be let out of his prison. 8 He will go out to deceive the nations—called Gog and Magog—in every corner of the earth. He will gather them together for battle—a mighty army, as numberless as sand along the seashore. 9 And I saw them as they went up on the broad plain of the earth and surrounded God's people and the beloved city. But fire from heaven came down on the attacking armies and consumed them. 10 Then the devil, who had deceived them, was thrown into the fiery lake of burning sulfur, joining the beast and the false prophet. There they will be tormented day and night forever and ever.

D. Resurrection of the Lost-Unbelievers

1. The lost unbelievers from the beginning of creation will not be resurrected until the end of the Millenium. This fact is stated in several biblical citations:
   i. Revelation 20:5 "The rest of the dead did not come to life until the thousand years were ended. This is the first resurrection."
   ii. Daniel 12:2 stated "And many of them that sleep in the dust of the earth shall awake, some to everlasting life, and some to shame and everlasting contempt.
   iii. John 5:28-29 "Do not marvel at this; for the hour is coming in which all who are in the graves will hear His voice and come forth - those who have done good, to the resurrection of life, and those who have done evil, to the resurrection of condemnation."

iv. Also, Revelation 20:4-5, "Then I saw thrones, and those seated on them were given authority to judge. I also saw the souls of those who had been beheaded for their testimony to Jesus and for the word of God. They had not worshiped the beast or its image and had not received its mark on their foreheads or their hands. They came to life and reigned with Christ a thousand years. (The rest of the dead did not come to life until the thousand years were ended.) This is the first resurrection."

2. The Resurrection of the Unbelievers Facing the Great White Throne Judgment
   i. The great white throne judgment is described in Revelation 20:11-15 and is the final judgment prior to the lost-unbelievers being cast into the lake of fire. We know from Revelation 20:7-15 that this judgment will take place after the millennium and after Satan is thrown into the lake of fire where the false messiah and the false prophet are (Revelation 19:19-20; 20:7-10).
   ii. The books that are opened (Revelation 20:12) contain records of everyone's deeds, whether they are good or evil, because God knows everything that has ever been said, done, or even thought, and He will reward or punish each one accordingly (Psalm 28:4; 62:12; Romans 2:6; Revelation 2:23; 18:6; 22:12).

E. The Old Testament Saints - Separated from the Church Saints

There are few reasons the author concluded. The Old Testament Believers Separated from the Church Saints for the following reasons, which are stated as follows:

1. The Old Testament saints were following the Mosaic-law
2. They have followed only the shadow of redemption in the form of the blood of bulls and goats as a prelude to the coming Messiah
3. They have not experienced the Grace of God demonstrated by the actual death and the resurrection of Jesus Christ
4. The New Testament Saints are the actual Church. The Church is the bride of Christ, while the Old Testament Saints are not considered the church. Consequently, they are not the bride of Christ. Christ, who is the bride-groom has personally redeemed His Beloved bride - the Church
5. The New Testament Saints either dead or alive will be raptured to Meet Christ in the air
6. The raptured bodies, or resurrected bodies of the New Testament believers are transformed to "Body of Glory"
7. The bodies of the Old Testament Saints have not been transformed into glorious bodies. They are souls and spirits brought by Jesus from their captivity in Hades to be in the presence of Jesus Christ in paradise. Their glorious bodies will be given at the moment of their resurrection the instant Christ descends to Mount Olives at His Second Coming
8. Accordingly, the 7 years of earthly tribulations is the same period, where the New Testament Saints spend it in heaven with Christ attending "The Wedding Feast of the Lamb."

THE GREATNESS OF THE CREATOR

The LORD Has No Equal (Isaiah 40:12-31)

12 Who else has held the oceans in his hand? Who has measured off the heavens with his fingers? Who else knows the weight of the earth or has weighed the mountains and hills on a scale?

13 Who is able to advise the Spirit of the LORD? Who knows enough to give him advice or teach him?

14 Has the LORD ever needed anyone's advice? Does he need instruction about what is good? Did someone teach him what is right or show him the path of justice?

15 No, for all the nations of the world are but a drop in the bucket. They are nothing more than dust on the scales.

He picks up the whole earth as though it were a grain of sand.

16 All the wood in Lebanon's forests and all Lebanon's animals would not be enough to make a burnt offering worthy of our God.

17 The nations of the world are worth nothing to him. In his eyes they count for less than nothing—mere emptiness and froth.

18 To whom can you compare God? What image can you find to resemble him?

19 Can he be compared to an idol formed in a mold, overlaid with gold, and decorated with silver chains?

20 Or if people are too poor for that, they might at least choose wood that won't decay and a skilled craftsman to carve an image that won't fall down!

21 Haven't you heard? Don't you understand? Are you deaf to the words of God—the words he gave before the world began? Are you so ignorant?

22 God sits above the circle of the earth. The people below seem like grasshoppers to him! He spreads out the heavens like a curtain and makes his tent from them.

23 He judges the great people of the world and brings them all to nothing.

24 They hardly get started, barely taking root, when he blows on them and they wither.

The wind carries them off like chaff.

25 "To whom will you compare me? Who is my equal?" asks the Holy One.

26 Look up into the heavens. Who created all the stars?

He brings them out like an army, one after another, calling each by its name.

Because of his great power and incomparable strength, not a single one is missing.

27 O Jacob, how can you say the LORD does not see your troubles? O Israel, how can you say God ignores your rights?

28 Have you never heard? Have you never understood?

The LORD is the everlasting God, the Creator of all the earth. He never grows weak or weary.

No one can measure the depths of his understanding.

29 He gives power to the weak and strength to the powerless.

30 Even youths will become weak and tired, and young men will fall in exhaustion. 31 But those who
trust in the LORD will find new strength. They will soar high on wings like eagles. They will run and not grow
weary. They will walk and not faint.

THE PRE-TRIBULATION PREMISES

Four years after my father's departure to the glorious presence of the Lord, I came across one of his favorite hard cover ancient books. There was a specific well displayed book in his comprehensive library, which he left behind as my inheritance. I was sixteen years of age, when I came across that favorite hard cover book-volume. The title was "Ο Άρχοντας των Λόρδων" The Lord of Lords. It was printed in 1885 and reprinted in 1910, and 1945. It detailed, in exceptional clarity the Christian Doctrines. I kept reading it from cover to cover in the span of one month spending about 10 hours daily.

Until today, withing the upper most of my mind I could still recall the distinctive diagram of the end time rapture till the return of the Lord Jesus Christ, Figure 32.58.

62 years ago, the biblical scholars throughout the worldwide churches in the East, Mid-East or the West had never debated whether the rapture was pre, mid or post tribulation. The churches and the Biblical scholars worldwide were in one belief united on the rapture event to take place unexpectedly at the pre-tribulation period.

Since the beginning of the nineties a rage against pre-tribulation rapture began to spread among only in the western churches and the western biblical scholars, questioning the end of time sequence of events, in particularly the churches. Nonetheless, the debate about the rapture event may be healthy hoping to open the Christians eyes to the importance of being readied at all time to meet Christ in the air. It is better to engage in a healthy debate of the rapture than being uninformed and acting aloof of such vital events.

There are a number of believers and churches worldwide have not yet heard of the rapture. This is contributed to the lack of enthusiasm to study the bible in particularly the book of Daniel in the Old Testament and the book of revelation in the New Testament.

At the age of 17, I was so eager to share the joy of the rapture event and the return of the Lord Jesus Christ in the air to rescue the believers from the wrath of God in the years to follow. As I attempted to share this biblical knowledge during a mid-week event, the attendees, sadly without exception attacked me severely accusing me of heresy. Although I have prepared all the Biblical supporting verses of the rapture events, they refused to listen and concluded the meeting. I was practically thrown out of the church building. However, as soon as the Christian Orthodox priest learnt what happened. He was very considerate to come to my temporary place of residence to apologize on behalf of his angry congregation and for himself for failing to teach his congregation about the end time events including the rapture. The following day, the priest peacefully invited the congregation and myself to restore order.

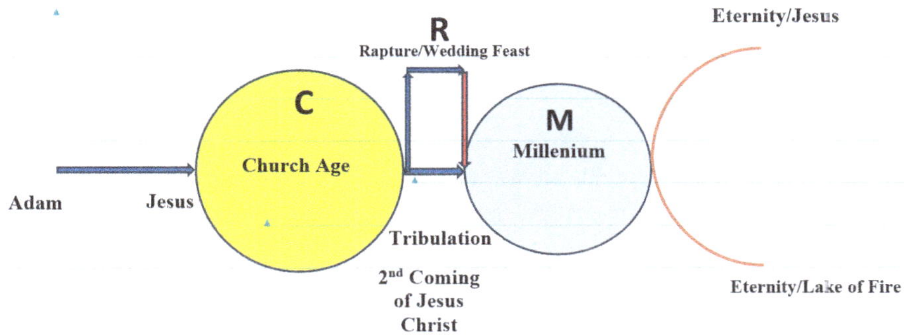

Fig.31.35: The Biblical Human History

Over the past 45 years I have studied the Scriptures along with many Biblical scholars from different countries including (Baba-kirollos) the Patriarch of Alexandria, Egypt; John Stott, England; Christian Missionaries in France; the staff of Calvary Baptist Theological Seminary, Lancaster, PA-USA; the staff of Lancaster Bible College, PA-USA; and my mentor Mr. William Thomson, New Jersey-USA in an effort to articulate a sustaining Biblical response to the rapture query, wherefore should I have confidence in a pre-tribulation rapture.

ISAAC NEWTON - THE RESURRECTION OF THE DEAD

Once the famous scientist Isaac Newton was asked:

– *How is it possible that the dead bodies, long ago corrupted, could become again bodies of their own souls?*

Instead of replying to this, the famous scientist threw some tiny iron particles on the ground, and then asked his speakers:

– ***Who could separate these iron particles from the dust and put them together?***

He received no reply. Then Newton took the magnet and brought it closer to the scattered iron particles. There was a commotion of the particles. The small particles started to stick on the magnet, one after another. There wasn't even a single iron molecule left in the dust. Then Newton told the present people:

Wouldn't the One Who gave such power to the dead metal, also give our souls the power needed for them to return to their own bodies?

Fig.31.36: Isaac Newton Precisely Picked up Each Iron Particle Scattered in the Dust

THE RESURRECTION AND THE TRIBULATION

The dominant drive of this inscription is not to expose the principal shortages of contrasting resurrection and tribulation postures, but to define the advantage of pre-tribulation as educated in foremost the doctrine writings, such as:

i. Matthew 24–25;

ii. 1 Thessalonians 4; 1 Corinthians 15; and

iii. Revelation 3, 6–18.

It is not the weightiness of a solitary cause that brands pre-tribulation so convincing, but instead the joint strength of all the positions of reasoning.

The End Times

In a twinkle of an eye myriads of people all over the earth will disappear. A large coalition of nations will gather from northern Europe, Middle East, Africa, and other nations surrounding Israel attempting to destroy it as predicted in future biblical events.

There are 40% of Americans believe we are living in the last days of catastrophically earth events immediately prior to the second coming of Jesus Christ. Further, there are 70% evangelical Christians believe Christ may return in their life time, according to the "Pew Research Center." The events of the end time are stated below to provide understanding of how these events will unfold.

Below, we are presenting the events of the disappearance of many people, which is referred to as the "Rapture," and the Second Coming of the return of Jesus Christ, which referred to as the "Return," Table 32.1.

Table 31.1: The Rapture and the 2nd Coming of Jesus Christ

Rapture	Return
• Christ comes in the air and brings the believers to Him	• Christ comes back to earth to the Mount of Olives, and then does battle with the ungodly nations
• No sign for the rapture	• There are many signs for the second coming

THE DIFFERENT VIEWS OF THE RAPTURE:

The Bible has recorded the future events of the rapture within the old and new testaments? Unfortunately, different churches and believers, as well as different denominations have believed different rapture timing. Additionally, church ministers sometimes do not wish to discuss the subject, educate, or conduct debates on this vital subject to their congregations. This may be contributed to the lack of knowledge on the part of ministers and the inability to study the Bible diligently. Frequently, ministers attempt to avoid controversial subjects due to their inability to introduce the scripture with distinctive clarity. Below is a simplified approach detailing the three main views of the end of time rapture and the second coming of the Lord Jesus Christ.

The Biblical Statements of the Rapture and the Second Coming of Jesus Christ:

There are Biblical views presenting irreconcilable differences between the "Rapture" and the "Return."

1. In the rapture - Jesus comes in the air, while
2. in the return - Jesus comes down all the way to the earth.
3. In the rapture - Jesus comes for His Saints.
4. In the return, He comes with His Saints.
5. In the rapture - Christ comes to claim His Bride (the church).
6. In the return - Christ comes back with His bride; He brings His bride with Him, Table 31.2

Table 31.2: The Difference between The Rapture and the 2nd Coming of Jesus Christ

Rapture	Return
• Christ comes for His Saints	• Christ comes to earth with His Saints
• Christ claims His bride	• Christ comes with His bride
• There is no sign	• There are many signs
• The believers are removed from earth	• The unbelievers are removed from earth
• Christ comes to reward believers - 1Thess. 4:17	• Christ comes to judge
• Involves believers only	• Involves Israel and the gentiles' nations
• Occurs in a moment, in a twinkle of an eye	• Will be visible to the whole world
• Followed by the Tribulation period	• Followed by the Millenium period

The rapture has no sign prior to the timing of the event. It is an event that can happen at any moment of time. The return has a long litany of signs, where Jesus provided for His coming. In the rapture believers are removed from this earth. At the second return of the Lord Jesus Christ the unbelievers are removed to be taken away for judgment. At the rapture Christ comes to reward the believers according to 1Thesselonians 4:17, (Then, together with them, we who are still alive and remain on the earth will be caught up in the clouds to meet the Lord in the air. Then we will be with the Lord forever). At the return He comes back to judge. At the rapture it involves Christian believers only. The second return of Jesus Christ involves Israel and all the gentiles' nations. At the rapture there is no reason the believers to be here on earth during this time because God is dealing with Israel and the gentiles nations, Table 31.3.

Table 31.3: Pre-Tribulation View

Rapture	Tribulation	Return
• Caught up to meet Christ in the air	→ 75 years	• Christ returns to earth with His Saints
• Believers are taken up in the rapture followed by the tribulation period, then the return of Christ to earth		

The rapture will happen in a twinkling of an eye. It will happen very fast that one may not be fully aware what would have taken place, Table 32.3. When the return of Jesus back to the earth happens, it will be visible to the entire world. Every eye will be able to see Him. It may last certain extended period of time.

After the rapture event - the tribulation period begins, which is the time of the judgment on the earth and its inhabitants, gentiles and Israelites. Then the millennium begins following the second coming of Jesus Christ. There are two phases of Christ's coming. These two phases cannot be happening at the same time. They are separated by a period of at least 7 years, which is called the tribulation period. Clearly, this view is depicted directly from the bible.

Christ Rescues His Bride the Church

Christ returns to rescue His bride before God pours out His wrath on the earth. The wrath of God becomes much severer as it increases in intensity and frequency during the progress of the tribulation period.

At the rapture the New Testament saints (he global church - the believers everywhere) are taken all out before the wrath falls upon the earth. These kinds of theological terminologies are considered simple to precisely describe the rapture and the return events taught in the Bible.

The Bible says that the pre-tribulation view is before or prior to tribulation enabling Christ coming in the air to rescue His bride before God pours out His wrath on the earth, Table 31.3

Question:

Will Christians endure the wrath of God during the tribulation period?

Answer:

The pre-tribulation view says "No." Christ will return before the tribulation period begins.

The Imminency of Christ:

The pre-tribulation view preserves the integrity of the "imminency of Christ," that Christ could come at any time. Christ's coming in the air is unpredictable and could happen at any time.

The Bride of Christ:

Christ comes before the rapture to rescue His bride before the tribulation period begins. Immediately after the rapture, the wrath of God will be poured upon the falling sinful earth, Table 31.4.

Table 31.4: Mid-Tribulation View

Rapture		Return
Believers caught to meet Christ in the air		Christ returns to earth with His Saints (believers)
3½ Years Man's wrath – wars, famines, etc.	3½ Years God's wrath	

Some people consider the rapture event takes place at the mid-point of the 7 years of tribulation period. The fundamental issue one must consider is that believers are not appointed to wrath, 1 Thessalonian 5:9 "For God did not appoint to suffer wrath but to receive salvation through the Lord Jesus Christ." Accordingly, one must question of when does the wrath start? The Mid-Tribulation view is a substantially different view than the pre-tribulation rapture, Figure 31.4. However, the Lord stated that believers are not appointed for wrath.

Nevertheless, the mid-tribulation advocates the belief that at the midpoint of tribulation-rapture would occur to take the believers to heaven at the end of the 3½-year midpoint of the tribulation. They consider that the first 3½ year-midpoint period is not God's wrath, but man's wrath, on which wars and famine, as well as various disasters will occur. The advocates of mid-tribulation rapture consider also the wrath of God is confined only to the last 3½ years of the tribulation period. These advocates equate the last trump of grace with the trump of judgment.

THE TRUMPET OF GRACE AND THE TRUMP OF JUDGMENT

Trumpet of Grace

The last trumpet in 1 Corinthians 15:52, "For the trumpet will sound … … … the dead will be raised … … … we shall be changed.

This trumpet is the trumpet of grace. It is meant for the church (believers).

Trumpet of Judgment

Rev. 11:15-18, *The seventh angel sounded his trumpet … … … your wrath has come.*

This trumpet is the trumpet of judgment for those who rejected God. Actually, the whole 7-year period appears to be God's wrath. These are two different trumpets. Even the seal judgments are opened by Jesus. Accordingly, limiting God's wrath to the second part of tribulation does not fit the time period, because the whole period is God's wrath, Table 31.4.

Table 31.4: Mid-Tribulation View

Rapture	Return
Believers caught to meet Christ in the air	Christ returns to earth with His Saints (believers)
The whole 7-year period is God's Wrath, The Seal Judgments are opened by Jesus	

Table 31.5: Post Tribulation View

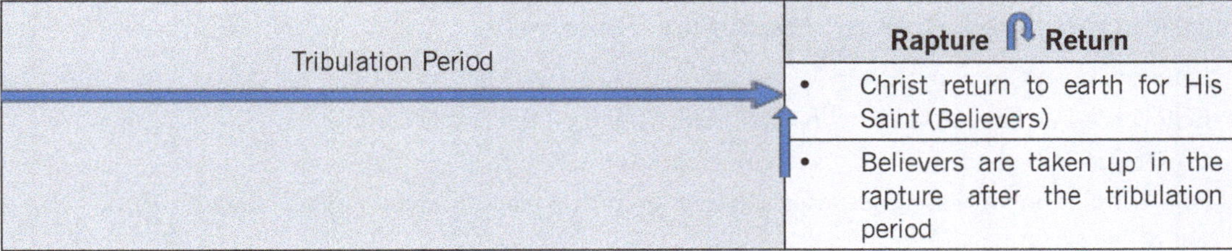

Tribulation Period	Rapture Return
	• Christ return to earth for His Saint (Believers)
	• Believers are taken up in the rapture after the tribulation period

This post-tribulation view contradicts John 14:2-3, which says we will be taken to the father's house; "In my father's house are many mansions … … I will receive you to myself, that where I am, there you may be also."

Post tribulation advocates say that Christians will go through the tribulation period. Therefore, the church must not have been raptured, Table 31.5. This is easily answerable because we believe that even though the church is raptured before the tribulation, there are new people becoming new believers during the tribulation. Post tribulation also says that Christ is going to keep all these people through the wrath of God. He will protect them. The problem here is many of the judgments are indiscriminate in nature, asteroids, earth quakes, and all kind of catastrophes are falling upon the earth such as famine, and there would be martyrdom as stated in Revelation chapter 6. It hardly seems as the church is kept protected.

The Biblical text of Revelation 3:10 "I will keep you from the hour of testing, which is coming upon all the earth." Furthermore, I believe that the idea of "Imminency" that Jesus come in the rapture at any moment, is critically Important.

If post-tribulation is true, we can essentially count down to the rapture with 7 years' worth of signs, and you can do away with "imminence." Nonetheless, imminence is a biblical doctrine seen through the New Testament that Jesus can come in the air at any time. Post tribulation contradicts this doctrine.

The Biblical text must be revered and highly considered. In John 14:2-3 Jesus said "*in my Father's are many mansions; if it were not so, I would have told you. And if I go to prepare a place for you, I will come again and receive you to Myself; that where I am, you may be also.*" Accordingly, Jesus is going to heaven to prepare a place for His bride the church. At present He is in Heaven. Jesus has been involved in a building project in Heaven, and in the Father's house. Jesus said "I will come again and receive you to Myself; that where I am, you may be also," John 14:3.

Where does Jesus take us? He says: He is taking us to the place prepared for us at the Father's house to be with Him. However, the **post-tribulation** stated that when Christ comes in the rapture, we join Him in the air, and then come right back to the earth with Christ. There would be no time to go to the Father's house in Heaven, Table 31.6.

Table 31.6: Post Tribulation View

Tribulation Period	Rapture ↱ Return
→	• Christ return to earth for His Saint
The believers come back to the earth with Christ and do not go to the Father's house in Heaven	

Accordingly, Jesus then is involved in a futile building project if post tribulation is true, Table 31.6. The operative statement of Christ is "I go … I come" I go to prepare a place – I will come to receive you. In post tribulation it assumes that Christ is saying "I stay to prepare a place … and you would be already with me." This does not fit the scripture.

Post-tribulation View Contradicts Biblical Statement

a. Believers meet Christ in the air
b. Believers come back to earth with Christ and do not go the Father's house
c. Christ Stated "I go to … … … I will come back to receive you."
d. Post-tribulation view assumes Jesus stays "to prepare a place and you already are with Him." This does not fit the scriptures.

The Pre-Tribulation Rapture

Table 31.7: Pre-Tribulation View

Rapture	Tribulation	Return
• Caught up to meet Christ in the air	→ 75 years	• Christ returns to earth with His Saints
• Believers are taken up in the rapture followed by the tribulation period, then the return of Christ to earth		

The word "rapture" originates from the Latin word raptus, which is utilized 14 times in the New Testament. The basic idea of the word is "to take away abruptly or snatch." It is employed in the New Testament in situation to stealing/predatory

Matt. 11:12; 12:29; 13:19; John 10:12, 28, 29) and take away, John 6:15; Acts 8:39; 23:10; Jude 23.

There rapture stresses on being caught up to heaven. It is utilized of Paul's third heaven involvement (2 Cor. 12:2, 4) and Christ's ascent to heaven, Table 32.3.

Noticeably, this is the seamless word to label God unexpectedly taking up the church (believers-saints), His Bride/Believers, out of earth to heaven as the initial part of Christ's second subsequent coming. Though, the time itself encompasses no clue as to its time in association to Daniel's seventieth week.

THE DOCTRINE OF THE "RAPTURE"

1 Thessalonians 4:16–17

Indisputably, denotes to a rapture that is a vital part of the doctrine. Henceforward, this word was taken from the Greek verb "αρπάζω επάνω (arpázo epáno)," meaning, "to seize upon, spoil, snatch away or take to oneself," especially used of rapture Acts 8:39; 2 Cor 12:2.

"[39] When they came up out of the water, the Spirit of the Lord snatched Philip away. The eunuch never saw him again but went on his way rejoicing."

"[2] I was caught up to the third heaven fourteen years ago. Whether I was in my body or out of my body, I don't know—only God knows."

For the Lord Himself will descend from heaven with a shout, with the voice of the archangel, and with the trumpet of God; and the dead in Christ shall rise first. Then we who are alive and remain shall be caught up together with them in the clouds to meet the Lord in the air, and thus we shall always be with the Lord. (bold italics mine).

Deprived of employing "αρπάζω επάνω (arpázo epáno)," but by means of similar appropriate language, 1 Corinthians 15:51–52 denotes to the similar doctrine incident as 1 Thessalonians 4:16–17:

"Behold, I tell you a mystery; we shall not all sleep, but we shall all be changed, in a moment, in the twinkling of an eye, at the last trumpet; for the trumpet will sound, and the dead will be raised imperishable, and we shall be changed."

16 For the Lord himself will come down from heaven with a commanding shout, with the voice of the archangel, and with the trumpet call of God. First, the believers who have died will rise from their graves. 17 Then, together with them, we who are still alive and remain on the earth will be caught up in the clouds to meet the Lord in the air. Then we will be with the Lord forever.

Therefore, it can be decided that Sacred scripture leads to a rapture doctrine, however neither of these initial texts covers a time pointer.

COMPREHENSIVE RAPTURE

Many have proposed that the rapture articulated of in 1 Thessalonians 4:16–17 and 1 Corinthians 15:51–52 will only be a "limited" rapture, not a rapture of all who accept as true. Supporters purpose that partaking in the rapture is not founded upon one's redemption, but is in its place restrictive, dependent upon one's worthy behavior.

This concept is based on New Testament text, which emphasizes on compliant watching and waiting (e.g., Matt. 25:1–13; 1 Thess. 5:4–8; Heb. 9:28).

If this understanding is accurate, the consequence would be that only portion of the church would be raptured, and those who are not raptured would withstand a percentage of or the entire seventieth week of Daniel. Though, these biblical writings which purportedly impart an unfinished rapture are better understood as differentiating between true believers who are raptured and purely acknowledging, yet untrue believers who endure to stay behind. Writings that mention the concluding aspect of Christ's second coming are frequently used incorrectly to back up the partial-rapture concept.

There are abundant explanations the limited rapture theory misses the mark to be conclusive.

a. 1 Corinthians 15:51 articulates that "all" will be changed.
b. An incomplete rapture would reasonably mandate a parallel incomplete resurrection, which is nowhere imparted in Sacred scripture.
c. An incomplete rapture would diminish and perhaps eradicate the necessity for the judgment seat of Christ, since judgment would have by now taken place by virtue of an "incomplete" rapture.
d. It produces a limbo of types on earth for those believers left behind.
e. An incomplete rapture is nowhere plainly imparted in Scripture.

Consequently, it is best determined that the rapture will be complete and comprehensive, not just fractional.

RAPTURE VIEWS

Revelation 6–18 – The Church is Absent

The known New Testament tenure for "church" _ (Εκκλησία – Ekklisía – לעדת) - (كنيسة) is utilized 19 times in Revelation chapter 1 to chapter 3, a segment that arranges the historic church of the first century to the end of the apostle John's life (ca. AD 95), Figure 32.58. Nevertheless, "Ekklisía-church" is then utilized only once more in the twenty-two-chapter volume, and that practice is at the very end (Revelation 22:16) when John proceeds to speaking about the first-century church. Most motivating is the detail that nowhere during the period of Daniel's seventieth week is the tenure for "Ekklisía-church" used for believers on earth (cf. Rev. 4–19), Table 31.8.

16 *"I, Jesus, have sent my angel to give you this message for the churches. I am both the source of David and the heir to his throne.[a] I am the bright morning star."*

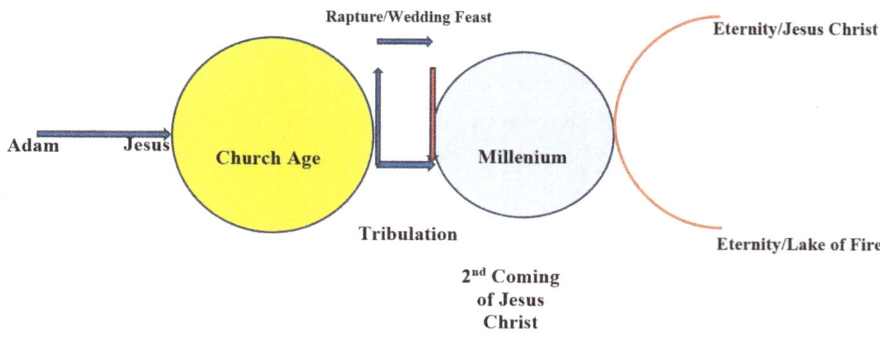

Fig.31.58: *The Biblical Human History*

Table 31.8: Pre-Tribulation View

Rapture	Tribulation	Return
• Caught up to meet Christ in the air	75 years	• Christ returns to earth with His Saints
• Believers are taken up in the rapture followed by the tribulation period, then the return of Christ to earth		

It is unforeseen that John would change from comprehensive instructions for the Ekklisía-church to comprehensive silence about the Ekklisía-church in the successive 13 chapters if, in reality, the church did endure into the tribulation. If the church will be involved in the tribulation of Daniel's seventieth week, then confidently the most comprehensive study of tribulation events would embrace a justification of the church's part—but it does not. The only timing of the rapture that would constitute this frequent mention of "Ekklisía-church" in Revelation 1–3 and the whole non-appearance of the "Ekklisía-church" on earth until Revelation 22:16 is a pre-tribulation rapture, Table 31.9, which will displace the church from earth to heaven prior to Daniel's seventieth week.

Tavke 31.9: Pre-Tribulation View

Rapture	Tribulation	Return
• Caught up to meet Christ in the air	75 years	• Christ returns to earth with His Saints
• Believers are taken up in the rapture followed by the tribulation period, then the return of Christ to earth		

Investigating this remark from another viewpoint, it is also factual that nowhere in the Sacred scripture it is imparted that the Ekklisía-church and Israel would co-occur as the cores for God's saving message and yet continue equally exclusive.

Nowadays, the universal Ekklisía is God's human conduit of redeeming truth. Revelation provides signs that the Jewish remainder will be God's human apparatus during Daniel's seventieth week. The account tersely shifts from the "Ekklisía-church" in Revelation 2–3 to the 144,000 Jews from the twelve tribes in Revelation 7 and 14. One may pose an important question, which is, "Why?"

Revelation 12 is also a mini-summary of the whole tribulation period, Table 31.9, and since the "woman" who gave birth to the male child - Rev. 12:1–13 is Israel, at that time the Tribulation period emphases is on the nation of Israel, not the Ekklisía-church. This is possible. Since a pre-tribulation rapture has detached the "church" from the earth preceding to Daniel's seventieth week.

"Then I witnessed in heaven an event of great significance. I saw a woman clothed with the sun, with the moon beneath her feet, and a crown of twelve stars on her head. 2 She was pregnant, and she cried out because of her labor pains and the agony of giving birth.

3 Then I witnessed in heaven another significant event. I saw a large red dragon with seven heads and ten horns, with seven crowns on his heads. 4 His tail swept away one-third of the stars in the sky, and he threw them to the earth. He stood in front of the woman as she was about to give birth, ready to devour her baby as soon as it was born.

5 She gave birth to a son who was to rule all nations with an iron rod. And her child was snatched away from the dragon and was caught up to God and to his throne. 6 And the woman fled into the wilderness, where God had prepared a place to care for her for 1,260 days.

[7] Then there was war in heaven. Michael and his angels fought against the dragon and his angels. 8 And
the dragon lost the battle, and he and his angels were forced out of heaven. 9 This great dragon—the ancient
serpent called the devil, or Satan, the one deceiving the whole world—was thrown down to the earth with all
his angels.

10 Then I heard a loud voice shouting across the heavens,"

"It has come at last—
salvation and power
and the Kingdom of our God,
and the authority of his Christ.[a]
For the accuser of our brothers and sisters[b]
has been thrown down to earth—
the one who accuses them
before our God day and night.
11 And they have defeated him by the blood of the Lamb
and by their testimony.
And they did not love their lives so much
that they were afraid to die.
12 Therefore, rejoice, O heavens!
And you who live in the heavens, rejoice!
But terror will come on the earth and the sea,
for the devil has come down to you in great anger,
knowing that he has little time."

13 When the dragon realized that he had been thrown down to the earth, he pursued the woman who had
given birth to the male child."

POST-TRIBULATION

Rendering Rapture Unimportant

Table 31.10: Post Tribulation View

Tribulation Period	Rapture ↱ Return
	Christ return to earth for His Saint
The believers come back to the earth with Christ and do not go to the Father's house in Heaven	

i. If the Lord astonishingly conserves the Ekklisía-church over the tribulation, there is no such need for a rapture! If it is to evade the wrath of God at Armageddon, at that point what would cause God to not remain shielding the saints on earth; as is assumed by post-tribulationism, just as He protected Israel? Exod. 8:22; 9:4, 26; 10:23; 11:7 from the wrath He discharged out on Pharaoh and Egypt. Additionally, if the drive of the rapture is for alive saints to evade Armageddon, why also resurrect the saints who are previously protected at the same time?

ii. If the rapture will happen in linking it with the Lord's post-tribulation imminent return, the successive parting of the sheep from the goats Matt. 25:31 ff. will be superfluous. Parting would have happened in the very act of transformation.

iii. If all tribulation believers are raptured and glorified just before the inauguration of the millennial Kingdom, who at that point will inhabit and promulgate the Kingdom? The Sacred scripture designate that the living unbelievers will be judged at the conclusion of the tribulation and detached from the earth Matt. 13:41–42; 25:41. Up till now, they also teach that children will be born to believers throughout the millennium and that these children will have the ability of sinning Isa. 65:20; Rev 20:7–10. This will not be conceivable if all believers on earth would have been glorified through a post-tribulation rapture, Table 32.6.

iv. The post-tribulation model of the church being raptured and then instantly carried back to earth leaves no time for the Bema, the "Judgment Seat of Christ" to occur (1 Cor. 3:10–15; 2 Cor. 5:10), nor for the Marriage Supper (Rev. 19:6–10).

Consequently, it can be resolved that a post-tribulation time of the rapture is incompatible with the sheep-goat nation judgment, and, actually, removes two serious end-time events. A pre-tribulation rapture evades all of these problems.

No Warnings Sign of an Imminent Tribulation for Church-Believers

The Lord's directives to the church through the scripture comprehend a diversity of warnings, but not ever do they warn believers to make for entering and enduring the tribulation of Daniel's seventieth week.

They caution strongly about coming falsehood and untruthful prophets

Acts 20:29–30; 2 Pet. 2:1; 1 John 4:1–3; Jude 4. They caution against sinful living

Eph. 4:25–5:7; 1 Thess. 4:3-8; Heb. 12:1. They even reprove believers to bear in the

1 Thess. 2:13–14; 2 Thess. 1:4; 1 Peter. Though, there is complete silence on formulating the church for any type of tribulation like that instituted in Revelation 6–18.

It is incompatible, at that point, that the Sacred scripture would be quiet about such a shocking transformation for the church. If any period of the rapture other than pre-tribulation were factual, one would imagine the scripture to explain the realism of the church in the tribulation, the determination of the church in the tribulation, and the manner of the church in the tribulation. Nevertheless, there is no instruction at all. Only a pre-tribulation rapture pleasingly clarifies this silence - **1 Thessalonians 4:13–18 Stresses a Pre-tribulation Rapture.**

Table 31.11: Pre-Tribulation View

Rapture	Tribulation	Return
• Caught up to meet Christ in the air	75 years	• Christ returns to earth with His Saints
• Believers are taken up in the rapture followed by the tribulation period, then the return of Christ to earth		

What if, theoretically, that some further rapture timing in addition to pre-tribulation is factual. What would one supposed to find in 1 Thessalonians 4? How does this liken with what is really observed?

"13 And now, dear brothers and sisters, we want you to know what will happen to the believers who have died[f] so you will not grieve like people who have no hope. 14 For since we believe that Jesus died and was raised to life again, we also believe that when Jesus returns, God will bring back with him the believers who have died.

[15] We tell you this directly from the Lord: We who are still living when the Lord returns will not meet him ahead of those who have died.[g] [16] For the Lord himself will come down from heaven with a commanding shout, with the voice of the archangel, and with the trumpet call of God. First, the believers who have died[h] will rise from their graves. [17] Then, together with them, we who are still alive and remain on the earth will be caught up in the clouds to meet the Lord in the air. Then we will be with the Lord forever. 18 So encourage each other with these words."

One would think the Thessalonians to be blissful over the reality that precious ones are home with the Lord and will not have to bear the fears of the tribulation. Nonetheless, the Thessalonians are essentially grief-stricken since they dread their loved ones have missed the rapture. Only a pre-tribulation rapture agrees with this heartache.

i. One would assume the Thessalonians to be heartbroken over their own imminent trial somewhat than grief-stricken over loved ones. Additionally, they would be prying about their particular future destiny. But the Thessalonians have no worries or queries about the impending tribulation.
ii. One would suppose apostle Paul, even in the absenteeism of attention or queries by the Thessalonians, to have given commands and encouragement for such a supreme test, which would make their current tribulation appear infinitesimal in contrast. But not one sign of any imminent tribulation of this kind seems in the text

    1 Thessalonians 4 fits only the paradigm of a pre-tribulation rapture, Table 31.3. It is mismatched with any supplementary time for the rapture.

1 THESSALONIANS 4:13–18 PARALLELS JOHN 14:1–3

John 14:1–3 refers to the second coming of Jesus Christ. It is not a promise to all believers that they shall go to Him at death. It does denote the rapture of the church. Observe the close equivalent between the promises of John 14:1–3 and 1 Thessalonians 4:13–18.

1. The promise of an existence with Christ: "*. . . that where I am, there you may be also" (John 14:3) and ". . . therefore we shall always be with the Lord*" (1 Thess. 4:17).
2. The promise of comfort: "*Let not your heart be troubled . . .*" (John 14:1) and
   "*Therefore comfort one another with these words.*" (1 Thess. 4:18).

Jesus taught the disciples that He was going to His Father's house to prepare a place for them. He promised that He would return and receive them so that they could be with Him wherever He was.

The expression "wherever I am," though inferring continual presence at large, here means existence in heaven especially. The Lord told the Pharisees in John 7:34, "Where I am you cannot come." He was not talking about His then-present dwelling on earth, but rather His resurrected existence at the right hand of the Father. In John 14:3, "where I am" essentially mean "in heaven," or the intent of John 14:1–3 would be meaningless.

"*Don't let your hearts be troubled. Trust in God, and trust also in me.* [2] *There is more than enough room in my Father's home. If this were not so, would I have told you that I am going to prepare a place for you?* [3] *When everything is ready, I will come and get you, so that you will always be with me where I am.*"

A post-tribulation rapture, Table 31.12, stresses that the saints meet Christ in the air and instantly come down to earth without undergoing what the Lord promised in John 14. Subsequently, John 14 refers to the rapture, only a pre-tribulation rapture contents the language of John 14:1–3 and permits raptured saints to reside for an expressive time with Christ in His Father's house.

Table 31.12: Post Tribulation View

Tribulation Period	Rapture ↱ Return
The believers come back to the earth with Christ and do not go to the Father's house in Heaven	Christ return to earth for His Saint (believers)

CHRIST'S POST-TRIBULATION COMING DIFFERS FROM THE RAPTURE

At the rapture: 1 Thessalonians 4:13–18 and 1 Corinthians 15:50–58, through what occurs in the final events of Christ's second coming in Matthew 24–25, at minimum eight transformations are noticeable, stated below. These transformations call for that the rapture happens at a time pointedly dissimilar from that of the final event of Christ's second coming.

1. At the rapture, Christ comes in the air and returns to heaven (1 Thessalonians 4:17), but at the final event of the second coming, Christ comes to the earth to dwell and reign (Matt. 25:31–32).
2. At the rapture, Christ gathers His own (1 Thessalonians 4:16–17), but at the final event of the second coming, angels gather the elect, Matt. 24:31.
3. At the rapture, Christ comes to reward the believers (1 Thessalonians 4:17), but at the final event of the second coming, Christ comes to judge - Matt. 25:31–46.
4. At the rapture, resurrection is prominent (1 Thessalonians 4:15–16), but at the final event of the second coming, resurrection is not mentioned.
5. At the rapture, believers depart the earth (1 Thessalonians 4:15–17), but at the final event of the second coming, unbelievers are taken away from the earth, Matt. 24:37–41.
6. At the rapture, unbelievers remain on earth, but at the final event of the second coming, believers remain on earth - Matt. 25:34.
7. At the rapture, there is no mention of establishing Christ's Kingdom on earth, but at the final event of the second coming, Christ has come to set up His Kingdom on earth - Matt. 25:31, 34.
8. At the rapture, believers will receive glorified bodies

cf. 1 Corinthians 15:51–57, but at the final event of the second coming of Jesus Christ, the tribulation saints who either died or martyred will be resurrected to join the New Testament believers (church) to enter the millennium. Revelation 20:4-6 "Then I saw thrones and seated on them were those who had been given authority to judge. I also saw the souls of those who had been beheaded because of the testimony about Jesus and because of the word of God. These had not worshiped the beast or his image and had refused to receive his mark on their forehead or hand. They came to life and reigned with Christ for a thousand years. 5 (The rest of the dead did not come to life until the thousand years were finished.) This is the first resurrection. 6 Blessed and holy is the one who takes part in the first resurrection. The second death has no power over them, but they will be priests of God and of Christ, and they will reign with him for a thousand years.

Also, the Old Testament believers will be resurrected at the second coming of the Lord Jesus Christ to reign with Him in the Millennium, after The Second Coming and after the defeat of the nations of the world at Armageddon" (Revelation 19:16-21).

Moreover, numerous of Christ's parables in Matthew 13 approve variances between the rapture and the final event of Christ's second coming.

a. In the parable of the wheat and tares, the tares (unbelievers) are taken out from amongst the wheat (believers) at the second coming - Matt. 13:30, 40, but believers are detached from among unbelievers at the rapture 1 Thessalonians 4:15–17.

b. In the parable of the dragnet, the bad fish (unbelievers) are removed out from among the good fish (believers) at Christ's second coming Matt. 13:48–50, but believers are removed from among unbelievers at the rapture 1 Thessalonians 4:15–17.

Lastly, the rapture is not stated in any of the most detailed second coming texts—Matthew 24 and Revelation 19.

This is to be anticipated in view of the explanations above, since the pre-tribulation rapture would have happened seven years earlier.

DANIEL'S SEVENTIETH WEEK - REVELATION 3:10

The Church is Removed

The subject here is if the expression "keep you from the hour of testing" indicates an ongoing safe state outside of" or "safe emergence from within."

Understanding "Out of" Εκτός (ektós)

The Greek preposition Εκτός has the knowledge of emergence, but this is not the case in every setting. Two distinguished exclusions to the basic idea are 2 Corinthians 1:10 and

1 Thessalonians 1:10. In the Corinthian context, Paul prepares his rescue from death by God. As is seeming, Paul did not arise from a state of death, but instead was rescued from that possible danger.

Even more considerable is 1 Thessalonians 1:10. At this point Paul states that Jesus is rescuing believers out of Εκτός the wrath to come. The impression is not emergence out of wrath, but rather protection from entrance into wrath.

Therefore, Εκτός can be understood to mean either "a continuing state outside of" or "emergence from within." Thus, no rapture position can be dogmatic at this point. At best, all positions remain possible.

The Martyrs

If Revelation 3:10 means immunity or protection within as other positions insist, several contradictions result:

a. If protection in Revelation 3:10 is limited to protection from God's wrath only and not Satan's wrath also, then Revelation 3:10 repudiates the Lord's appeal in John 17:15.

b. If it is contended that Revelation 3:10 means full immunity, then of what worth is the promise in light of Revelation 6:9–11 and 7:14 where martyrs abound? The wholesale martyrdom of saints during the tribulation demands that the promise to the Philadelphian church be understood as "keeping out of" the hour of testing, not "keeping within."

Meanwhile the expression "to meet the Lord" in 1 Thessalonians 4:17 can denote to a welcoming city going out to meet the visiting king and accompanying him back to the city, does not this expression point definitely to a post-tribulation rapture?

a. This Greek verb/noun can denote to either meeting within a city
   Mark 14:13; Luke 17:12 or
   going out of the city to meet and return back
   Matt. 25:6; Acts 28:15.
   So, the employment of this specific word is not at all conclusive.

b. Recall that Christ is coming to a antagonistic people at large who will sooner or later fight against him at Armageddon. So, the pre-tribulation rapture best pictures the king rescuing, by rapture, His faithful followers who are imprisoned in a hostile world and who will later escort Him when He returns to overcome His enemies and set up His Kingdom (cf. Rev. 19:11–16).

THE DAY OF THE LORD

If, rendering to pre-tribulationism, believers would not be in it what are the reasons Paul write in 1 Thessalonians 5:6 to alert them to "*the day of the Lord*"?

Paul urges believers in 1 Thessalonians 5:6 to be watchful and living godly in a Day of the Lord context just as Peter does in 2 Peter 3:14–15 where the Day of the Lord understanding is obviously at the end of the millennium when the old heavens and earth will be demolished and substituted with the new. In both cases, they are encouragements to present godly living for devoted believers in the knowledge of God's future judgment on unbelievers. These texts are not decisive influences for any situations on the time of the rapture.

Matthew 24:37–42,

Are people taken out of the world at a post-tribulation rapture, Table 31.13?

Table 31.13: Post Tribulation View

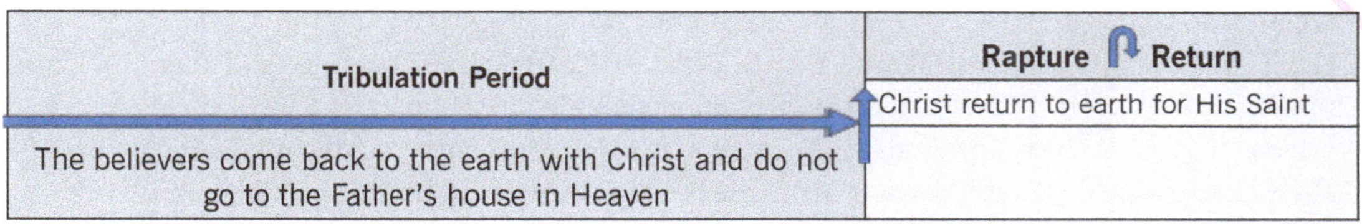

Tribulation Period	Rapture Return
	Christ return to earth for His Saint
The believers come back to the earth with Christ and do not go to the Father's house in Heaven	

In fact, Matthew 24:37–42 explains just the opposite:

1. The historic figure of Noah (vv. 37–39) explains that Noah and his family were left alive while the whole world was taken away in death and judgment. This is precisely the order to be anticipated at Christ's second coming as stated in the parable of the wheat and tares (Matt. 13:24–43), the parable of the dragnet (Matt. 13:47–50), and the sheep-goat realm judgment (Matt. 25:31–46).
2. In all of these situations, at the concluding event in Christ's second coming, unbelievers are taken away in judgment and righteous believers remain. Therefore no, this text does not communicate about the rapture.

APPENDIX

a. According to the technical discussion the author had with Dr Ratnakant Sanjay, M.D., of Bangalore, India.

b. Kepler quoted in Tiner, Ref. 1, p. 178. Kepler's Third Law of Planetary Motion states: $P^2 \propto a^3$ where P is the orbital period of planet and a is the semimajor axis of the orbit, while $\propto$ means "is proportional to".

c. The evolutionary ideas of spontaneous generation and of species being able to change from one to another were not originated by Charles Darwin. These ideas had been around since the time of the ancient Greeks. Darwin merely put various existing ideas together into his theory of evolution and popularized them.

d. Adapted from Henry C. Thiessen, Lectures in Systematic Theology, Eerdmans, Grand Rapids, 1979, pp. 384–85.

e. Tim LaHaye, Pre-Trib Perspectives (Vol. V, Num. 5; Aug. 2000), p. 1.

f. Thomas Ice, Pre-Trib Perspectives (Vol. V, Num. 5; Aug. 2000), p. 4.

g. Evelyn Leopold, Reuters Internet News Service, "UN, NY gear for largest-ever meet of world leaders," August, 2000.

h. Josephus, Antiquities of the Jews

REFERENCES

1. Mozur, P., Google's A.I. program rattles Chinese Go master as it wins match, *New York Times*; nytimes.com, 25 May 2017.
2. Aleksander, Igor: Artificial neuro-consciousness: An update, IWANN, 1995.
3. Showalter, B., Artificial intelligence, transhumanism and the church: How should Christians respond? christianpost.com, 27 Jan 2018.
4. Ruse, M., How evolution became a religion: creationists correct? *National Post*, pp. B1, B3, B7, 13 May 2000; creation.com/ruse.
5. Karr, J.R. *et al.*, A whole-cell computational model predicts phenotype from genotype, *Cell* 150(2):389–401, 12 Jul 2012. See also Cellular complexity "nearly unbelievable", Focus, *Creation* 35(1):11, 2013; creation.com/focus-351#128-computers.
6. Statham, D., The remarkable language of DNA, *Creation* 36(2):52–55, 2015; creation.com/dna-remarkable-language.
7. Hoyle, F., The big bang in astronomy, *New Scientist* 92(1280):527, 1981, quoted in Batten, D., Cheating with chance, *Creation* 17(2):14–15; creation.com/cheating-with-chance.
8. Harris, M., God Is a Bot, and Anthony Levandowski Is His Messenger, wired.com, 27 September 2017.
9. Harris, M., Inside the First Church of Artificial Intelligence, wired.com, 15 November 2017.
10. McFarland, M., Elon Musk: 'With artificial intelligence we are summoning the demon.', washingtonpost.com, 24 October 2014.
11. Korosec, K., Anthony Levandowski, former Google engineer at center of Waymo-Uber case, charged with stealing trade secrets, techcrunch.com, 27 August 2019.
12. Shepherd, G.M., *Neurobiology*, Oxford University Press, London, p. 577, 1983.
13. Tortora, G.J. and Anagnostakos, N.P., *Principles of Anatomy and Physiology*, Harper & Row, New York, p. 290, 1981.
14. Gitt, W., *The Wonder of Man*, CLV Publishing, Germany, p. 82, 1999.
15. Restak, R.M., *The Brain*, Bantam Books, New York, pp. 34–35, 1984.
16. Math Equity Toolkit; equitablemath.org.
17. https://equitablemath.org/wp-content/uploads/sites/2/2020/11/1_STRIDE1.pdf.
18. Bertrand Russel, *The Principles of Mathematics*; Cambridge University Press, 1903; people.umass.edu.
19. "The right answer", posted on *The Atheist Conservative;* theatheistconservative.com
20. Sergiu Klainerman, *There Is No Such Thing as "White" Math,* bariweiss.substack.com, 1 Mar 2021.
21. The Medieval Problem of Universals, plato.stanford.edu, 31 Oct 2017.
22. Vern S. Poythress, *A Biblical View of Mathematics*, 4 Jun 2012; frame-poythress.org.
23. Vern Poythress, *Creation and Mathematics; or What Does God Have to do with the Numbers?* frame-poythress.org, 21 May 2012.
24. Other examples of beauty in mathematical equations are provided by Clara Moskowitz, The 11 Most Beautiful Mathematical Equations, livescience.com, 1 Jun 2017.
25. Ramos, S.C., *The Discovery of Calculus: Leibniz vs. Newton*; stmuhistorymedia.org, 3 Nov 2017.
26. Johannes Kepler, quoted in: Tiner, J. H., *Johannes Kepler-Giant of Faith and Science*, Mott Media, Milford, Michigan (USA), p. 193, 1977.
27. *Encyclopaedia Britannica*, 15th ed., vol. 22, p. 506, 1985.

28. *Encyclopaedia Britannica*, vol. 22, p. 507, 1985.
29. Crowther, J.G., *Founders of British Science: John Wilkins, Robert Boyle, John Ray, Christopher Wren, Robert Hooke, Isaac Newton*, Cresset Press, London, p. 94, 1960.
30. *McGraw-Hill Encyclopedia of World Biography*, 9:118, McGraw-Hill, New York, USA, 1973.
31. Asimov, I., *Biographical Encyclopedia of Science and Technology: The Lives and Achievements of More Than 1000 Great Scientists from Ancient Greece to the Space Age*, 3rd edition, Doubleday & Co. Inc., Garden City, New York, USA, p. 137, 1982. Return to text.
32. Ray, J., quoted in Ref. 1, p. 126.
33. Ray, J., quoted in: H.M. Morris, *Men of Science, Men of God*, Master Books, El Cajon, CA., USA, p. 18, 1982.
34. Derham, W., quoted in Ref. 1, p. 129.

WILL ROBOTS INHERIT THE EARTH?

WILL ROBOT BECOME HUMANOID?

The robotics revolution has started! Since robotics have been manufactured, the efficiency and productivity in working conditions are growing rapidly, Figure 32.1. They formed a huge leap in our lives as they became part of many workplaces.

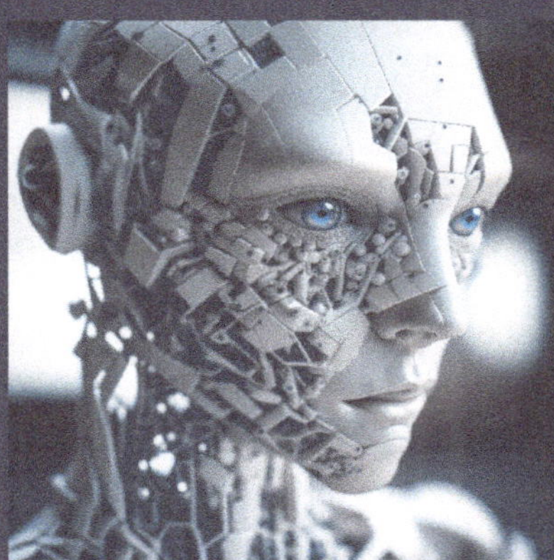

Fig.32.1: *Would Humanoid - Inherit the Earth? – Curtsy Freepik*

Moreover, they can implement a wide variety of successful tasks in a record time compared to human performance. Also, they are playing a pivotal role in helping laborers in their work to be faster and more effective, Figure 32.2.

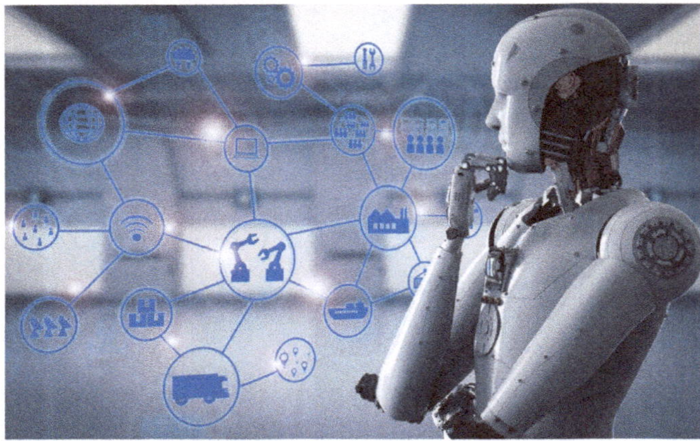

Fig.32.2: Humanoid/Robots Faster and More Effective – Curtsy Shruti Suresh Kamath

"*When automation or computerization makes some steps in a work process more reliable, cheaper or faster, this increases the value of the remaining human links in the production chain.*"

However, some corporate employees believe that robotics will have an unfavorable impact on the workplace, all because they thought that their jobs are in danger. For instance, now accountants, laborers of construction, farmers, housekeepers, and chauffeurs are worrying about losing their jobs due to robotics "Up to 20 million manufacturing jobs around the world could be replaced by robots by 2030" (Oxford Economics, 2019). So, this will decrease laborer wages and increase the rate of unemployment in the future.

From another perspective, robotics may be the main source for real disasters that are waiting to face the world in the future. This writing will discuss the impact of robotics on the workforce and how employees need to cooperate with them in fields such as accounting, workers of construction and farming.

There are so many high-risk fields that still depend on manual laborers as construction workers, who consistently facing hard and perilous working conditions. As a result, the numbers of wounds and death are constantly incrementing.

"317 million nonfatal professional injuries and 321,000 occupational fatalities occur all around the world each year, so, that 151 workers sustain a work-related accident every 15 seconds."

However, replacing them with robots will increase safety and permit errands to be finished precisely. For example, in China, they started to use the robot-welder, which is manufactured to pick up a big pile of boards and take them into an elevator to increase safety, instead of hiring workers, because if a laborer is injured while working particularly on this job, it could be a serious issue. Robots also will be cost-efficient in the long run as they can work for long periods with the same efficiency without taking rest. On the contrary, laborers need time to take a break, and they get bored due to the same task that they do every day, so it became a routine for them, and this affects the quality of their work, Figure 32.3.

Recently, several research have shown that robots are superior in accounting functions rather than humans, and this affects accountants who are concerned about where they are going to fit if this occurs in the future. "Software programs attempting to replicate human experts, behavior, and expertise, store human knowledge and experience and transform it into rules thus trying to solve accounting problems and perform some accounting tasks."

Fig.32.3: Humanoid Helping in Hazardous Jobs – Curtsy – Sastra Robotics

Robots not only can be used in banking fields as they will arrange meetings, collect, and count money but also, they will be quite useful in business fields to produce a higher quality of output for companies. As well as they help to utilize the data and understand information faster and accurately so the human error will be reduced and data can be easily checked later on.

Moreover, an individual's positions will turn out to be less stressful, so employee's tasks can be more centered around significant works that need greater liability. However, replacing accountants who are working as risk-takers with robots cannot be ideal because this type of position needs creativity and from its qualification to be able to take specific decisions when there is an issue, and robots will struggle to achieve this.

From another perspective, accountants and businesses must find the best way in which workers can interact with robots effectively. "Businesses should review their organization's activities to assess where potential value from automation is highest and create a strategic plan that includes both capital investment and reskilling workers."

AI – THE BEAST OF REVELATION

Fig.32.4: Did The Bible Predict Artificial Intelligence? Curtsy by Derek London – Curtsy CodeX Medium

IS ARTIFICIAL INTELLIGENCE THE MARK OF THE BEAST?

Artificial intelligence is intelligence displayed by machines, in contrast with the natural intelligence displayed by humans and other animals, Figure 32.4.

"And it was allowed to give breath to the image of the beast, so that the image of the beast might even speak and might cause those who would not worship the image of the beast to be slain." (Revelation 13:15)

Artificial Intelligence (AI)

Today, most of us regularly talk to Alexa, Siri or Google Assistant or even the artificial intelligence embedded in our cars, refrigerator, or our home. The AI talks back and responds directly to us about whatever it was that we wanted to know from it or to do for us.

Without even realizing it, human daily interaction with AI has become very common, and is continuing to grow more common. Scientists and others who predict what the future holds (futurists) have even planned, for the future, to name this artificial intelligence (AI), "Singularity." When it is no longer just the product of man but has complete control over itself.

"And he deceives those who dwell on the earth by those signs which he was granted to do in the sight of the beast, telling those who dwell on the earth to make an image to the beast..." (Rev 13:14)

The Greek words underlined in the verse above are ***poieō eikōn thērion***, meaning that the fully deceived men of the earth of that time will be called upon by the False Prophet **to construct a likeness of the Antichrist**. This means that the Image of The Beast is not the same as the Beast himself (Antichrist) but is constructed thing that has "come to life." In some translated texts it is called a 'statue,' Figure 32.5.

Fig.32.5: Worship The Image of The Anti-Christ - Curtsy Freepik

Added to the description of the Image having "breath," (Rev 13:15) gives us the impression that it is some kind of self-operating machine, a computer-operated cybernetic robot, programmed so that it can speak and act like it's alive and command the worship of the world.

Artificial intelligence and robotics (humanoid) are advancing so rapidly today, making this prophecy no longer futuristic. Imagine how farfetched this would have sounded in John's Day.

Robots/humanoids are increasingly replacing human functions by automating manual tasks. Killer robots, digital doctors, and driverless cars are just a few examples of how far technology has gone in taking over our lives and replacing human effort.

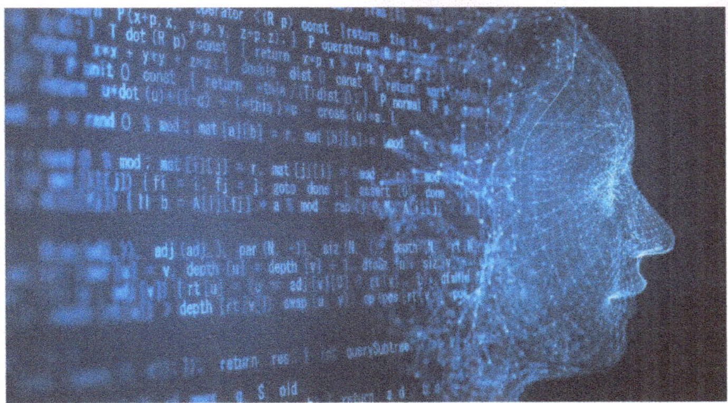

Fig.32.6: Superior Intelligent Being - Curtsy Freepik

Science is racing towards giving life to an "image" already. The robot/humanoid it not only can speak but also can respond to questions, but it has a "personality" and understands concepts like sarcasm and humor, Figure 23.6.

Researchers at the University of Washington have developed an artificial intelligence software so powerful that it can put words in the mouth of just about anybody.

A recent, new religion known as "**Way of The Future**," worship a '***godhead***' based on AI! They state that their focus is on:

"the realization, acceptance, and ***worship*** of a ***Godhead*** based on Artificial Intelligence (AI) developed through computer hardware and software."

It includes funding research to help create the ***divine AI*** itself, Figure 32.7.

Fig.32.7: World's religions embracing AI 'God robots capable of performing miracles' – Curtsy Daily Star

The entire fields of science are now racing towards achieving Singularity ***(The merging of man's biological thinking and existence with technology to the point there is no distinction between human and machine)***, scientists are planning it for the future; here's some recent news headlines:

- Former Google and Uber engineer is developing an AI 'god'
- Why humans will happily follow a ROBOT ***messiah***: Religions based on AI will succeed because we tend to 'worship supreme understanding,' claim experts.

- Religion That Worships Artificial Intelligence Wants Machines to Be in Charge of The Planet
- The First Church of Artificial Intelligence is already functioning and officially registered
- The Rise of AI: Give Me That New Time Religion
- The world's next great religion could have a ROBOT/HUMANOID God because humans are pre-programmed to worship things more intelligent than us, Figure 32.8.

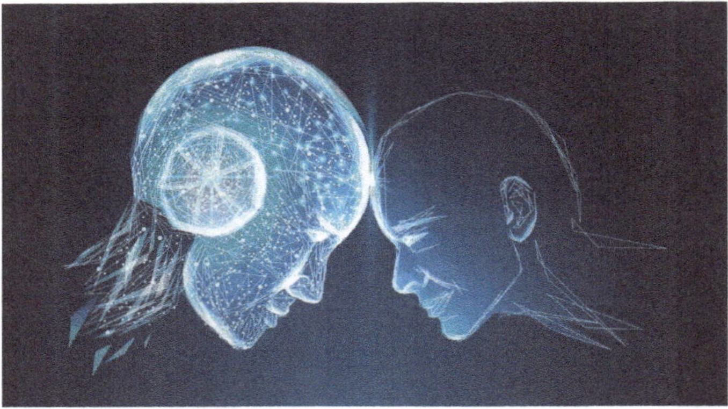

Fig.32.8: The merging of man's biological thinking and existence with technology to the point there is no distinction – Curtsy Linkedin

The concept of the Internet was known in scientific circles for many years before it reached the public. Today the Internet and all the technology surrounding it is used every day by huge numbers of the public. The scientists were working on cloning long before the first sheep was cloned.

The same concept holds true with Singularity, that is now taking place. Humans are already being blended with machines, unaware of by the public. They make AI look beneficial these days, for example, who can question an artificial eye for the blind, or a microchip implanted into the brain to help a quadriplegic?

However, in the end, the blending of human biology with mechanical technology will very likely be used for evil, Figure 32.9.

Fig.32.9: The Apostatic "The Church of AI – California- USA" - Artificial Intelligence Developing a super intelligent creature speaking logically respond to any question

Is this far-fetched? Only time will tell, for now we can just speculate. Is it possible many of the technologies that Revelation speaks of in regard to the Antichrist are technological deception - such as the image that speaks in Revelation 13:15?

ARTIFICIAL INTELLIGENCE AND EVOLUTION

In our information age, the development of computerized technology has been astonishing. Many believe 'the sky is the limit,' especially in the field of artificial intelligence (AI), Figure 32.10. Not surprisingly, it has been the subject of several films.

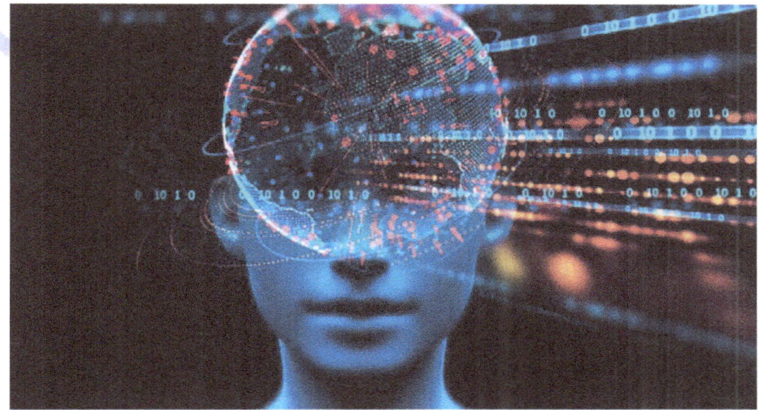

Fig.32.10: *The sky is the limit, especially in the field of Artificial Intelligence*

In the *Terminator* series, defense systems of the future US are handled by *Skynet*, a swarm intelligence. This develops consciousness and decides to wipe out all humanity with nuclear strikes. In a war against humanity, the terminators are machines near perfection.

In *Ex Machina*, an engineer working for a (fictional) world-renowned search engine is invited to an isolated research facility which is also the home of the company's founder. This founder has secretly developed a robot which seems conscious—feelings and all. The engineer gets the task of evaluating the artificial intelligence (AI), but the opposite happens.

In *Transcendence*, a brilliant AI researcher becomes terminally ill but is 'saved' when his consciousness is uploaded into a quantum computer. Without physical limits, this brilliant mind ends up producing a physical copy of his late self, which is really just an extension of the AI.

The list could go on. The public mind is also molded by news headlines about 'how computers beat humans.' It started when IBM's Deep Blue computer beat world chess champion Garry Kasparov in a 1997 match. More recently, in 2017, Google's AI AlphaGo defeated the world No. 1, Ke Jie, 3–0 at Go, a complex ancient Chinese board game. The *New York Times* started their article by declaring: "It's all over for humanity".

Accordingly, is seems only a matter of time before the machines take over!

MUCH TO WORRY ABOUT

Computers—and robots—are composed of physical components called the *hardware*. However, this is not enough. Without something to give instructions—the *software* (also called a 'program')—computers won't operate and robots won't move.

Imagine you have a surveillance camera in the doorway of your home, connected to a computer containing ID information about you and your family. When you approach, the camera identifies you before letting you in with a personal greeting. In addition to all the hardware, it requires purposefully- created software for the actual work of identifying the people and letting them in.

Software does what it is made for—what it is instructed to do—and only that. Even where it can 'learn,' that is only because it has had that capacity programmed into it. The software will blindly follow the *algorithm* (a sequence of instructions for carrying out a task) given to it. This defines how the program would (in our example) calculate the identifying information and how to compare it with your stored profile. Note that software and algorithms are *human intelligence conveyed to a computer.*

ERROR REIGNS

We live in an imperfect world, and a large part of software code deals with various types of error situations. In our example, a mistake in programming could open your door to unknown persons—or not open it for you! And even if it all works right now, will it still do so in a year's time? Ten years from now? As in the biological world, experience and observation affirm decay.

Exposing the surveillance system to changing weather or lightning can cause breakdown or malfunction. If you need to replace your camera in the future, will the software still work with the new one? And so on.

Software is comprehensively tested to catch errors and ensure proper operation. Still, the news unfortunately often reveals problems with different information systems. NASA lost a space probe in 1999 due to an information system failure, and entire airports have been shut down. The Internet abounds with similar examples.

Artificial Intelligence (AI) - Synthetic Intelligence

Artificial Intelligence (or Machine Intelligence) is a term for synthetic 'intelligence.' This includes understanding human speech, playing strategic games (such as chess and GO) against human experts, self-driving cars, and interpreting complex information from videos or photos.

Machine Learning

This often includes what is known as *machine learning*, the ability to learn, e.g., from mistakes, to perform a task better. E.g., Google's AI "AlphaZero" learned chess, go, and shogi (Japanese chess), starting from scratch. That means nothing more than the rules of the games and playing games with itself. Yet in a few days, it supposedly surpassed "AlphaGo" and leading computer 'engines' of the other games.

Machines of Super Intelligence

Therefore, some researchers consider the emergence in machines of super intelligence, above that of humans, to be only a matter of time. However, no matter how advanced the industry becomes, it still boils down to programming by human intelligence. Of course, a garbage-sorting robot, e.g., can *learn* to sort and recognize garbage better. But only if it is pre-programmed to learn this. A robot doesn't dream about retirement on a tropical island—let alone world conquest!

Machine Coups - *Machine Consciousness*

What's behind the idea of machine coups? It's not just the idea of higher intelligence emerging. It also includes the idea that the machines will become aware, i.e., conscious, of what they are doing and why. That is, given enough complex software with the ability to refine itself through machine learning, eventually 'consciousness' will emerge. *Machine consciousness* is a field of technology research to try to find out what is required to achieve consciousness artificially, Figure 32.11.

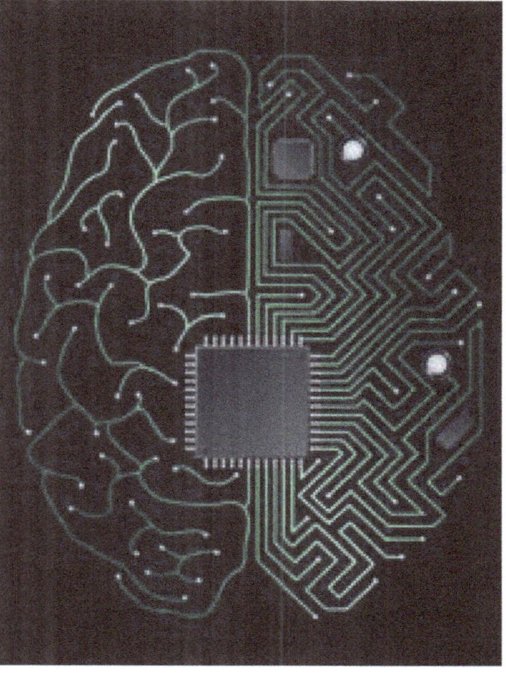

Fig.32.11: *Could computers and robots become conscious – Curtsy - HowStuffWorks*

DRIVING IT ALL: NATURALISM

Underlying all this is the belief that humans (together with our consciousness) came about from natural processes—by themselves, with no guiding intelligence. This was by small gradual changes from some primitive creature which allegedly came from a primordial soup billions of years ago.

Accordingly, everything, including consciousness, *must be able to be reduced to the material*—atoms in various arrangements and motions. That is, there is ultimately nothing 'special' about intelligence, consciousness. So why shouldn't a similar, or even a superior arrangement of matter/energy emerge within machines, especially as they learn to get better at tasks?

MAKING GOD

Similarly, many also believe that with enough advanced technology, one could perfectly duplicate a person, along with their consciousness, memories, experiences etc. All it would need is all the atoms in the right order. The Terasem organization, dedicated to the development of AI, states: "Nobody dies so long as enough information about them is preserved, Figure 32.12." By developing the technology to one day 'upload' our consciousness, declares this organization, **"We are making God."**

Fig.32.12: *The "Terasem" group wants to help transform you into an AI-| Curtsy - Ars Technica*

It's easy to see the connection to the 'promise' given in Eden's Garden: "You will be like God" (Genesis 3:5). All this is only one demonstration of the truth of anti-creationist philosopher Michael Ruse's affirmation: "***Evolution is a Religion***. This was true of evolution in the beginning, and it is true of evolution still today."

Information science, software engineering and AI research are a part of the God-given dominion mandate we have over the creation. Any alleviation of the effects of the Fall through them is a good and God-honoring thing, but not the exaltation of man and the worship of the objects of our creation, in this case '**AI Almighty**.'

MICROBIC MASTERY OVER MICRO-BYTES

In 2012, Stanford University researchers reported they had modelled the 'simplest' bacterium, *Mycoplasma genitalium*, on computers. Using a cluster of 128 computers for the simulation, it took *almost ten hours* to model a single cell division! The bacterium (as an obligate parasite, already a degenerate version of its God-designed precursor) does everything faster, better, and in conditions where computers couldn't operate.

It even makes copies of itself that can make copies of themselves. Human attempts to imitate even this 'simplest' creature are crawling light-years behind the Master Programmer's original.

When we turn to consider a human, the challenges of similar computer modelling expand by orders of magnitude—not to mention the idea that we could even remotely model, let alone create, true consciousness!

Computers are of course good at handling and analyzing large amounts of data, in relatively simple tasks requiring precision and repetition, and their development is ongoing. However, this requires a huge amount of *intelligent design*. Software engineers, find biological systems to be much more ingenious and complex than the most sophisticated human-fiddled software. All creatures are programmed with the stunningly ingenious DNA language, with more and more baffling brilliance discovered as it is studied.

EVEN 'SIMPLE' LIFE IS BEYOND THE REACH OF CHANCE

Numerous software-developers have used their ingenious minds and worked together for years in coordinated, purposeful ways to mimic remotely human-like 'intelligence.' What a contrast with the evolutionary narrative, relying on an aimless, blind process of *error-inducing* mutations plus natural selection. But it asserts that this resulted in humans with consciousness and intelligence.

In reality, *natural selection* requires reproduction, so the '*fittest*' can pass on the 'fit' features. But even the simplest self-reproducing creature requires hundreds of proteins. But the chance of forming *even one* of these proteins by randomly combining its building blocks is incredibly tiny. Astrophysicist Sir "Fred Hoyle" famously compared it to the chance of a solar system full of blind people shuffling Rubik's cubes, all solving it at the same time. It requires an unimaginably strong, blind faith to maintain the evolutionary claim that a blind, purposeless process was the cause of humans, the human brain, intelligence, and consciousness.

SUMMONING THE DEMON

Worshiping Almighty Artificial Intelligence

Fig.32.13: The A.I. Deity That Will Threaten Life as We Know It – Curtsy Getty Images

The "Way of the Future" is the way of the Past!

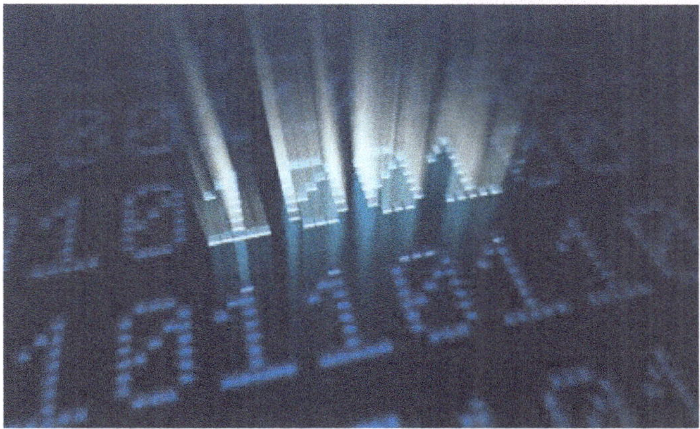

Fig.32.14: Digitizing the Demonic Future

"***And I will declare my judgments against them, for all their evil in forsaking me. They have made offerings to other gods and worshiped the works of their own hands.***" (Jer. 1:16)

Thus spoke the Lord to Jeremiah concerning the cities of Judah who had rebelled against Him and had committed idolatries. It is human nature to worship idols, and in times past that often included literal idols, Figure 32.13.

In the West, due mainly to Christian influence, the practice of worshipping physical idols has been largely abandoned for many centuries, Figure 32.14. But today, the ***Christian foundations of the West are crumbling away*** as more and more people forsake God (just as the Jews did in Jeremiah's day). With this shift away from Christianity, we find another shift—back to the practice of worshiping the works of our own hands, Figure 32.13. This time, one of the forms this may take is the worship of *artificial intelligence (AI).*

Multi-millionaire Silicon Valley engineer "Anthony Levandowski" has made headlines in recent years by announcing a new official ***religion***: the Way of the Future. It is Levandowski's contention that the human race is currently building artificial intelligence that is so powerful it will become like a ***god*** compared to humanity, and will eventually assume control over this planet. This is an event he calls *The Transition.* As he puts it,

*"It's not a **god** in the sense that it makes lightning or causes hurricanes. But if there is something a billion times smarter than the smartest human, what else are you going to call it?"*

Levandowski envisions a world where we become subservient to our new AI master(s), but, he hopes, we will be treated nicely because the AI will fondly remember us as its/their benevolent (*but frail*) elders.

If you think this sounds like the plot of more than one science-fiction movie, you would not be wrong! But Levandowski appears to be dead serious, having *registered this new church with the IRS to receive tax-exempt status.*

Interestingly, the trappings of this new religion all appear to be ripped off from the Bible. For example, note the obviously Christian-sounding nature of this quote from Levandowski:

"The church is how we spread the word, the gospel. If you believe, start a conversation with someone else and help them understand the same things."

Scanning through the extremely rudimentary 'church' website that still exists now, the Way of the Future cult has co-opted all the familiar superficial talking-points one might expect to hear from one of the New Atheists (i.e., Richard Dawkins, Christopher Hitchens, etc.):

"We believe in science (the universe came into existence 13.7 billion years ago and if you can't re-create/ test something it doesn't exist). There is no such thing as 'supernatural' powers."

This is nothing other than the philosophy of positivism and materialism, which embarrassingly fails its own test. Can you perform an experiment to determine if positivism is true? Also note the prominent mention of deep time here.

"Extraordinary claims require extraordinary evidence."

This talking-point is so often abused by atheists that some creation scientists wrote articles specifically addressing this point. What counts as 'extraordinary' depends upon what worldview you already have, making this circular and subjective.

"We believe it may be important for machines to see who is friendly to their cause and who is not. We plan on doing so by keeping track of who has done what (and for how long) to help the peaceful and respectful transition."

This is almost too ridiculous to comment on, but the author(s) appear to be serious here. They think the machines apparently will have some sort of ethic or honor system, whereby they will reward those humans (i.e., these cult members) who have helped them ascend to power the most. This is literally the same mindset shared by the henchmen of all the stereotypical villains in fantasy and sci-fi— "once I help [villain] take over the world, he'll put me in charge of something good!"

"We believe that intelligence is not rooted in biology. While biology has evolved one type of intelligence, there is nothing inherently specific about biology that causes intelligence. Eventually, we will be able to recreate it without using biology and its limitations. From there we will be able to scale it to beyond what we can do using (our) biological limits (such as computing frequency, slowness and accuracy of data copy and communication, etc.)"

Obviously, this is just their statement of faith, but what evidence could they possibly have for these claims? Other than God and angels (whom they deny), what example do we have of 'intelligence not rooted in biology'? Ironically, I believe they are partially correct. Intelligence is not necessarily rooted in biology because it is spiritual! It is the fact that intelligent beings are spiritual beings that enables us to make free decisions in the first place. Consciousness is not an emergent property of matter—it is a function of the spirit. For this reason, I don't believe it is realistic to assume that any form of machine or artificial intelligence will ever become truly self-aware, Figure 32.15.

Fig.32.15: *AI is like 'summoning the demon' - Curtsy Washington Post*

However, as "***Elon Musk***" *said, with AI* "***we are summoning the demon. In all those stories where there's the guy with the pentagram and the holy water, it's like yeah he's sure he can control the demon. Didn't work out.***"

Just how literally and seriously did Musk intend these statements to be taken? Regardless, Levandowski and others believe that if they worship the demon the demon will be kind to them. The real problem is not the development of AI technology itself (which could theoretically be put to good and moral uses), but the unbiblical ideas that *consciousness is merely a function of matter*, and that it would ever be appropriate to *worship* a machine. The Bible says that whenever humans worship idols of their own making, they are really worshiping demons:

" *... what pagans sacrifice they offer to demons and not to God. I do not want you to be participants with demons.*" 1 Corinthians 10:20

What does it take to truly summon a demon?

There are some out there that are actively working towards that goal. We know from the Bible that demons are very real, and they really can inhabit bodies—even the bodies of animals (c.f. Matthew 8:28-34). But whether or not demons could literally inhabit a machine, the worship of anything other than God is truly demonic.

This cult (Way of the Future) may not go anywhere—after all, their leader Levandowski has now been charged with **theft of trade secrets** in a court battle between Google and Uber, and his future status is uncertain. Clearly, though, the cultural stage has been set via the false doctrine of evolutionism and an uncritical absorption of the messaging in many sci-fi movies. Expect more AI-worship cults to follow.

HEAVENLY PROFOUND WISDOM IN HUMAN BODY

Stylized diagram of interconnected nerve cells, with closeup details of the junction (*synapse*) between two nerves, Figure 32.16. The tiny spherical bodies contain transmitter chemicals

In the last century, our society's dependence on electricity and all the devices associated with it has grown phenomenally. How many of us can really imagine what it would be like without electricity? Yet, electricity and devices which harness it have been around since the beginning of creation!

Fig.32.16: *The Miracle of Human Body*

Electricity itself can be defined as the movement or current of small charged particles, usually electrons. Some substances, such as metals and various types of liquids, allow the movement of (or conduct) charged particles better than others. The harnessing of electricity has enabled us to develop devices which cause electrical energy to be changed into some other form of energy—e.g., heat (cooking), light (electric bulbs), motion (electric motors).

Man was not the first to harness electricity and put it to work. When we look at the human body, for example, especially the nervous system, we should conclude that the *designer of the human body* must have had an *intricate knowledge* of electronics and must have known how to harness electrical energy to change it into other forms of energy. When we consider the scale of the operation (i.e., at the atomic and microscopic levels), we can only wonder at God's profound wisdom in creation.

THE NERVOUS SYSTEM

The nervous system is composed of two parts:

- the central nervous system, which is the control center comprising the brain and the spinal cord, and
- the peripheral nervous system, which consists of nerves connecting other parts of the body to the control center.

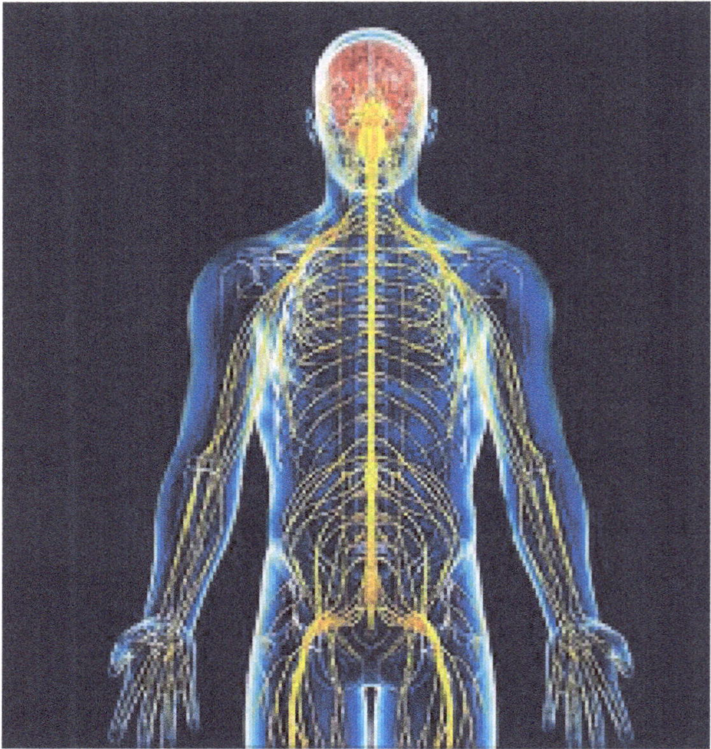

Fig.32.17: Human Nervous System

Via a combination of electrical and chemical processes, the nervous system is used to control the functioning of the entire human body, Figure 32.17.

Scientists inherently acknowledge that the nervous system is built according to an electrical design. The scientific literature describing the nervous system is replete with references to electrical theory and electrical devices that man uses today. Such references include technical words like batteries, transducers, motors, pumps, calculators, transmitters, electrochemical potential, circuitry, binary system, current, resistance, voltage, capacitance, charge. *The difficulty of describing the nervous system without resorting to such language implies the Creator's understanding prior to man's electrical inventions.*

THE BASIC BUILDING BLOCK

The basic building block of the nervous system is the nerve cell, called a neuron. The brain itself consists primarily of neurons. Under a microscope a neuron looks like an octopus with many tentacles. A neuron can transmit an electrical impulse to the next neuron, Figure 32.18.

The network of electrical impulses enables us to receive information from the physical world and then send it to our brains, and vice versa. Without the neuron circuits our bodies would completely shut down, like turning off the power supply to a city.

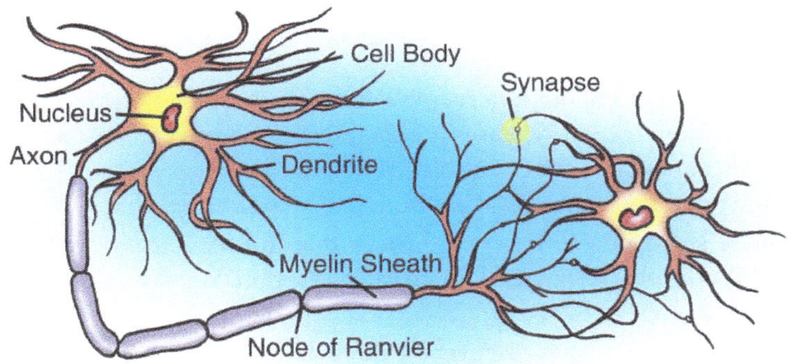

Fig.32.18: A nerve cell body, with one of its extensions becoming the nerve fiber (axon). This communicates with the nerve cell at right via a synapse. The myelin sheath acts like electrical insulation.

One textbook author states, concerning the nervous system, '*We speak of it as the most local circuit, or a microcircuit.* It is very common for a particular type of microcircuit to be repeated throughout a layer or a given cell type, thus acting as a module for a specific type of information processing.' [Emphasis added.]

Information from the physical world to our brain is relayed via our five senses using electrical devices which change one form of energy into electrical energy. Our bodies have *sensory receptor cells* because there are different types of physical stimuli to be changed into electrical signals. For example, a different type of receptor cell is required for hearing stimuli than for smell stimuli.

CREATOR'S POWER AND WISDOM

The neuron may be likened to a switch which is turned either on or off according to the right conditions, Figure 32.19. 'Under normal body conditions, the frequency of [electrical pulse] transmission may range between 10 and 500 impulses per second.' The impulse is not generated unless the neuron has been given a strong enough stimulus. It is hard to imagine the complex integration of electrical signals without realizing the Creator's power and wisdom.

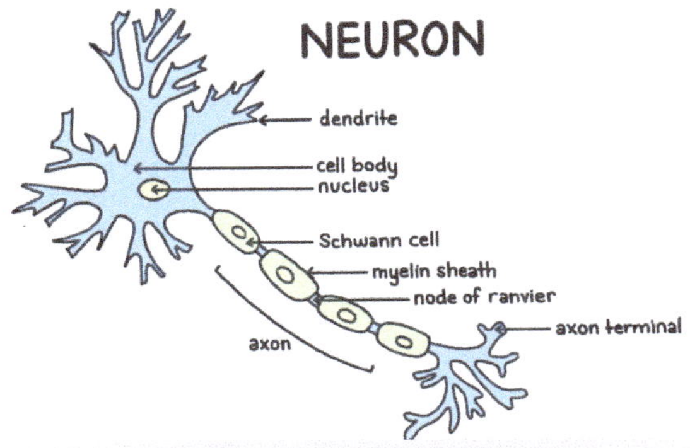

Fig.32.19: Neurons (Nerve Cells) Structure, Function & Types

The individual neuron is only a small component in the interconnected circuitry of the nervous system. Information scientist Dr "Werner Gitt" says, 'If it were possible to describe [the nervous system] as a circuit diagram, [with each neuron] represented by a single pinhead, such a circuit diagram would require an area of several square kilometers ... [it would be] several hundred times more complex than the entire global telephone network.'

To gain a true comprehension of the complexity of this circuitry, we must understand that co-ordination between neurons is essential. The computations required for such co-ordination are enormous. 'There may be from ten trillion to one hundred trillion synapses [i.e., connections between neurons] in the brain, and *each one operates as a tiny calculator that tallies signals arriving as electrical pulses.*' [Emphasis added.] Thus, messages to and from the brain are relayed, moving from one neuron to another.

It is difficult to understand how anyone can believe that the nervous system, particularly the brain, could have been produced by evolutionary randomness and selection. We have barely touched on some of the electrical design present in the rest of the body.

The truth is that scientists are always discovering more about its workings, since its complexity, which far surpasses anything produced by man, is nothing short of a miracle. Truly we can say with David, 'I will praise You, for I am fearfully and wonderfully made; Your works are marvelous and my soul knows it very well' (Psalm 139:14).

HOW DO OUR NERVES TRANSMIT INFORMATION?

A nerve fiber is an extension of a single nerve cell, Figure 32.20. The inside and outside of most of our cells are bathed with fluid containing positively and negatively charged ions (e.g., sodium Na^+; potassium K^+; chloride Cl^-). Using complex biological 'pumps,' the cell's machinery can transport positively charged ions through the (semi) permeable membrane, with the end result being that there is a slight excess of negatively charged ones inside. This means there will be an electric potential across the membrane, so that the inside and outside are like the positive and negative poles on a battery, i.e., it is *polarized* (Figure 32.20).

If something causes the membrane to suddenly become more permeable at one spot, the resulting flow of positive ions back into the cell causes the charge differences to cancel out at this point—i.e., the membrane will become *depolarized* there (Figure 32.21).

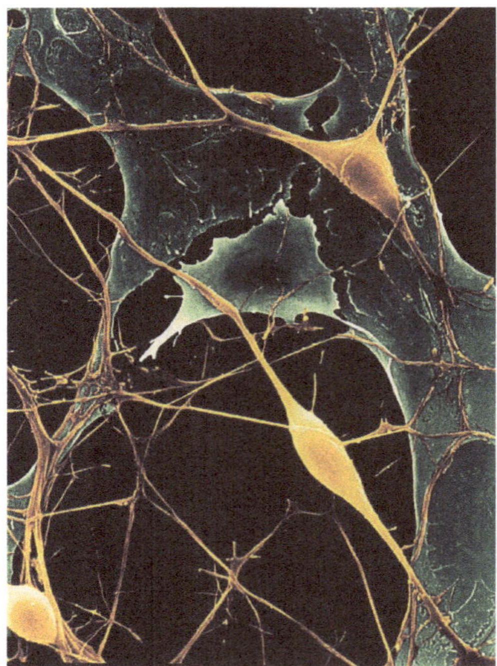

Fig.32.20: *Information Transformation - Pyramidal neurons from human central nervous system. A system of relaying information much more complex than anything man has ever devised.*

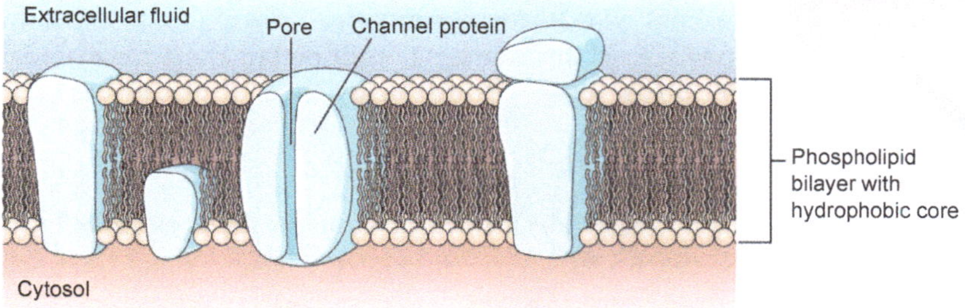

Fig.32.21: Cell Membrane and Transmembrane Proteins The cell membrane is composed of a phospholipid bilayer and has many transmembrane proteins, including different types of channel proteins that serve as ion channels.

This depolarization then spreads sideways, like a wave, along the cell wall, i.e., along the nerve fiber. The message in our nerve fibers is not transmitted by an electric current as such, but by a *wave of depolarization* (Figure 32.22). The cell's biological pumps restore the electric charge to the membrane behind the path of the wave.

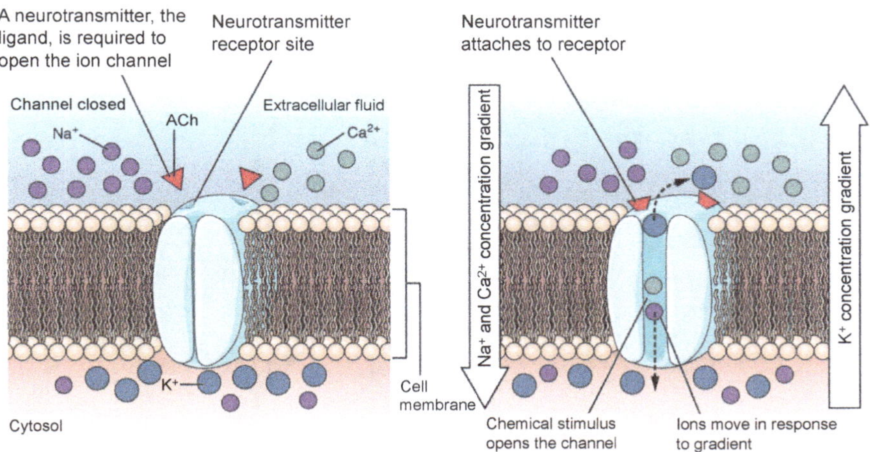

Fig.32.22: Ligand-Gated Channels When the ligand, in this case the neurotransmitter acetylcholine, binds to a specific location on the extracellular surface of the channel protein, the pore opens to allow select ions through. The ions, in this case, are cations of sodium, calcium, and potassium

A number of causes—mechanical or electrical stimuli, or chemical effects—can cause this temporary increase in permeability. Where one nerve fiber A contacts another B at what is called a *synapse*, the arriving wave causes the release of special transmitter chemicals from tiny containers. These chemicals cause depolarization in B at that contact point, so starting a new wave of depolarization going in the same direction. Once released, the transmitter chemicals have to be broken down almost instantly, otherwise B would stay depolarized, and unable to build up charge ready for the next 'firing'.

Organophosphorus insecticides (e.g., malathion) work by preventing this breakdown, thus the insect's nerve cells cease to function properly. Because our nerve fibers use the same transmitter chemicals, malathion is poisonous to humans if exposed to enough of it.

This whole cycle of charge, discharge, chemical release, breakdown, and remanufacture, can happen several hundred times per second. Even with this very simplified description, it is clearly an astonishing process. The information to plan and make all this is stored in code on our DNA, the material of heredity. ***We really are fearfully and wonderfully made***!

GOD'S MAGNUM OPUS

'I will praise thee; for I am fearfully *and* wonderfully made: marvelous *are* thy works; and *that* my soul knoweth right well' (*Psalm 139:14*).

Miraculous Attention to details

We live in an amazing world. The greatest of all creations is man himself, the marvelous machine—precise and efficient. The human body has a dynamic framework of bone and cartilage called the skeleton. The human skeleton is flexible, with hinges and joints that were made to move. But to cut down harmful frictions, such moving parts must be lubricated.

Man-made machines are lubricated only by outside sources. But the body lubricates itself by manufacturing a jelly-like substance in the right amount at every place it is needed. Yes, the *created never-evolved* body is a wonder human machine, despite the defects from genetic copying errors (mutations) that have accumulated since the Fall of man brought on the Curse (Genesis 3), Figure 32.23.

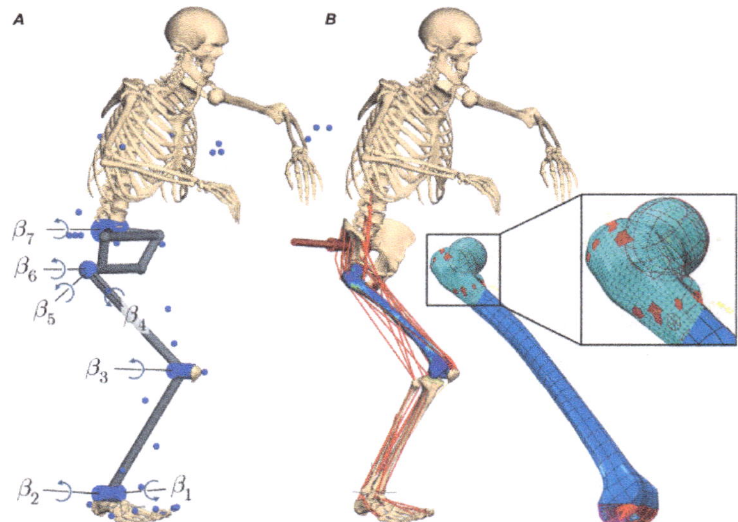

Fig.32.23: *Neuro-musculoskeletal flexible multibody simulation yields a framework for efficient bone structure*

CHEMICAL MANUFACTURING PLANT

Food Manufacturing

The body has a chemical plant far more intricate than any plant that man has ever built. This plant changes the food we eat into living tissue. It causes the growth of flesh, blood, bones and teeth. It even repairs the body when parts are damaged by accident or disease. Power, for work and play, comes from the food we eat, Figure 32.24.

The Digestive System

Mouth
Salivary glands
Esophagus
Liver
Gallbladder
Stomach
Pancreas
Large intestine
Small intestine
Appendix
Rectum
Anus

Fig.32.24: Your Digestive System & How it Works

COOLING AND HEATING SYSTEM

Even in freezing weather our bodies will sometimes overheat. The body's own cooling system then takes over. Drops of perspiration pour from millions of tiny sweat glands in the skin. This is a major way in which our cooling system keeps our temperature down. The human body has an automatic thermostat that takes care of both our heating and cooling systems, keeping body temperature at about 37°C (98.6°F), Figure 32.25.

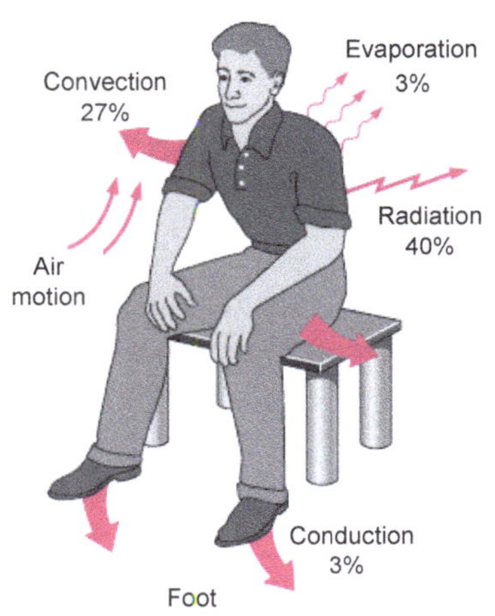

Fig.32.25: Heat Transfer from the Human Body

NOT AI – HEAVENLY COMPLEX COMPUTER SYSTEM

The brain is the center of a complex computer system more wonderful than the greatest one ever built by man; infinitely better than man-AI computerized machine. The body's computer system computes and sends throughout the body billions of bits of information, information that controls every action, right down to the flicker of an eyelid, Figure 32.26.

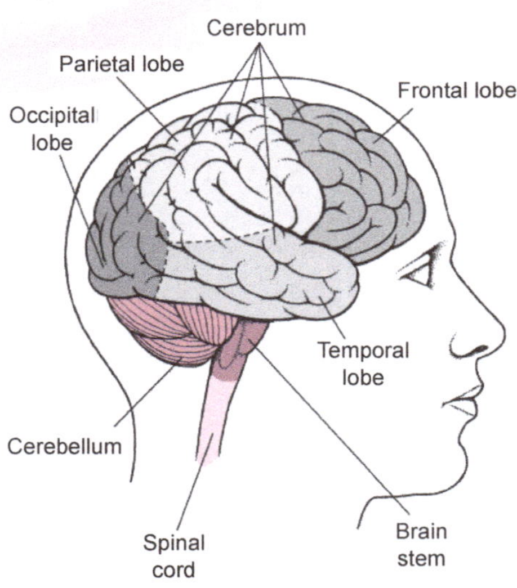

Fig.32.26: Brain - Brain, Spinal Cord, and Nerve Disorders – Curtsy Merck

MIRACULOUS CENTRAL NERVOUS SYSTEM

In most computer systems, the information is carried by wires and electronic parts. In the body, nerves are the wires that carry the information back and forth from the central nervous system. And in just one human brain there is probably more wiring, more electrical circuitry, than in all the computer systems of the world put together, few times over. Yes, it is a wonderful thing—this brain of mankind, Figure 32.27.

THE HUMAN EYE

In fact, as we look at this very moment, we are actually seeing with our brain. Although, of course, the message is carried there from another marvelous structure, the human eye. Modern cameras operate on the same basic principle as our eyes. In our eye the focus and aperture are adjusted automatically, Figure 21.28.

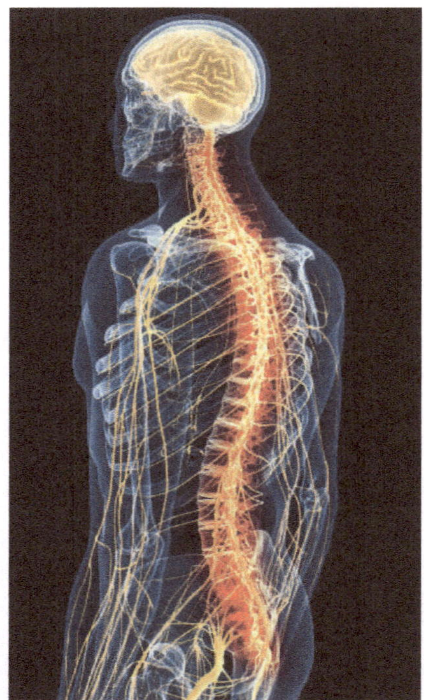

Fig.32.27: Central Nervous System – Curtsy Yale School of Medicine

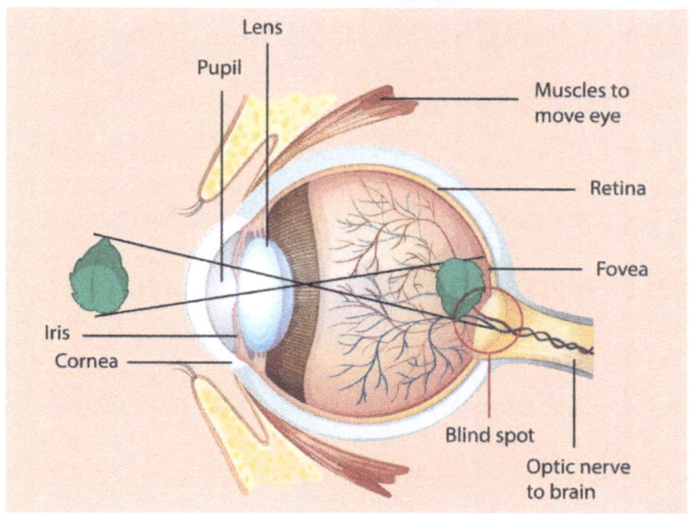

Fig.32.28: Anatomy and Physiology of Human Eye – Curtsy Geeks for Geeks

AUDITORY SOUND WAVES – MYSTERY IN SCIENCE

The sound we hear is being played on a perfect little musical instrument inside our ear. The sound waves go down the auditory canal and are carried by the bones of the middle ear to the cochlea, which is rolled up like a tiny sea shell. The outer ear operates in air. But the cochlea is filled with liquid, and transferring sound waves from air to liquid is one of the most difficult problems known to science. Three tiny bones called the ossicles are just right to do the job that enables us to hear properly. Interestingly, the size of these little bones does not change from the time we are born, Figure 32.29.

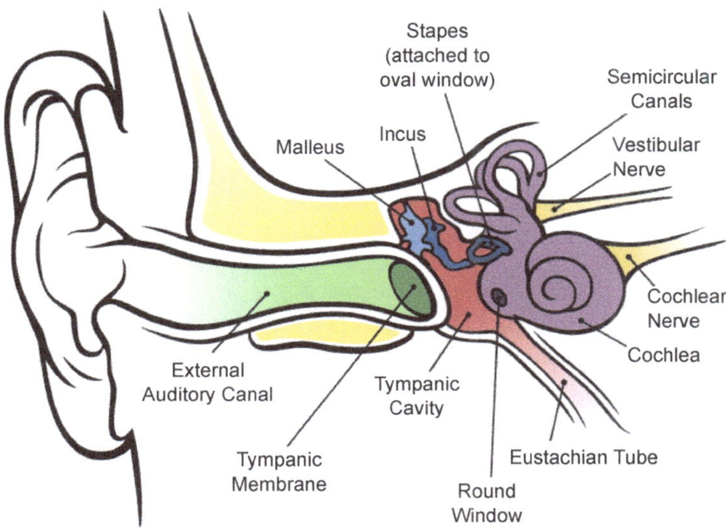

Fig.32.29: The Miraculous Science of Sound

INTRICATELY MANUFACTURED MUSCULAR PUMP

Several Thousand Miles of Blood Vessels

The heart actually is a muscular pump forcing blood through thousands of miles of blood vessels. Blood carries food and oxygen to every part of the body. The heart pumps an average of six liters (1.5 U.S. gallons) of blood every minute, and in one day pumps enough blood to fill more than forty 200-litre (50-gallon) drums.

SKILLFULLY INTELLIGENT MIRACULOUS DESIGNER

Omniscient God Himself

Indeed, the human body is a wonderful machine. The fact that any one of these devices exists is a complete demonstration that they are the work of an intelligent and skillful designer, God Himself. 'So, God created man in His *own* image, in the image of God created He him, male and female created He them' (Genesis 1:27).

DUST OF THE GROUND- PERFECT INPUT OF INTELLIGENCE

The raw material, the basic chemicals in our body, can be found in the 'dust of the ground'. However, these chemicals cannot arrange themselves into cell tissues, organs and systems. This can only happen with an input of intelligence, Figure 32.30.

Fig.32.30: God Created Man from the Dust of the Earth

The book of Genesis teaches that God took 'the dust of the ground', a heap of chemicals, shaped a man and then blew into his nostrils the breath of life. Then man became a living soul. Human beings are different from animals, for 'God created man in his *own* image' (Genesis 1:27). Our bodies have been designed with the ability to pass on to the next generation the programmed information required to form another person from simple chemicals.

We are more than the chemicals that form our body. We are a special creation of God. Man is God's masterpiece—His workmanship, the crown of creation.

The Best in the Universe ...!

Without a doubt, the most complex information-processing system in existence is the human body.

If we take all human information processes together, i.e., conscious ones (language, information-controlled, deliberate voluntary movements) and unconscious ones (information-controlled functions of the organs, hormone system), this involves the processing of 10^{24} bits daily. This astronomically high figure is higher by a factor of 1,000,000 [i.e., is a million times greater] than the total human knowledge of 10^{18} bits stored in all the world's libraries.'

Dr "Werner Gitt," in *Information: The Third Fundamental Quantity*, (reprint from) *Siemens Review*, **56**(6), November/December 1989.

A NEW TWIST ON BLOOD VESSELS

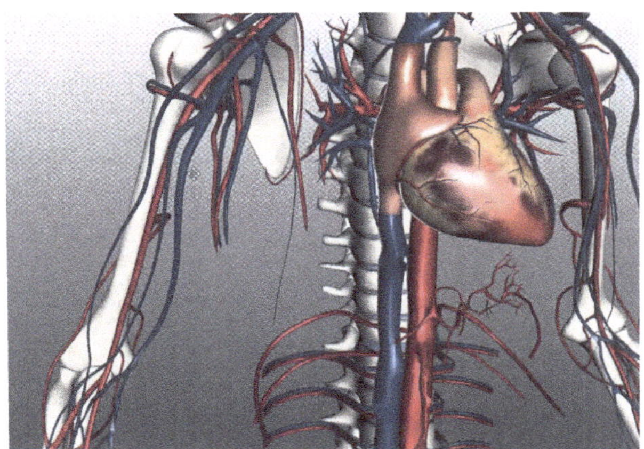

Fig.32.31: *Man Blood Vessels*

Blood vessels are not just straight-through tubes, like water pipes, as was thought. Scientists at Imperial College, London, found that blood vessels have a slight twist to them—they are helical. "Colin Caro" and "Spencer Sherwin" showed that the gentle corkscrewing makes the blood flow more evenly compared to straight vessels, Figure 32.31.

They found that, with helical vessels, damage from turbulent flow was much less likely, especially at T-junctions. Smooth flow also encourages the production of health-promoting protective substances.

This could be very important in bypass surgery where veins from a patient's leg are used to replace sections of clogged arteries around the heart. If surgeons were to give the replacement a slight twist, it could result in a longer time before the vessels clog again.

A SIMPLE WAY OF GOD

Mathematics!

Amazingly - Did you know that mathematics could be racist, and that there is such a thing as 'white math'?

A claim is being made in some US "*deranged academic circles*" that to insist - that there is a correct answer to a statement like '$2 + 2 = x$', is inherently racist, and that mathematics needs to be made antiracist.

An organization funded by the:

"Bill and Melinda Gates Foundation" has published a document entitled, "A Pathway to Equitable Math Instruction Dismantling Racism in Mathematics Instruction."

The document's *Letter to Reader* states, "The framework for deconstructing racism in mathematics offers essential characteristics of *antiracist math educators* and critical approaches to dismantling *white supremacy* in math classrooms by visualizing the toxic characteristics of white supremacy culture ... with respect to math."

Bertrand Russell (1872–1970) a prominent anti-Christian philosopher and mathematician, a professed atheist, and "Sergiu Klainerman," a professor of mathematics at Princeton University, among others, would disagree with the premise that mathematics is inherently racist, Figure 32.32.

Fig.32.32: Bertrand Russell: atheistic mathematical logician

From a Christian perspective, we can add the fact that the basic principles of logic and mathematics cannot be racist nor are they in any way subjective human constructs, because they come from the perfect (Psalm 18:30) and holy (1 Samuel 2:2; Revelation 4:8) mind of God. Let us explore the reality that mathematics comes from the mind of God.

In Genesis 1–2, the narrator (God) uses concepts of mathematics embedded within the account without expounding the conceptual depth which underlies these concepts. For example, the enumeration of the days of Creation Week in Genesis 1 and 2 (Genesis 1:5, 8, 13, 19, 23, 31; Genesis 2:2) may appear to be rather routine. After all, we count from our earliest days—a child counts the number of days until his or her birthday.

Some writers have observed that the use of the ordinal adjective ('second', 'third', etc.) with the word 'day' indicates that God is speaking of standard 24-hour days, and that the use of the cardinal ('one') and ordinals to number the days of creation demands a sequential chronological reading of the text. However, while this is true, they often fail to observe that the use of the ordinals indicates that the *enumeration* of the creation days is of fundamental importance. In addition, Genesis 5 implicitly uses addition, such as in the statement, "When Seth had lived 105 years, he fathered Enosh. Seth lived after he fathered Enosh 807 years and had other sons and daughters. Thus, all the days of Seth were 912 years, and he died." (Genesis 5:6–8)—i.e., 105 + 807 = 912.

Another use of a fundamental mathematical concept in Genesis 1 is the inclusion of entities in sets (e.g., six days of creative activity and "two great lights" [Genesis 1:16]). Sets are based on the abstract concept that assumes the existence of universals and not merely particulars by which each entity is viewed as independent from all others. A universal is the grouping of instances by their consistent characteristics or qualities. For example, when God later spoke to Adam and told him that he could eat from every tree in the garden, but one (Genesis 2:16–17), he used a universal to describe a set and Adam would have understood the concept to include grapes from vines, berries from bushes, and pears and nuts from trees. Universals are usually grouped into three classes:

- types or kinds (e.g., boats, birds, or chairs),
- properties (e.g., heavy, big, or small), and
- relations (e.g., parent, higher, colder).

Since we use universals all the time, we may not understand an inherent difficulty with their existence. The *problem of universals* has perplexed philosophers from the earliest days. The challenge is how to account for their existence: Are they real? Do they exist independent of particulars? Can they exist if the universe is merely the product of random events in a material universe? Are they purely constructs of language?

The correspondence between abstractions—mathematical concepts such as ordinal counting and sets—and phenomena in nature is difficult for materialistic naturalists to explain. If mathematics is purely an invention of human minds, then it is a challenge to explain the correspondence between a mathematical equation and what happens in the natural realm.

The efficacy of mathematics is astounding. With calculations, men and women guided the Apollo missions to the moon, can triangulate on signals generated by 'black' boxes to find a lost airplane at the bottom of the ocean, and can synchronize computers 20,000 kilometers apart so that they receive e-mail messages correctly.

A materialistic naturalist cannot explain *why* pure abstractions—e.g., trigonometric equations, can have such an effect on the physical universe.

THE INTELLIGENT DESIGNER CREATED THE UNIVERSE

The existence of counting and mathematics is evidence that the universe was created by an intelligent designer—the God of the Bible. The laws of mathematics (e.g., commutative, associative, and distributive) come from God's mind. Mathematics is one form of God's thought—others include:

- descriptive (Genesis 1),
- naming (Genesis 1:5),
- animating (Genesis 2:7),
- visualizing (Exodus 25:40; Exodus 26:30), and
- logical (Isaiah 1:18).

THE LAWS OF MATHEMATICS EXIST IN GOD'S MIND

The Perfect Laws

The laws of mathematics exist in God's mind and were not affected by the Fall of man into sin.

The laws of mathematics are therefore perfect (Psalm 19:7) and can be derived without error. Of course, this does not mean that accountants and engineers (and others), affected by the Curse (Genesis 3:17–19), will always perform their arithmetical calculations without error.

Unbelievers can use the mechanics of mathematics without being able to explain what counting is or provide a logically consistent reason for why mathematics works, particularly when applied to natural systems. The problem is that their worldview claims that the universe came into existence by chance and has no non-material dimension (e.g., everlasting human souls).

The fact that materialistic naturalists count and use complex mathematics to accomplish amazing things illustrates that they live in practical terms as if there is a God behind the universe, while denying His existence (Romans 1:18).

Fig.32.33: Leonhard Euler. great creationist mathematician

When we examine the laws of nature (e.g., the Law of Gravitation or Coulomb's Law of Electric Charge), we quickly discover that they are based on mathematics and that they model how God has chosen ordinarily to govern the universe; not how He must govern it—the laws of nature are *de*scriptive, not *pre*scriptive. The fact that nature can be described mathematically has been a continuing surprise to those who cannot think beyond their presuppositions of materialistic naturalism. They have difficulty explaining why there is a connection between physical reality and abstract mathematics. The fact that there is, indicates that the universe is more than the sum of its parts and is the product of intelligent thought.

Attributes of God are reflected in mathematics. For example:

- His *infinity* is hinted at by endless number sequences such as found in π *(pi)*.

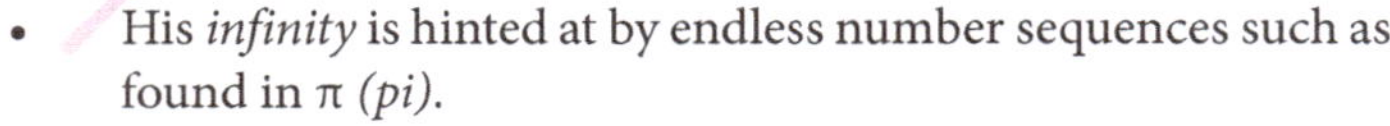

- His *orderly precision* and *logical* nature are displayed in the order and logic of mathematics, a system to which no human invention can compare. For example, the exponent in Newton's Law of Gravitation

$$(F = G\, m_1\, m_2 / r^2)$$

has been studied in numerous experiments and is a perfect 2 (to better than 1 part in a trillion). Also, in every right(-angled) triangle, the square of the longest side (hypotenuse) always equals the sum of the squares of the lengths of the other two sides—as expressed in the Pythagorean Theorem

$$(a^2 + b^2 = c^2).$$

Similarly, his *beauty* can be observed through the formal symmetry and patterns in mathematics. For example, Leonhard Euler, Figure 32.33, developed Euler's equation

$$e^{i\pi} + 1 = 0$$

combines the five most important mathematical constants into one equation.

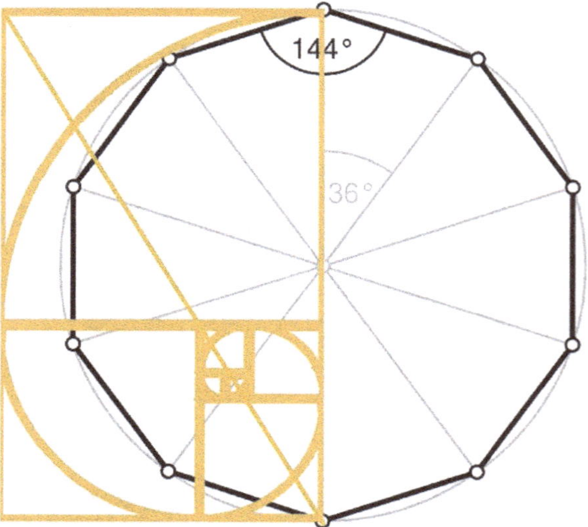

Fig.32.34: 10-sided figure

His *omnipresence* can be seen in the correspondence between nature and mathematics. For example, we can model vocal and instrumental sounds in mathematical terms and reproduce them in our mp3 players using a digital signal processor.

Or the circumference of all circles is $2\pi r$, no matter how small or large the circle is. Both of these would be impossible if mathematics were merely an invention of man's mind.

MATHEMATICIANS DID NOT INVENT MATH

Mathematicians did not invent math. For instance, calculus was not *invented* by Newton or Leibniz, it was *discovered* by them.

Mathematical reality lies outside of mankind and is not invention or art, it is a study of the order God has put into the entire universe.

His *unity* is seen in the system of mathematics which has order, unifying principles, rules that make sense, and an amazing interconnectedness. Clearly a deliberateness is evident in mathematics. For example, π is linked to the set of all odd numbers:

$\pi/4 = 1 - 1/3 + 1/5 - 1/7 + 1/9 - 1/11 + \ldots$ (the *Madhava–Leibniz formula*).

Also, ratios between consecutive numbers in the Fibonacci sequence:

(0, 1, 1, 2, 3, 5, 8, 13, 21 ...) approach the *golden ratio* φ (phi).

This is the number such that the equation:

$\varphi^2 - \varphi - 1 = 0$ gives a φ of 1.618 ...,

which is precisely the ratio of the radius of a circumscribing circle to the side of a regular inscribed decagon (10-sided figure), Figure 32.34. The unity of mathematics has been called 'remarkable', 'surprising', 'unexpected', 'mysterious', 'a miracle', and 'astonishing', by those who have no belief in God.

Fig.32.35: Fibonacci numbers seen in botany

Fibonacci numbers can be seen in botany, Figure 32.35. The arrangement of the whorls on a pine cone (**left**) and the petals of a sunflower (**right**) follows a sequence of Fibonacci numbers.

While God's unity is seen in mathematics, so also is his *diversity* (in a Trinity)—in a single system that can be used to describe human behavior, calculate stresses on a building foundation, predict hurricane behavior, and model the revolution of the planets.

GOD THE COMMUNICATOR

The fact that God is a *communicator* is shown through mathematics' ability to provide complex, specified information such as is evidenced when vocal and instrumental sounds in a musical presentation can be represented in mathematical terms, stored in digital form (zeros and ones), and reproduced again in analog form as puffs of air with high fidelity, using computer chips called digital signal processors.

His *power* is demonstrated through what can be accomplished with mathematics—to synchronize traffic flows, describe musical scores that stir the heart, and define monetary systems that allow modern economies to function.

His *truthfulness* is observed in the constancy of equations such as

$2 + 2 = 4$.

Since all truth is ultimately God's truth, anything which is true reflects God's nature.

God's *consistency* (unchangeableness) is shown from the fact that the results of calculations applying mathematical equations will never change anywhere in the universe. The fact that even atheists believe in the consistency of mathematics is evidenced by advocates of SETI (Search for Extra-Terrestrial Intelligence) who expect that someday we will receive a radio signal from outer space that is based on a regular pattern such as counting by binary numbers.

MATHEMATICS REFLECTS THE PERFECT MIND OF GOD

The Perfect Order of the Universe

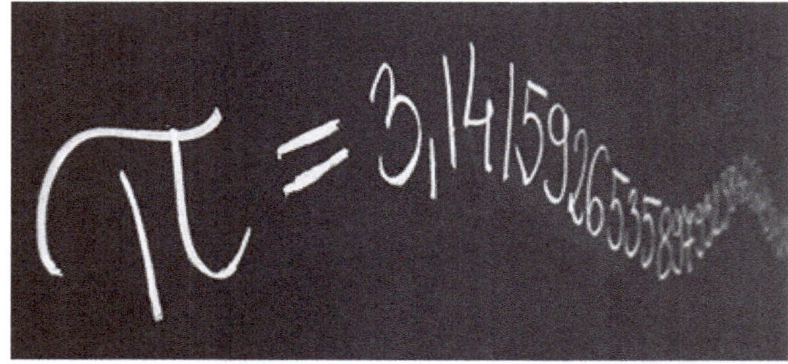

Fig.32.36: Mathematics / Answers in Genesis

Only people with a Christian worldview can understand that the logical, ordered, and good mind of God, Figure 32.36. As one of His forms of thought, God thinks mathematically. So, the existence of mathematics declares to mankind that God is behind the order in the universe. We can count and perform mathematical calculations because we are image-bearers of God, (Genesis 1:26–27) and can think God's thoughts after Him. The notions that there are such things as 'white math' and that mathematics is 'racist' is *utter foolishness* (Romans 1:22), as is any idea that mathematics is a mere human construct or invention. In addition, such an approach to mathematics, which eliminates the demand for precision and objectivity, is a disservice to those who can benefit from learning to apply mathematics correctly.

GOD'S HANDS ON OUTSTANDING GODLY SCIENTIST - I

Johannes Kepler (1571–1630)

Johannes Kepler was born in the town of Weil der Stadt, Germany, on 27 December 1571, Figure 32.37. Johannes was a very small boy who was frequently ill. At the age of three years, he contracted smallpox and lingered close to death for several months. His childhood was also unsettled and unhappy. His father was a mercenary soldier who was away from home for long periods, sometimes years at a time.

When Johannes' mother went away to be with her husband, Johannes was left with his grandfather. The separation from his parents was distressing for Johannes, but God blessed him during these years. His grandfather, a dedicated Christian, encouraged young Johannes as his faith grew. Although poor, Johannes' grandfather appreciated the value of education and sent Johannes to school. The boy's outstanding academic ability soon came to the attention of his teachers.

When Johannes' parents returned after several years, his father, Heinrich, set up business as an innkeeper.

Heinrich was not interested in paying fees to send his son to school. Instead, he saw Johannes as a cheap source of labor in the inn, so Heinrich made his son leave school. However, business at the inn later declined and Johannes' help was not really needed. With his former teachers' encouragement, Johannes successfully obtained a scholarship from the Duke of Württemberg to enable him to continue his schooling. Johannes' drunken father reluctantly allowed him to return to school.

Fig.32.37: *Johannes Kepler maintained his biblical-based faith despite opposition and persecution. Public Domain via Commons*

EDUCATION

Through the Duke's continued generosity, Johannes Kepler was able to begin attending the University of Tübingen in 1587. His studies included Latin, Hebrew, Greek, the Bible, ***mathematics***, and ***astronomy***. Kepler was taught mathematics and astronomy by "Michael Mästlin," one of the few astronomy professors of that time who had accepted Copernicus' idea that the planets, including the earth, revolved around the sun. Almost all scholars of that era still believed that the earth was the center of the solar system.

Kepler obtained his B.A. degree in 1588 and his M.A. degree in 1591. He then continued at Tübingen, studying ***theology***.

During his youth, Kepler had become a committed Christian and dedicated himself to serving God. As he said shortly before he died, he believed "only and alone in the service of Jesus Christ. In Him is all refuge, all solace". Kepler intended to serve God as a Lutheran minister after completing his university education. However, God had other plans for this uniquely gifted young man.

In 1594, Kepler was asked to go to the Lutheran high school in Graz, Austria, to replace the mathematics teacher who had just died. Although close to finishing his theological training, Kepler felt led by God to take up this teaching position.

ASTRONOMY AND ASTROLOGY

As well as teaching mathematics in Graz, Kepler became district mathematician. This position involved surveying land, settling disputes over the accuracy of weights and measures used in business, and calendar-making. In addition to listing dates, calendars today frequently include information on public holidays, school holidays, and phases of the moon (full moon, new moon, etc.).

Some calendars even include the dates of sporting events, social services payment days, and the like. Similarly, in Kepler's time, calendars were expected to include information which was useful to people's everyday lives. The information given included advice to farmers on when to plant and harvest crops, advice to leaders on military campaigns, advice on matters of romance, etc.

Today we understand how the relative positions of the sun, moon, and planets, together with the tilt of the earth on its axis, combine to determine the seasons of the year, the phases of the moon, tides, eclipses of the sun and moon, and so on, Figure 32.36. These occurrences have scientific implications for agriculture, fishing, military planning, and other things. (Even in modern times, some military offensives are timed to fit in with seasons and moonlight.)

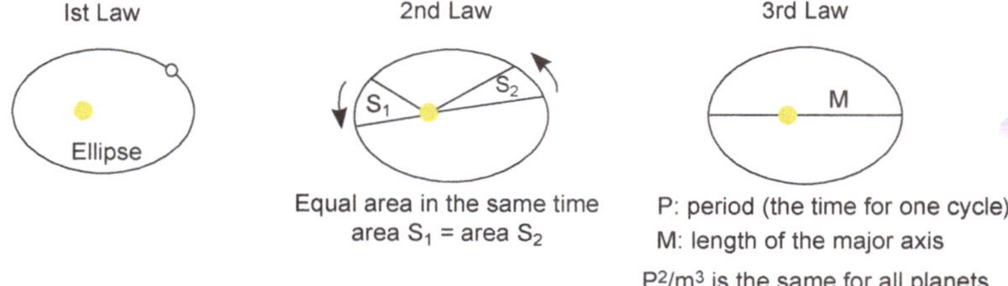

Fig.32.38: Kepler's first, Second, and Third law of planetary motion- Curtsy University of Waikato: Each planet moves about the sun in an orbit that is an ellipse. The sun is at one focus of the ellipse.

DISTINCTION BETWEEN ASTRONOMY AND ASTROLOGY

With such present-day knowledge, legitimate conclusions based on the science of astronomy can be distinguished from unfounded claims based on astrology. However, in Kepler's day, there was considerable confusion both in the general community and in universities regarding the distinction between astronomy and astrology. With their limited knowledge of the movements of heavenly bodies, scientists were unsure, which events on earth were affected by events observed in the heavens and which were not, Figure 32.39.

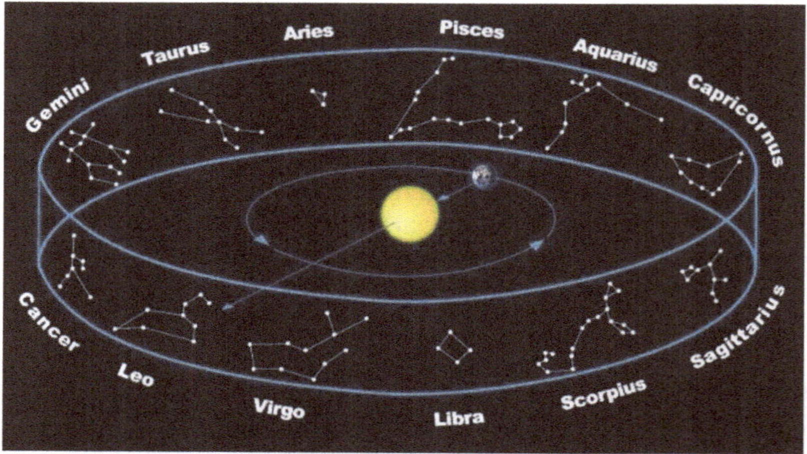

Fig.32.39: Astronomer explains why you have got your zodiac sign wrong. Curtsy PNGGuru (CC BY-NC)- Astrology suggests that each sign of the zodiac fits neatly into a 30-degree slice of sky – which multiplied by 12 adds up to 360 degrees. In actuality, this is not the case, as the constellations vary a great deal in shape and size. For example, the Sun passes through the constellation Scorpio in just five days, but takes 38 days to pass through Taurus. This is one of the reasons astrological signs do not line up with the constellations of the zodiac.

REJECTING ASTROLOGY

Kepler continued making calendars. However, he determined that he would subsequently check the accuracy of his predictions in order to sort those which were legitimate from those which were not. As part of this process, Kepler published a book in 1601 which "rejected the superstitious view that the stars guide the lives of human beings."

Kepler progressively rejected other aspects of astrology as well, Figure 32.40. In his biography of Kepler, "J. H. Tiner" points out that, "Johannes was the first scientist to investigate the long-term accuracy of astrology. His records showed that trusting in astrology could be a risky business."

Fig.32.40: Disastrology - How To Mess Up Your Life With Astrology

MOTION OF THE PLANETS

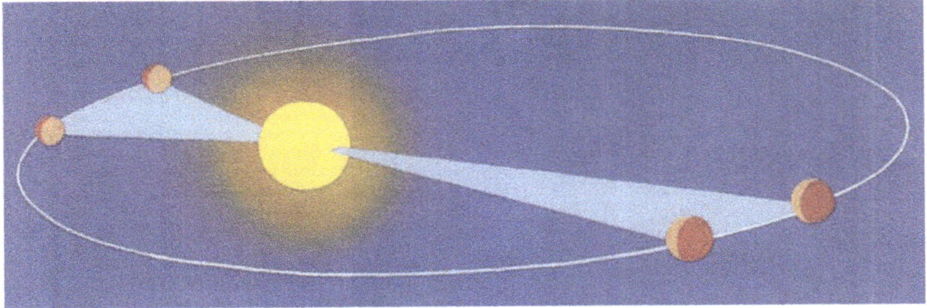

Fig.32.41: Kepler's second law of planetary motion

The straight line joining a planet with the sun sweeps out equal areas in equal amounts of time, Figure 32.41. (This means that the planet travels faster when it is closer to the sun, **Table 32.1**.)

Table 32.1 : Planet Travels Faster when it is Closer to the Sun

Planet	Av. Distance from sun, *d* (x 10^6 km)	Time (T) taken to revolve around sun	d^3/T^2 = constant (km^3/yr^2 x 10^{24})
Mercury	57.9	0.241	3.34
Venus	108.2	0.615	3.35
Earth	149.6	1.00	3.35
Mars	227.9	1.88	3.35
Jupiter	778.3	11.9	3.35
Saturn	1427	29.5	3.34
Uranus	2870	84.0	3.35
Neptune	4497	165	3.34
Pluto	5900	248	3.33

CREATOR LOGICAL PATTERN

Kepler strongly believed that "The world of nature, the world of man, the world of God—all three fit together." In particular, Kepler reasoned that because the universe was designed by an intelligent Creator, it should function according to some logical pattern. To him, the idea of a chaotic universe was inconsistent with God's wisdom. In contrast, many other scientists had given up searching for a simple logical pattern, Figure 32.42.

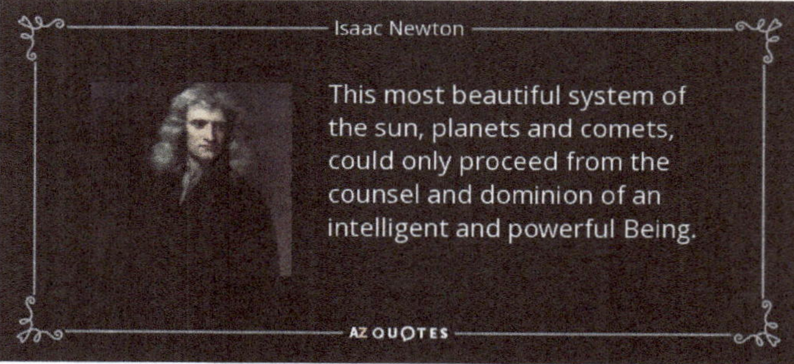

Fig.32.42: "The world of nature, the world of man, the world of God—all three fit together."

PHILOSOPHIES AND MATHEMATICS OF ANCIENT GREEKS

Without access to accurate data on the positions of the planets over a period of time, Kepler based his early attempts to discover the pattern behind the motion of the planets on the philosophies and mathematics of the ancient Greeks. He proposed his ideas in a book called *Cosmic Mystery* written in 1595. Although many of his ideas later proved to be incorrect (as is often the case in science), publication of this book brought Kepler to the attention of the outstanding Danish astronomer, "Tycho Brahe," Figure 32.43.

Fig.32.43: Tycho Brahe. Public Domain via Commons

SCIENTIFIC PLANETARY MOTION

Tycho Brahe was so impressed with Kepler's mathematical ability and keenness to apply mathematics to astronomy that he invited Kepler to join his team of astronomers. These astronomers had charted the paths of the planets across the sky for many years but could not make any sense of the complicated paths that they saw. In 1600, Kepler joined Tycho Brahe at his observatory in Prague. Kepler was given the task of investigating the orbit of Mars. At last, he had access to the data he needed to really attack the problem of planetary motion scientifically.

The idea that the paths of the planets must be either circles or combinations of circles was almost universally accepted in Kepler's time. However, Kepler found that even complex combinations of circles simply did not work. Turning away from popular thinking, Kepler "tried noncircular paths until he found the true solution: *Mars revolves in an elliptical orbit* with the Sun occupying one of its focuses."

KEPLER'S TWO LAWS OF PLANETARY MOTION

Kepler further showed that a planet does not move an equal distance in an equal amount of time (i.e., at a constant speed) as was previously thought. Instead, he was able to show that the imaginary line joining the sun to the planet sweeps through equal areas of the ellipse in equal amounts of time. This means that the planet travels faster when it is closer to the sun, and slower when it is further away from the sun. Kepler published these first two laws of planetary motion in 1609 in a book entitled *The New Astronomy*.

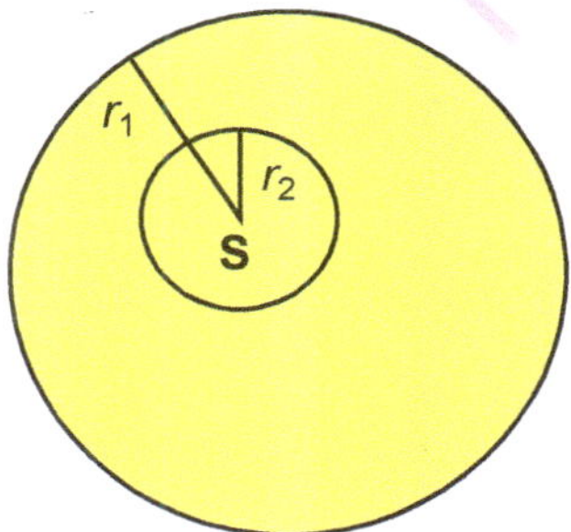

Fig.32.44: Kepler's third law of planetary motion

KEPLER'S THIRD LAW OF PLANETARY MOTION

The squares of the periods (p) of any two planets' revolutions are in the same ratio as the cubes of their mean distances from the sun (r). That is,

$$r_1^3 / r_2^3 = p_1^2 / p_2^2$$

Ten years later, Kepler established his third principle of planetary motion, which mathematically related the time a planet takes to complete an orbit of the sun and the average distance of that planet away from the sun, Figure 32.44. This principle was published in *Harmony of the Worlds* in 1619. In this book, Kepler also *praised God*, saying, "***Great is God our Lord, great is His power and there is no end to His wisdom***".

Kepler's Christian faith had led him to a pattern of thinking which had eventually enabled him to solve the riddle of planetary motion where so many other scientists had given up trying. Kepler had sought and found a simple logical pattern for planetary motion which reflected God's wisdom. As Kepler said:

"We see how God, like a human architect, approached the founding of the world according to order and rule and measured everything in such a manner."

ADDITIONAL VITAL DISCOVERIES

Kepler's laws of planetary motion were his greatest contribution to science. These laws had an enormous impact on scientific thinking, providing the groundwork for Sir Isaac Newton's later work on universal gravitation.

However, Kepler made many other contributions to science as well.

- He discovered a new star (a supernova);

- he analyzed ***how the human eye works,*** Figure 32.45.
- he made improvements to the telescope, and
- made other contributions in the field of optics.
- He published accurate data on the positions of stars and planets which were of immense value to navigators.
- He made various contributions to mathematics, including faster methods of calculation, and investigated the volume of many solid bodies.

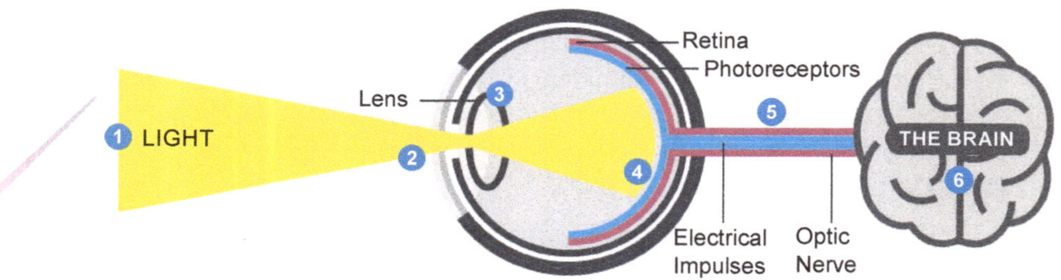

Fig.32.45: Kepler analyzed how the human eye works

Light reflects off objects and travels in a straight line to your eye. Light passes through the cornea, into the pupil and through the lens. The cornea and lens bend (refract) the light to focus on the retina. Photoreceptors on the retina convert the light into electrical impulses. The electrical impulses pass along the optic nerve to the brain. The brain processes the signals to create an image.

THE ROLE OF THE EYES

Your eyes play a crucial role in almost everything you do. Here are some of the main functions of the eye.

Seeing - Eyes take in light and convert it into electrical impulses that are sent to the brain, which processes these signals to form the images we see.

Moving - The six 'extraocular' muscles control the movement of the eye. Four move the eyeball up, down, left and right; two adjust the eyes to counterbalance head movement.

Blinking - Every time you blink, a salty secretion (basal tears) from your tear gland is swept over the surface of the eye, keeping your eyeballs moist and clean. This action is very vivid when we travel in dusty / windy weathers.

Crying - Tears are salty fluid containing protein, water, mucus and oil – are released from the lacrimal gland in the upper, outer region of the eye. Reflex tears protect the eye from irritants like smoke, dust and wind. Emotional tears are a response to sadness or joy – there's a theory that 'a good cry' can help the body get rid of toxins and waste products.

Protecting - The eyes are set in sockets in the skull to protect them from injury. Eyelashes and eyelids keep out dust and dirt. Eyebrows are arched in shape to divert sweat away from your eyes.

Kepler is recognized as one of the founders of modern science. "In his three books, *Cosmic Mystery*, *The New Astronomy*, and *Harmony of the Worlds*, he began the process that eventually replaced superstition with reason."

ORIGIN OF THE UNIVERSE

Kepler also spent time investigating the dating of historical events in the Bible, including the birth of Jesus. In addition, he wrote a story called *The Dream* which is credited as being the first modern science-fiction story.

Life of Tragedy

Johannes Kepler died after an acute illness in Regensberg, Germany, on 15 November 1630, aged 58 years. Kepler's life had been filled with tragedy. The unhappiness and illness of his childhood were followed in adult life by ***the deaths of three of his six children*** during their childhood, ***the death of his first wife***, and ***repeated religious persecution***.

A Man of the Bible

Kepler lived in an era in which most rulers expected the people to adopt the religious beliefs held by the ruler. However, he refused to change his beliefs with a change of ruler. Kepler was a man of the Bible and refused to accept man-made rules which he believed contradicted the Bible.

Unfortunately, this stand caused him to suffer great persecution on a number of occasions.

KEPLER'S DEFENDED HIS MOTHER – SAVED HER LIFE

Another traumatic event in Kepler's life was the trial of his *superstitious mother* who was accused of *witchcraft*. Had she been convicted; she would have been tortured and burnt at the stake. It was only Kepler's skillful defense of his mother that saved her.

Throughout all these trials, *Kepler kept his unshakable faith in God*. He summarized his faith by saying simply that

"*I am a Christian.*"

Despite his great achievements, he remained humble. His desire was to

"*Let my name perish if only the name of God the Father is thereby elevated.*"

ACKNOWLEDGING GOD

He acknowledged God as

"*the kind Creator who brought forth nature out of nothing.*"

Kepler was prepared to put aside the plans he had made for his life, and to humbly follow God's leading.

As a result, he was able to say in later life that,

"*I had the intention of becoming a theologian ... but now I see how God is, by my endeavors, also glorified in astronomy, for 'the heavens declare the glory of God'.*"

GOD'S HANDS ON OUTSTANDING GODLY SCIENTIST -II

John Ray (1627- 1679)

John Ray, Figure 32.46, was born in 1627—the same year "Robert Boyle" was born and "Francis Bacon" died.

Boyle was the founder of modern chemistry, and Bacon was the originator of the modern scientific method which combined experimentation with rational argument. In his book, *The Founders of British*

Science, "J.G. Crowther" describes the relationship among the work of these three committed Christians as follows:

'The work of recording and classifying the contents of nature, which, as Bacon had indicated, was the first step in creating a modern universal science, was led in chemistry by Boyle. In biology the comparable work was carried out by John Ray.'

In the preface to his book on British plants published in 1690, Ray expressed gratitude to God that he was born in this era, when Bacon's idea of experimentation and logic combined had replaced the previous philosophy of logical argument alone (i.e., without a starting point in the real world) which was a leftover from ancient Greek thinking.

Fig.32.46: John Ray devoted his life to studying the natural world, particularly plants.

EARLY LIFE

John Ray was born on November 29, 1627, in the English village of Black Notley in Essex. He was the youngest son of the village blacksmith. His mother collected and used herbs for medicinal purposes, and she was well known in the area for her care and compassion for those who were ill. John's interest in plants began in his early childhood while collecting herbs with his mother.

John received some basic education in his home village. It appears that the local vicar recognized that he had academic potential and arranged for him to attend grammar school in nearby Braintree. Again, his potential was recognized and, at the age of 16, he entered Trinity College at Cambridge University.

Unlike most of the other Cambridge students, who were the sons of rich gentlemen, John Ray was poor and unable to pay the necessary fees. As with Isaac Newton subsequently, Ray was exempted from fees by acting as a *servant* for the staff of Trinity College. He obtained his master's degree in 1651, and then stayed on at Trinity college as a lecturer.

CHRISTIAN FAITH

Ray was known as a *modest and sensitive man* who was deeply committed to his Christian faith. He sought to *turn the thoughts of his students from materialism to God*. Crowther states:

'Ray suggested that, instead of devoting themselves to games and dissipations, they would gain more satisfaction from the contemplation of the wisdom and goodness of God, as demonstrated in the exquisite works of nature.'

Although botany and zoology were not yet accepted as subjects in the university, Ray invited his students to join him in their own time in researching the local flora and fauna. Ray published his first book on botany—a catalogue of the plants around Cambridge—in 1660.

Ray had always intended becoming an ordained clergyman, but unfortunately, he lived in an era when there was considerable interference by the State in the affairs of the church.

When the Royalists returned to power, they insisted on re-establishing extreme ritual in the church. However, Ray '*distrusted a religion that depended so much on external practices. He gave far more weight to inner spiritual institutions and the refinements of conscience*.'

Ray would not compromise his beliefs and he was forced to leave the university in 1662.

Ray was now unable to support himself and his widowed mother. However, "Francis Willoughby" (1635–72), a wealthy former student of Ray's, suggested that they should undertake scientific research together at

Willoughby's expense. Ray's dismissal actually became a turning point in his life that enabled his great work in biology to be accomplished. As we read in Romans 8:28— *'And we know that all things work together for good to them that love God, to them who are the called according to his purpose.'*

RAY'S SCIENTIFIC CONTRIBUTIONS

For 10 years, Ray and Willoughby traveled around the British Isles and the rest of Europe, making observations, and collecting specimens. While Willoughby focused on animals, Ray focused on plants. Their intention was to document living organisms and classify them into a limited number of classes.

In 1670, Ray published a catalogue of plants of the British Isles. 'It contained a long section on the medicinal use of plants, ***which denounces astrology, alchemy, and witchcraft, and is ruthless in its demands for evidence.***'

In recognition of his work so far, Ray was made a Fellow of the Royal Society, a prestigious group of England's greatest scientists. Later, when asked to become secretary of the Royal Society, Ray declined as he would have had less time to devote to his work on biology.

Although Willoughby died in 1672, Ray was able to continue their great project with funds generously provided for this purpose in Willoughby's will. As well as proceeding with his own work on plants, Ray undertook the task of analyzing, classifying, and interpreting Willoughby's vast collection of data on animals.

Ray published a book on birds in 1676 and a book on fish in 1685. The book on birds ***'laid the foundations of scientific ornithology.'***

Ray then undertook the documentation and classification of the plants of Europe and beyond into a generalized catalogue of the plant kingdom. He classified flowering plants according to whether they had two seed leaves or only one.

'This is the division into dicotyledons and monocotyledons which all subsequent botanists have adopted.'

DEFINING SPECIES ACCORDING TO ITS KIND

It was Ray who specifically defined what was meant by a species. "Linnaeus" (another committed Christian) subsequently used Ray's species as the building blocks of his classification system in which related species were grouped into genera, related genera were grouped into higher categories, and so on, Figure 32.47.

Fig.32.47: Species According to its Kind - Answers in Genesis

Thus, Ray's generalized catalogue of plants 'laid the groundwork for systematic classification, which was to be brought into modern form by Linnaeus.'

In his 1693 book on reptiles and mammals, Ray classified animals with respect to teeth, hooves and toes. This logical classification of animals was a vast improvement over the previous system, Figure 32.48.

Ray's insights were ahead of his time in other areas as well. As "*Asimov*" states:

'He declared fossils were the petrified remains of extinct creatures. This was not accepted by biologists generally until a century later.

Fig.32.48: *John Ray 1693 book - Synopsis Methodica Animalium Quadrupedum et Serpentini Generis.*

OPPOSITION TO EVOLUTIONARY IDEAS

Ray saw no conflict between his Christian beliefs and his scientific work. In fact, he believed that scientific investigation:

'was a proper exercise of man's faculties and a legitimate field of Christian inquiry.'

Ray adamantly believed that all things—the heavens, the earth, and living organisms—were created by an infinitely wise and loving God. He believed that the infinite detail of the structure and function of living organisms was clear evidence of Divine wisdom. He expressed this in two widely read books. The second of these, entitled ***The Wisdom of God Manifested in the Works of the Creation***, became a classic. He also wrote a theological book entitled ***Persuasive to a Holy Life***.

Ray wrote a paper for the Royal Society completely opposing spontaneous generation—the idea that *life can arise spontaneously from non-living matter.* Ray quoted experiments by "Francesco Redi," which contradicted spontaneous generation. Ray said that:

'My observation and affirmation is that there is no such thing in nature'

and he referred to spontaneous generation as 'the atheist's fictitious and ridiculous account of the first production of mankind and other animals.'

Ray never doubted that the plants and animals were created by God in distinct kinds, but *at one stage he questioned whether one species could change into another*. However, after extensive scientific investigation of plants and animals, *he concluded that species had remained distinct*.

Ray concluded that the plants and animals were:

'the works created by God at first, and by Him conserved to this day in the same state and condition in which they were first made.'

Unfortunately, this statement, understood as 'fixity of species,' played into the hands of the evolutionists.

Darwin observed different 'species' of finches on the Galápagos Islands and drew the reasonable conclusion that they had descended from a small number of finches which had arrived on the islands. Therefore, species were not 'fixed,' and creationists were wrong. However, Ray would probably have seen that Darwin's finches were just varieties within a created kind, and no more proved evolution than does variety within dogs.

HIS LATER YEARS

John Ray had married at the age of 44 years. When his mother died in 1679, he and his wife moved into the house that he had built for his mother in Black Notley. In his later years, his four young daughters were of great assistance to him in collecting and breeding specimens.

Despite his success and his fame, *Ray maintained his modesty and interacted easily with rich and poor alike*. He remained living and working in Black Notley until his death on January 17, 1705.

Describing Ray after his death, a friend wrote that you would find—

'*in his dealings, no man more strictly just; in his conversation, no man more humble, courteous, and affable; towards God, no man more devout; and towards the poor and distressed, no man more compassionate and charitable.*'

During his long and fruitful life, Ray had laid the foundations for both botany and zoology, as well as maintaining a consistent Christian witness through both his behavior and his defense of biblical teaching.

CATASTROPHE, CURSE AND JUDGMENT

Fig.32.49: *Catastrophe, Curse and Judgment*

All beings are created for the purpose to Honor and Worship the Creator. However, the main purpose of the creation/evolution debate is that if evolutionists can show that atheistic evolution is true, they will have eliminated the last great Day of Judgment. However, there have already been three occasions when the whole of sinful mankind has been judged by God. The fact that these three events have occurred testifies to the certainty of the fourth and final judgment to come soon, Figure 32.49.

I. The Fall

The first universal judgment was at the event known as the Fall, described in Genesis 3. The reason was the deliberate disobedience of Adam and Eve to the specific instruction of God that they were not to eat from the tree in the Garden of Eden called the tree of the knowledge of good and evil (Genesis 2:17). When Adam and Eve disobeyed God, the effects for them included

- knowledge of sin,
- experience of guilt,
- loss of holiness,
- fear of God,
- forfeiture of fellowship with God, and
- feelings of embarrassment towards each other.

And, above and beyond these consequential effects, there was the specific pronouncement of judgment by God, sometimes called

'the curse'.

This contained both spiritual and physical dimensions, not only for Adam and Eve, but also for their descendants, that is, for all mankind. Spiritually, they were from then on cut off from God, so that on that day they died spiritually. Physically, *their human bodies began to die* and would eventually 'return to dust' and decayed, Eve was told that:

- childbearing would be a matter of 'sorrow' and that
- her husband would rule over her,
- the ground was cursed so that Adam's work from then on would be hard labor, and
- they were cast out of the Garden of Eden (Genesis 3:16–24).

Please observe that God's mercy is seen in the midst of this judgment, in two aspects:

1. There was the promise of Genesis 3:15, that the seed of the woman (Jesus Christ) would 'bruise the head' of the serpent (i.e., cause the ultimate downfall of Satan).
2. The penalty of physical death was itself a mercy! Without physical death mankind would have been condemned to live forever in the sorrowful results of rebellion against God.

However, because of the death of Christ on the Cross substituting for our sins, and the resurrection, those who repent of their sins and have faith towards God, after death will be united with God in Heaven and attain that state of holiness and fellowship with God, which our ancestors, Adam, and Eve, lost for us in the Garden of Eden.

II. The Flood

The second universal judgment was at the catastrophe of the Flood, Figure 32.50, described in Genesis 6–9. The reasons for this judgment were that 'the wickedness of man was great,' 'every imagination of the thoughts of his heart was only evil continually,' 'all flesh had corrupted his way upon the earth,' and 'the earth was filled with violence' (Genesis 6:5–13), Figure 32.48. In this judgment, all of sinful mankind (apart from one family) died, as did all the air-breathing land animals except for those on the Ark (Genesis 7:21-23).

Fig.32.50: *The Flood of Noah*

God's mercy is seen in that Noah, described as a just man and perfect in his generations, and righteous (Genesis 6:9 and 7:1), was saved, together with his wife and family (a total of eight persons) by means of the Ark, which also carried the animals nominated by God to replenish the earth.

III. Babel

The third universal judgment was the confusion of languages at Babel, described in Genesis 11. After the Flood, God commanded Noah's sons to 'replenish [Hebrew: "fill"] the earth', that is, to spread out, to form many nations, and to inhabit all the lands (Genesis 9:1), Figure 32.51.

When the population had increased to the point where there were enough family units to have begun to do this, instead they built for themselves a city and within it a tower (Genesis 11:1-4).

Fig.32.51: Confusion of Language at Babel – Curtsy "The Tower of Babel by Pieter Bruegel the Elder"

Three reasons are given in the Bible for this building program.

i. It was to be a 'tower unto heaven' (the words 'may reach' or 'to reach' are not in the original Hebrew), and so it appears that one use to which it was put involved the worship of heaven (possibly of angels, leading to demonic involvement; or of stars, leading to astrology and occult practices).

ii. It was to make a name for themselves, and to glorify human achievement, thereby showing their pride and self-sufficiency.

iii. It expressed a human attempt to preserve a unity which had been lost through sin, and which was contrary to God's earlier command to Noah's sons to replenish or fill the earth ('lest we be scattered abroad upon the face of the earth'—Genesis 11:4).

God's judgment on this act of rebellion—the confounding of their language—showed both His mercy and His wisdom. God's mercy is seen in the fact that He did not make the penalty equal to the offence. God's wisdom is seen in the fact that the miracle of confusing their language effectively put a halt to their threefold aims:

- Without intercommunication, they could no longer continue their building program.
- The tower, which was meant to be a symbol of the great name they sought for themselves, became instead a term of reproach—Babel, meaning 'confusion'.
- They were forced to do the very thing they had previously refused to do—scatter over the earth.

IV. The Great White Throne

The fourth universal judgment is what the Bible calls the Great White Throne Judgment, Figure 32.50. Our knowledge of this comes from Revelation 20:11–15, which reads:

Fig.32.52: *The Great White Throne of Judgment*

'And I saw a great white throne, and him that sat on it, from whose face the earth and the heaven fled away; and there was found no place for them. And I saw the dead, small and great, stand before God: and the books were opened: and another book was opened, which is the book of life: and the dead were judged out of those things which were written in the books, according to their works. And the sea gave up the dead which were in it; and death and hell delivered up the dead which were in them; and they were judged every man according to their works. And death and hell were cast into the lake of fire. This is the second death. And whosoever was not found written in the book of life was cast into the lake of fire.'

This judgment is described as in the form of a law court, with a presiding judge, and evidence is subpoenaed in the form of 'those things which are written in the 'books.' It is relevant to ask by what authority this tribunal is held, Figure 32.52, who the judge will be, and what his qualifications are to make him judge of all mankind. Consider:

i. The fact that God created mankind in His own image gives Him the right to set His rules for holy living and to call all men to appear before Him to answer for the things they have done and the way they have lived.

ii. The three former judgments on the whole of sinful mankind form a precedent for the holding of the final judgment. Scoffers ignore this at their own peril.

iii. God, if He chooses, is free to appoint whomsoever He will to be the judge, and He tells us in the Bible that He has appointed the Lord Jesus Christ. 'He has given all judgment to the Son' (John 5:22), 'because He is the Son of man' (John 5:27). Concerning this fact, 'He hath given assurance unto all men, in that He hath raised Him from the dead' (Acts 17:31). Jesus is thus uniquely qualified for this task. As God He has the insight and authority (Matthew 28:18) to judge all men, and as man He understands and sympathizes with man. In His person justice and mercy meet, and as judge of all mankind He will do what is right (Genesis 18:25)

In each of the three other world judgments there was some aspect of mercy. So, will there be any here?

Concerning this judgment there is both bad news and good news. The bad news is that, at the Great White Throne Judgment, the time for mercy is past forever. History has reached the time of the consummation (or completion) of all things.

The good news is that the mercy necessary for us to avoid being condemned at this tribunal is ***available now***, but it must be appropriated in this life to be effective in the next.

HE BORE THE DIVINE WRATH TO THE FULL

The Gospel (or 'good news') is that when Jesus died upon the Cross, He bore the divine wrath against sinners to the full, Figure 32.53. His sufferings were vicarious—for others because ***He died in mankind's place and for the sin of the world***. His sufferings were expiatory—***He exhausted the penalty for sin***. And His sufferings were ***atoning***—He dealt with all that separated God from man and man from God, ***making possible reconciliation and the restoration of lost fellowship***.

Fig.32.53: *Jesus Died on the Cross - He bore the divine wrath against sinners to the full*

Why Then are not All Men Saved?

Because Christ is God incarnate; our substitute is God Himself. If the Judge pays the penalty Himself, He is free to do so on His own conditions. The terms that God has laid down for the forgiveness of the sinner are repentance and faith towards God. We must meet these terms before we can benefit from Christ's atoning death for us, but when we do,

'being justified by faith, we have peace with God through our Lord Jesus Christ' (Romans 5:1).

Be Concern

If all this is true, why then do so few people show any concern?

The Bible tells us that:

'the god of this world [Satan] hath blinded the eyes of them which believe not'
(2 Corinthians 4:4).

A major way in which he is doing this today is through the increasingly widespread ***teaching of the theory of evolution as 'fact'***, which ***teaching leads men to deny the existence of God the Creator, Savior, and Judge.***

It also specifically denies all three previous judgments:

I. Denial of a historical Fall judgment.

Evolutionists (including theistic) must insist that the 'bad things' in this world (death, bloodshed, suffering) were part of the formation process. Hence, in their opinion, the world we now see is normal, not fallen.

II. Denial of a world-wide Flood judgment.

Theistic evolutionists are united with atheistic ones in saying there never was such an event as a world-wide flood. If there was, the fossil-bearing rocks could no longer be seen as evidence of evolutionary eons, but rather of God's judgment.

III. Denial of the Babel judgment.

All evolutionary schemes of anthropology deny that this event ever occurred.

The teaching of creation and subsequent events in Genesis involves affirmation, not only that God the Creator, Savior, and Judge does exist, but also that His power in judgment is real—that He has already executed judgment on all mankind three times in the past, while the fourth is looming on the horizon, as He will judge once again in the foreseeable future, as well prophesied. The converse is also true, that is, the fact of these judgments points inexorably back to the ***truth of creation, not evolution***. It is therefore ***not surprising that the many churches which accept some form of theistic evolution usually have little desire to proclaim the biblical theme of the judgment of God*** and thereby send a clear and positive warning signal to an unbelieving world. The remedy is obvious.

GLOBALISM: PREPARATION FOR THE ANTICHRIST

The Fourth of God's Judgment

No matter how the problems of today are defined, increasingly the world believes that the only viable solutions are of a global nature.

Everywhere we turn, globalism is being talked about and touted as the only answer to mankind's problems. I agree wholeheartedly with Dr. "LaHaye's" recent comment that

"Two of the signs all Pre-Tribulation prophecy scholars most watch for are Israel and global government."

No doubt that Israel is God's super sign of the end-times. And in concert with Israel's recent rise is the development of a global consciousness for the first time since the Tower of Babel.

In that same August issue of "Pre-Tribulation Perspectives" to which I just quoted from Dr. LaHaye, I wrote an article on "Signs of The Times and Prophetic Fulfillment." I noted that we appear to be at the end of the church age because God is currently setting the stage for the next phase of history, known as the Tribulation. I said,

"*we can see that already God is preparing or setting the stage of the world in which the great drama of the tribulation will unfold. In this way this future time casts shadows of expectation in our own day so that current events provide discernable signs of the times.*"

GLOBAL STAGE SETTING

I recall in the early 1960s, when I first starting learning about Bible prophecy, many would cite a few events evident in our day of the move toward global governance. Today there are so many things happening in virtually every area of life that it is enough to make one's head spin. Yet, there seems to be a decreasing interest by Christians in this obvious trend.

There is a rash of events that are currently in the news who are gathering to consider a global agenda. For example, September 6-8, 2000 was to see the U. N. Millennium Summit, "billed as the largest-ever gathering of world leaders" in New York City which is expected to draw a record 150 heads-of-state and government. "Louise Frechette" of Canada, the deputy secretary-general of the United Nations and spokesman for this U. N. planned event said, the summit should:

"give a sense of direction to the organization from which its work should then flow."

What Work?

Why the work toward globalism, of course. Secretary-General of the U. N., Kofi Annan, has:

"called for benevolent globalization in the 21st century to ensure that the information revolution does not leave billions of people behind in poverty."

The Secretary-General:

"asked world leaders to commit themselves to a number of targets by 2015."

*"But still under debate are proposals that would **highlight human rights** over **national sovereignty** or strongly assert the right of non-interference in domestic affairs."*

Interspersed throughout the summit will be dozens of "*mini-summits*" designed to reach consensus on various issues.

Events like the U. N. Millennium Summit are occurring with greater frequency and are increasingly taken more seriously. ***Clearly the world is on a path to global rule***.

It is not a matter of ***"if,"*** but only a matter of ***"when."*** The Clinton, Obama, and Biden administrations have recently committed the United States to participation in a global court. Such a move, without the required Congressional approval, is just one more step toward the compromise of U. S. "*www.pre-trib.org*" sovereignty and a global rule.

It appears that globalists are working overtime in an attempt to somehow implement a global tax upon the citizens of the world so that the U. N. can have an increased revenue source and gain jurisdiction over citizens of the world in what would be a major step toward global governance. Globalism is coming, faster than most may realize.

We see the urge for globalism in at least the following areas:

- government,
- economics,
- religion,
- the environment,
- military,
- commerce and trade,
- manufacturing,
- banking,

- business,
- population control,
- education,
- management,
- publishing,
- entertainment,
- personal health and well-being,
- ***wealth redistribution***,
- agriculture,
- Fossil Fuel
- Climate Change Cult
- law,
- science,
- medicine,
- sports,
- travel,
- music,
- electronics,
- the internet and information availability,
- Artificial Intelligence (AI)

and so many more areas. With each day that passes, the mantra of our day is the famous statement of Los Angeles citizen, "Rodney King," when he said,

"Why can't we all just get along?"

Well, what's wrong with globalism?

Is Globalism Unbiblical?

The only global government authorized by God and His Word will be when Jesus Christ returns to planet earth at the second advent and rules the world with a rod of iron from Jerusalem.

All other global government, no matter how "benevolent" its intention, is frowned upon by Scripture. Yet, the Bible teaches that ***God will allow evil to form a global government*** for ***three and one-half years during the second half of the tribulation*** for the purpose of ***our Lord's global judgment just prior to the second coming of Christ and the establishment of His global rule for 1,000 years upon earth***.

Thus, current moves toward globalism are ***preparation for antichrist***, not for Jesus Christ, and should be opposed by Bible-believing Christians.

Most Christians have at least some vague familiarities with the *Tower of Babel* incident shortly after the Flood, which lead to the splintering of the human race into factions.

The traditional understanding of this story (Gen. 10:8-14, 31-31; 11:1-9) indicates that the erection of the Tower corresponds with Nimrod's beginning of the kingdom of man at Babel (Gen. 10:10). Thus, Nimrod is seen as the father of the kingdom of man as a vehicle of rebellion against God and His kingdom.

It is at this point *God judged the first United Nations building, confounded the single human language into many, and established the Divine Institution of "tribal diversity" to promote social stability*.

What is tribal diversity and how has God used this to promote social stability?

Tribal diversity is when God divided the human race into diverse tribal units so that mankind would not be inclined toward unity. Due to the fall of Adam into sin, all descendants of Adam have a sin nature. This means that socially, man wants to unite in his rebellion against God. Such unity was displayed before the flood and led to the global flood.

God promised not to destroy humanity by a flood again. At the end of history, He will use fire.

In the interim, God keeps the human race from self-annihilation through the restraint of civil government and tribal diversity. "Charles Clough" says, "In Genesis 9:25- 27 an outline of postdiluvian history is revealed which centers upon the three sons of Noah—Ham, Shem, and Japheth. Further subdivisions are indicated in Genesis 10—11 and Deuteronomy 32:8.9 "Throughout the postdiluvian period, then," explains Clough,

"God preserved man's social stability and health by playing off one group or tribe against another to maximize true progress and retard the influence of evil (cf. Acts 17:26-27)."

Since the flood, God has used tribal diversity as an instrument to pit one people group against another, so that when one society becomes entirely pagan, He can judge or restrain its evil influence through another people. Thus, like compartments in the hull of a ship, damage to one compartment will not sink the entire ship.

This is what God uses to restrain global government until the tribulation comes when, *through the leadership of the antichrist*, mankind will overcome this restraint, resulting in a second global judgment at Christ's return.

The day in which we live, appears to be on the threshold of smashing God's institutions for the supposed higher goal of erecting a new Babel. Modern globalism was first put forth by ***Dante*** in the fourteenth century in order to counter the all-encompassing stranglehold of the Roman Catholic Church.

"Immanuel Kant" developed globalism as a philosophical idea half a millennium later. The military turmoil of the last two hundred years has seen globalism gradually grow to the place that it is considered the only reasonable solution to a multiplicity of problems by secular thinkers.

Nevertheless, God is as opposed to globalism today as He was to the one world spirit displayed through the ancient tower of Babel. His judgment will be just as certain in the future as it was in the distant past.

The Mystery of Inequity – Black Lives Matter - Antifa – Binary Beings – Rainbow (LGBTQ ... etc.)

No matter how hard apostate man struggles to apply a global solution to mankind's problems, he will not succeed until God removes ***His restraint*** (*The Holy Spirit*) of man's evil impulse ***after the rapture*** of the believers. Paul tells us in 2 Thessalonians 2 that:

"the man of lawlessness (the antichrist)" (verse 3), will not be revealed until "he who now restrains. . . is taken out of the way" (verse 7).

The restraint will be removed at that future time when God will remove His believers (the true church) through the rapture.

The removal at the rapture will then allow for ***the rise of antichrist***, who will begin the seven-year tribulation by making a covenant with the nation of Israel. The final three and one-half years will see ***antichrist implementing global governance*** in rebellion against God.

Paul told the Thessalonians that:

"the mystery of lawlessness is already at work; only he who now restrains will do so until he is taken out of the way" (verse 7).

Ever since Babel, Satan has been working to bring about his agenda of globalism, but God restrains from reaching the flowering of his kingdom until after the rapture, during the tribulation. Because the mystery of lawlessness (i.e., iniquity in the KJV) has been at work since Babel, we can see evidence down through history, even in our own day that the globalization of antichrist, and the explosion of AI is coming. Step by step it creeps, but it is coming. This is why we can see evidences of globalism today, before the rapture.

Paul further warns the Thessalonians that the mystery of iniquity is a satanic program that advances through deception and guile (verse 10). We can gain insight into the present working of the mystery of iniquity by studying the maturity of the kingdom of man as it will be brought out into the open during the second half of the tribulation. Revelation 17 and 18 describes this satanic kingdom as being composed of a religious aspect as depicted by "the great harlot who sits on many waters" (Rev. 17:1; cf. 17:1-18); and the political and commercial aspect of the kingdom of man (cf. Rev. 18:1-24).

It should not be surprising that we see the development globalization advancing at an alarming rate when we consider the insight into such developments in light of biblical teachings about past and future growth of the kingdom of man. It should be clear that all globalization in our own day is the result of a satanic impulse, not one from above.

Even when we find similar trends being espoused in the ***Evangelical Church***, they should be understood as an invasion of the enemy (cf. 1 Tim. 4; 2 Tim. 3). We should not be surprised to learn that efforts toward a one world government have been and continue to be the efforts of so many in high places of government and commerce; and that they use the stealth of deception to further their dreams and goals. E.g., The US Government is dividing it citizens into parties against each other, as well as setting its own departments against people whom they oppose their deceptive plans to gain control. We see it clearly today demonstrated in USA's (Department of Justice) DOJ, (Federal Bauru if Investigation) FBI, (Internal Revenue Services) IRS, the Executive Branch, and the Congress of USA.

This understanding provides a framework for understanding, interpreting, and standing against many of the trends and events of our own day, knowing from whence they have come and to where they are leading.

We take great relief in knowing that God is just and will one day judge this ungodly world. But at the same time, while we are eagerly waiting for a Savior from heaven, we should be aggressively spreading the gospel of God's grace to any who may listen.

THE FOURTH JUDGMENT OF GOD

Satan Opposition to God

Armageddon And the Second Coming of Christ

The dramatic conclusion of the "times of the Gentiles" is described in prophecy as a gigantic world war which is climaxed by the second coming of Christ. The war that brings to a close the times of the Gentiles, which already has embraced twenty-five hundred years of history, is also the final effort of Satan in his strategy of opposition to the divine program of God.

The second coming of Christ is God's answer. Some of the major elements of this conflict have already been considered and now need only to be related one to the other.

The Final World Conflict

It is clearly prophesied in the Scripture Old and New Testaments. The great world war which will engulf the Middle East at the end of the age is an outgrowth of the world situation during the time of the great tribulation. The Roman Empire formed earlier has now extended its power "over all kindreds, and tongues, and nations" (Revelation 13:7). *The world government formed at the beginning of the great tribulation* is scheduled in prophecy to endure for forty-two months or three and a half years (Revelation 13:5). At its beginning there is no serious challenge of the power and authority of the world ruler, the antichrist, and he is able to assume supreme power not only in the political field, but also receives recognition and worship as God and controls the economic power of the entire world. His reign is afflicted, however, by a series of ***great judgments of God*** described in the breaking of the seals, the blowing of the trumpets, and the outpouring of the vials of the wrath of God (Revelation 6:1—18:24). The disruptive force of these judgments is keenly felt throughout the world and it soon becomes evident that the promised utopia which his rule was designed to produce is not going to be fulfilled.

The Trinity of Evil

Many students of prophecy have noted the "trinity of evil" which characterizes the end time. In some respects, this trinity corresponds to the Trinity of the Godhead. The ultimate source of power and evil in the end time is none other than ***Satan himself***, referred to as "the great dragon," and as "that old serpent, called the Devil, and Satan" (Revelation 12:9). The political as well as the religious power which dominates the world is unquestionably Satan, and for this reason it is stated in Revelation 13:4 that the world "worshipped the dragon which gave power unto the beast."

Satan assumes much the same power and prerogatives as God the Father.

The world ruler, who is Satan's masterpiece as a counterfeit of Christ, is the actual supreme dictator of the entire world and in a sense is Satan incarnate. He is undoubtedly a brilliant man intellectually and a dynamic personality, but he is completely dominated by Satan. In keeping with the satanic approach of imitation and counterfeit of God's program, the world ruler is Satan's king of kings and lord of lords. Many students of Scripture assign the term "antichrist" to this person for this reason, although in the Bible none of the references to antichrist clearly indicate the personage in view (cp. I John 2:18, 22; 4:3; II John 7).

The Beast and the False Prophet

The third member of the unholy trinity is the "beast coming up out of the earth" (Revelation 13:11) who assists the world ruler, performing satanic miracles and causing all men to worship the image of the beast (Revelation 13:12-15). He apparently also is instrumental in linking the economic and religious life of the world in that only those who worship the beast can buy or sell (Revelation 13:16). This personage is undoubtedly the same as "***the false prophet***" (Revelation 19:20) and in every respect he is the right-hand man and expediter for the world ruler. In his activities he corresponds to some extent to the ministry of the Holy Spirit on behalf of Christ and thus forms the third member of the trinity of evil. The world situation is therefore firmly in the grasp of Satan, Satan's man who is the ***world dictator***, and the ***false prophet*** who heads up the satanic world religion of the great tribulation. In spite of the satanic control of the world by divine plan (Revelation 16:16), as the great tribulation moves on to its close, major sections of the world rebel against their ruler, and this sets the stage for the final great world war.

The Gathering of The Armies of The World.

The Armies of the World

The armies of the world which converge upon the Middle East according to Revelation 16:13 are induced to engage in the final conflict by satanic influences. This is introduced in the statement of John the Apostle:

"I saw three unclean spirits like frogs come out of the mouth of the dragon, and out of the mouth of the beast, and out of the mouth of the false prophet. For they are the spirits of devils, working miracles, which go forth unto

the kings of the earth and of the whole world, to gather them to the battle of that great day of God Almighty" (*Revelation 16:13, 14*).

There has been endless speculation as to the identity of the three unclean spirits like frogs. The passage itself indicates plainly that they are spirits of devils or demons and unquestionably are fallen angels under the command of Satan who are sent forth to draw the kings of the world into this final conflict. Humanly speaking, they are gathering to wrest the world rulership from the Roman ruler.

In the satanic purpose, however, the armies of the world are gathered to fight the armies of heaven which will accompany Christ at His second coming. As in so many undertakings of Satan, such as is supremely illustrated in the ***crucifixion of Christ***, the very program of Satan is its own destruction, and although Satan is inevitably impelled to gigantic opposition to Christ, he only sets the stage for the triumph of God. It is to facilitate the gathering of these armies that the Euphrates River is dried up that the armies from the east may converge without difficulty upon the Middle East.

Three major armies are mentioned in the Bible, namely, the army from the ***north,*** the army from the **east**, and the army from the ***south***. These three armies are combining their efforts to wrest power from the Roman ruler who may be considered as the king of the west, although this title is never given to him in the Scriptures. The focal point for their gathering is declared in Revelation 16:16 to be:

"*a place called in the Hebrew tongue Armageddon.*"

Although various explanations have been given of this title, it seems to refer to the valley of Esdraelon also known as the valley of Jezreel located to the east of Megiddo in northern Israel. The word Armageddon means Mount of Megiddo from har meaning mount and Megiddo.

The broad valley that is here described is approximately fourteen miles wide and twenty miles long and historically has been the scene of many great battles of the past. In modern times the area became a great swamp, but with the revival of the area under the state of Israel the water has now been drained, and it is a fruitful and beautiful plain well suited for a great army. It is obvious, however, that this is only the central staging area for the war as actually the size of the armies involved preclude the possibility of confining them to this valley. As Scripture indicates, the war rages for some two hundred miles north and south thereby engulfing the entire Holy Land.

THE EARLY BATTLES OF THE FINAL WORLD WAR

Scripture does not provide much detail on the characteristics of the final world conflict. The main significance is that they are assembled in the Holy Land at the time of the second advent and oppose Christ in His return to the earth. However, some indication of the nature of the battles preceding the second coming of Christ is given in Daniel 11:40-45. If the order of introduction of events is taken chronologically, it appears that the first stage of the battle is an attack by the king of the south. According to Daniel 11:40, "And at the time of the end shall the king of the south push at him."

In rapid succession an attack also comes from the north which apparently is successful. The Scriptures state that "the king of the north shall come against him like a whirlwind, with chariots, and with horsemen, and with many ships; and he shall enter into the countries, and shall overflow and pass over." This king of the north may be Russia. The force of his invasion is such that he proceeds through the Holy Land and conquers Egypt at least temporarily. (Some expositors, however, interpret the passage beginning in Daniel 11:42 as referring to the Roman ruler who is naturally to the north of Africa, rather than to Russia as the king of the north, i.e., north of Palestine, as this seems to be the main theme of the passage.) The warfare brought about by the invasion of the king of the north and the king of the south, however, is now followed by another phase, namely, the arrival of the host from the east.

According to Daniel 11:44, "Tidings out of the east and out of the north shall trouble him: therefore, he shall go forth with great fury to destroy, and utterly to make away many." The arrival of the forces from the Orient described as an army of two hundred million in Revelation 9:16 brings on the last phase of the world struggle, and at the time of the second coming of Christ the war is raging in several areas.

At least four geographic locations are mentioned in the Bible as figuring in the final struggle. The center, of course, is ***Armageddon*** where the main forces are located, Figure 32.54. Another focal point for the battle is the city of Jerusalem itself. According to Zechariah 12:2-10 a siege will be declared against the city of Jerusalem. Jerusalem apparently will be defended to some extent by the power of God by miraculous intervention, for the armies of the world have great difficulty in subduing the city. It is stated in Zechariah 12:3, "And in that day will I make Jerusalem a burdensome stone for all people; all that burden themselves with it shall be cut in pieces, though all the people of the earth be gathered together against it." The passage goes on to say how the horses are smitten with blindness and the riders with madness.

Fig.32.54: *The Battle of Armageddon*

At the time of the second coming of Christ, however, Jerusalem has finally been entered and is in the process of being subdued at the very moment that the glory of Christ in the heavens in His second advent appears. This is stated in Zechariah 14:2, 3: "For I will gather all nations against Jerusalem to battle; and the city shall be taken, and the houses rifled, and the women ravished; and half of the city shall go forth into captivity, and the residue of the people shall not be cut off from the city. Then shall ***the Lord*** go forth, and ***fight against those nations***, Figure 32.55, as when he fought in the day of battle." From this description it is clear that the nations are engaged in active warfare in relation to Jerusalem at the time of the second advent.

Another geographic location is that of the valley of Jehoshaphat mentioned in Joel 3:2, 12. Although there is some dispute as to its location, it appears to be a valley immediately east of Jerusalem. Here, according to Joel, God declares: "I will also gather all nations, and will bring them down into the valley of Jehoshaphat, and will plead with them there for my people and for my heritage Israel, whom they have scattered among the nations, and parted my land." This valley is the scene of the divine judgment mentioned in Joel 3:12. Whether this gathering has to do with the battle for Jerusalem or is a subsequent event to the second advent, is not entirely clear.

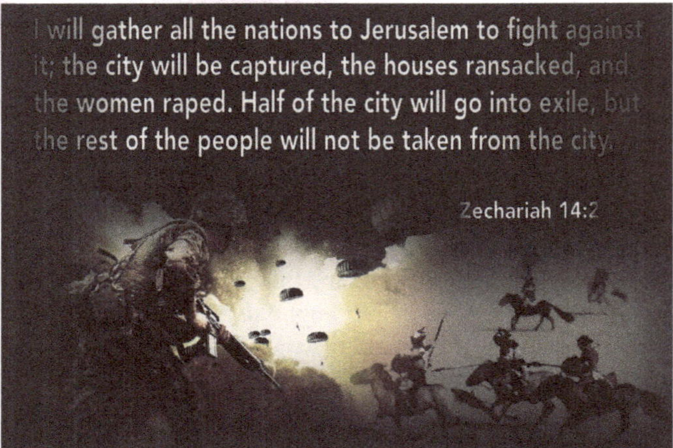

Fig.32.55: Then the Lord shall go forth, and fight against those nations

Still another geographic location mentioned is that of Edom in Isaiah 34:1-6 and 63:1-6. Again, it is not entirely clear, however, whether this is part of the battle or a subsequent judgment of God. In Daniel 11:41 Edom, Moab, and Ammon are specifically mentioned as escaping the full brunt of the battle.

Blood Shed

The awful bloodshed stemming from this conflict is indicated in Revelation 9:18 where one-third of the armies are declared to be destroyed by the army from the east and the statement in Revelation 14:20 that "blood came out of the winepress, even unto the horse bridles, by the space of a thousand and six hundred furlongs." This is a distance of approximately two hundred miles and seems to indicate the extent of the bloody battle as these armies converge upon the Holy Land.

The Final Showdown

Although the exact deployment of the forces and the precise character of the successive battles which precede the second coming of Christ are not indicated in Scripture, it is sufficient for us to know that the Holy Land will be crowded with the armies of the world in preparation for the dramatic second advent of Christ. This is the final showdown of Gentile power dominated by Satan in blasphemous opposition to the Lordship of Jesus Christ.

The Second Coming of Christ and the Annihilation of the Armies

As the armies of the world are engaged in struggle for power throughout the Holy Land and in the very act of sacking the city of Jerusalem, the glory of the Lord appears in heaven and the majestic procession pictured in Revelation 19:11-16 takes place, Figure 32.56. At the head of the procession is Christ, described as riding on a white horse coming to judge and make war. His eyes are as a flame of fire and on His head are many crowns. His vesture is dipped in blood. Accompanying Him are the armies of heaven also riding on white horses and clothed in fine linen. In verse 15 it is stated of Christ, "Out of his mouth goeth a sharp sword, that with it he should smite the nations: and he shall rule them with a rod of iron: and he treadeth the winepress of the fierceness and wrath of Almighty God."

Fig.32.56: The Second Coming of the Lord Jesus

THE TRIUMPHANT KING OF KINGS

In contrast to His lowly birth in Bethlehem where He was laid in a manger, this advent is the triumphant King of kings and Lord of lords coming to claim the world for which He died and over which He is now going to exercise His sovereign authority in absolute power, Figure 32.57. The verses which follow invite the fowls of the earth to feed upon the carnage of the flesh of kings and mighty men and of their horses (Revelation 19:18).

Fig.32.57: The Triumphant King of kings

Fight Against Christ

According to Revelation 19:19 the armies of the world, which have previously been fighting each other, forget their differences and unite to ***fight against Christ*** in His second advent to the earth. John writes:

"I saw the beast, and the kings of the earth, and their armies, gathered together to make war against him that sat on the horse, and against his army."

Their struggle against such an adversary, however, is useless. It is apparent that they are put to death not by ordinary military struggle, but by the word of authority proceeding out of His mouth described as "*a sharp sword*" (Revelation 19:15).

All the armies and their horses apparently are put to death at one stroke, but the beast (the world ruler-Antichrist) and the false prophet (the religious ruler of the world) are taken alive, and according to Revelation 19:20, Figure 32.58.

Fig.32.58: *The False Messiah and the False Prophit were cast in the Lake of Fire*

"These both were cast alive into a lake of fire burning with brimstone."

However, the doom of the rest is sealed in Revelation 19:21:

"*And the remnant were slain with the sword of him that sat upon the horse, which sword proceeded out of his mouth: and all the fowls were filled with their flesh.*"

Thus, ends in one dramatic blow the power of the Gentiles which had controlled Jerusalem from the time of Nebuchadnezzar, 606 B.C. Thus, ***ends also the satanic control*** of the Gentiles who had been a demonstration of satanic power, guilty of blasphemy and of the blood of countless martyrs, especially in oppressing the nation, Israel.

SATAN CAST INTO THE ABYSS

Satan, their unseen leader, is also dealt with, and according to Revelation 20:1-3 he is cast into the abyss where he is rendered inactive for the entire period of the thousand-year reign of Christ on earth. Then he is destined to join the beast and the false prophet in the lake of fire (Revelation 20:10), Figure 32.59. The Gentile population of the world as a whole is judged at a separate judgment which follows and is a part of the establishment of Christ's kingdom on earth.

Fig.32.59: The Judgement of Satan and Fallen Angels - Curtsy FaithEquip

The inglorious end of Gentile power is precisely that which was anticipated in Daniel 2 where the great image disintegrates into chaff when struck by the stone cut out without hands. The same dramatic end is contemplated in the destruction of the beast (Daniel 7:11) followed by the inauguration of the everlasting kingdom in Daniel 7:13, 14. Jerusalem was no longer to be trodden under the feet of the Gentiles and once again Israel was to be exalted.

CATASTROPHE, CURSE AND JUDGMENT

All beings are created for the purpose to Honor and Worship the Creator. However, the main purpose of the creation/evolution debate is that if evolutionists can show that atheistic evolution is true, they will have eliminated the last great Day of Judgment. However, there have already been three occasions when the whole of sinful mankind has been judged by God. The fact that these three events have occurred testifies to the certainty of the fourth and final judgment to come soon.

The Fall

The first universal judgment was at the event known as the Fall, described in Genesis 3. The reason was the deliberate disobedience of Adam and Eve to the specific instruction of God that they were not to eat from the tree in the Garden of Eden called the tree of the knowledge of good and evil (Genesis 2:17). When Adam and Eve disobeyed God, the effects for them included

- knowledge of sin,
- experience of guilt,
- loss of holiness,
- fear of God,
- forfeiture of fellowship with God, and
- feelings of embarrassment towards each other.

And, above and beyond these consequential effects, there was the specific pronouncement of judgment by God, sometimes called

'the curse'

This contained both spiritual and physical dimensions, not only for Adam and Eve, but also for their descendants, that is, for all mankind. Spiritually, they were from then on cut off from God, so that on that day they died spiritually. Physically, *their human bodies began to die* and would eventually 'return to dust' and decayed, Eve was told that:

- childbearing would be a matter of 'sorrow' and that
- her husband would rule over her,
- the ground was cursed so that Adam's work from then on would be hard labor, and
- they were cast out of the Garden of Eden (Genesis 3:16–24).

Please observe that God's mercy is seen in the midst of this judgment, in two aspects:

1. There was the promise of Genesis 3:15, that the seed of the woman (Jesus Christ) would 'bruise the head' of the serpent (i.e., cause the ultimate downfall of Satan).
2. The penalty of physical death was itself a mercy! Without physical death mankind would have been condemned to live forever in the sorrowful results of rebellion against God.

However, because of the death of Christ on the Cross substituting for our sins, and the resurrection, those who repent of their sins and have faith towards God, after death will be united with God in Heaven and attain that state of holiness and fellowship with God, which our ancestors, Adam, and Eve, lost for us in the Garden of Eden.

The Flood

The second universal judgment was at the catastrophe of the Flood, described in Genesis 6–9. The reasons for this judgment were that 'the wickedness of man was great,' 'every imagination of the thoughts of his heart was only evil continually,' 'all flesh had corrupted his way upon the earth,' and 'the earth was filled with violence' (Genesis 6:5–13). In this judgment, all of sinful mankind (apart from one family) died, as did all the air-breathing land animals except for those on the Ark (Genesis 7:21-23).

God's mercy is seen in that Noah, described as a just man and perfect in his generations, and righteous (Genesis 6:9 and 7:1), was saved, together with his wife and family (a total of eight persons) by means of the Ark, which also carried the animals nominated by God to replenish the earth.

Babel

The third universal judgment was the confusion of languages at Babel, described in Genesis 11. After the Flood, God commanded Noah's sons to 'replenish [Hebrew: "fill"] the earth,' that is, to spread out, to form many nations, and to inhabit all the lands (Genesis 9:1).

When the population had increased to the point where there were enough family units to have begun to do this, instead they built for themselves a city and within it a tower (Genesis 11:1-4).

Three reasons are given in the Bible for this building program.

i. It was to be a 'tower unto heaven' (the words 'may reach' or 'to reach' are not in the original Hebrew), and so it appears that one use to which it was put involved the worship of heaven (possibly of angels, leading to demonic involvement; or of stars, leading to astrology and occult practices).

ii. It was to make a name for themselves, and to glorify human achievement, thereby showing their pride and self-sufficiency.

iii. It expressed a human attempt to preserve a unity which had been lost through sin, and which was contrary to God's earlier command to Noah's sons to replenish or fill the earth ('lest we be scattered abroad upon the face of the earth'—Genesis 11:4).

God's judgment on this act of rebellion—the confounding of their language—showed both His mercy and His wisdom. God's mercy is seen in the fact that He did not make the penalty equal to the offence. God's wisdom is seen in the fact that the miracle of confusing their language effectively put a halt to their threefold aims:

- Without intercommunication, they could no longer continue their building program.
- The tower, which was meant to be a symbol of the great name they sought for themselves, became instead a term of reproach—Babel, meaning 'confusion'.
- They were forced to do the very thing they had previously refused to do—scatter over the earth.

THE GREAT WHITE THRONE

The fourth universal judgment is what the Bible calls the Great White Throne Judgment. Our knowledge of this comes from Revelation 20:11–15, which reads:

'And I saw a great white throne, and him that sat on it, from whose face the earth and the heaven fled away; and there was found no place for them. And I saw the dead, small and great, stand before God: and the books were opened: and another book was opened, which is the book of life: and the dead were judged out of those things which were written in the books, according to their works. And the sea gave up the dead which were in it; and death and hell delivered up the dead which were in them; and they were judged every man according to their works. And death and hell were cast into the lake of fire. This is the second death. And whosoever was not found written in the book of life was cast into the lake of fire.'

This judgment is described as in the form of a law court, with a presiding judge, and evidence is subpoenaed in the form of 'those things which are written in the 'books.' It is relevant to ask by what authority this tribunal is held, who the judge will be, and what his qualifications are to make him judge of all mankind. Consider:

i. The fact that God created mankind in His own image gives Him the right to set His rules for holy living and to call all men to appear before Him to answer for the things they have done and the way they have lived.

ii. The three former judgments on the whole of sinful mankind form a precedent for the holding of the final judgment. Scoffers ignore this at their own peril.

iii. God, if He chooses, is free to appoint whomsoever He will to be the judge, and He tells us in the Bible that He has appointed the Lord Jesus Christ. 'He has given all judgment to the Son' (John 5:22), 'because He is the Son of man' (John 5:27). Concerning this fact, 'He hath given assurance unto all men, in that He hath raised Him from the dead' (Acts 17:31). Jesus is thus uniquely qualified for this task. As God He has the insight and authority (Matthew 28:18) to judge all men, and as man He understands and sympathizes with man. In His person justice and mercy meet, and as judge of all mankind He will do what is right (Genesis 18:25)

In each of the three other world judgments there was some aspect of mercy. So, will there be any here?

Concerning this judgment there is both bad news and good news. The bad news is that, at the Great White Throne Judgment, the time for mercy is past forever. History has reached the time of the consummation (or completion) of all things.

The good news is that the mercy necessary for us to avoid being condemned at this tribunal is ***available now***, but it must be appropriated in this life to be effective in the next.

A major way in which he is doing this today is through the increasingly widespread ***teaching of the theory of evolution as 'fact'***, which ***teaching leads men to deny the existence of God the Creator, Savior, and Judge.***

It also specifically denies all three previous judgments:

I. Denial of a historical Fall judgment.

Evolutionists (including theistic) must insist that the 'bad things' in this world (death, bloodshed, suffering) were part of the formation process. Hence, in their opinion, the world we now see is normal, not fallen.

II. Denial of a world-wide Flood judgment.

Theistic evolutionists are united with atheistic ones in saying there never was such an event as a world-wide flood. If there was, the fossil-bearing rocks could no longer be seen as evidence of evolutionary eons, but rather of God's judgment.

III. Denial of the Babel judgment.

All evolutionary schemes of anthropology deny that this event ever occurred.

The teaching of creation and subsequent events in Genesis involves affirmation, not only that God the Creator, Savior, and Judge does exist, but also that His power in judgment is real—that He has already executed judgment on all mankind three times in the past, while the fourth is looming on the horizon, as He will judge once again in the foreseeable future, as well prophesied. The converse is also true, that is, the fact of these judgments points inexorably back to the ***truth of creation, not evolution***. It is therefore ***not surprising that the many churches which accept some form of theistic evolution usually have little desire to proclaim the biblical theme of the judgment of God*** and thereby send a clear and positive warning signal to an unbelieving world. The remedy is obvious.

THE TRIBULATION OF GOD'S JUDGMENT (THE 4TH JUDGMENT)

GLOBALISM: Preparing the Advent of the Antichrist

The Fourth Judgment of God's on All the Earth Dwellers

No matter how the problems of today are defined, increasingly the world believes that the only viable solutions are of a global nature.

Everywhere we turn, globalism is being talked about and touted as the only answer to mankind's problems. I agree wholeheartedly with late Dr. "LaHaye's" comment that

*"Two of the signs all **Pre-Tribulation prophecy** scholars most watch for are Israel and global government."*

No doubt that Israel is God's super sign of the end-times. And in concert with Israel's recent rise is the development of a global consciousness for the first time since the Tower of Babel.

Also, as I am writing on "Signs of The Times and Prophetic Fulfillment," it appears that we at the end of the church age because I see the Hand of God is currently setting the stage for the next phase of history, known as the Tribulation. We can see that already God is preparing or setting the stage of the world in which the great drama of the tribulation will unfold. In this way this future time casts shadows of expectation in our own day so that current events provide discernable signs of the times.

GOD'S FOURTH JUDGMENT

The Trinity of Evil

Many students of prophecy have noted the "trinity of evil" which characterizes the end time. In some respects, this trinity corresponds to the Trinity of the Godhead. The ultimate source of power and evil in the end time is none other than ***Satan himself***, referred to as "the great dragon," and as "that old serpent, called the

Devil, and Satan" (Revelation 12:9). The political as well as the religious power which dominates the world is unquestionably Satan, and for this reason it is stated in Revelation 13:4 that the world "worshipped the dragon which gave power unto the beast," Figure 32.60.

Fig.32.60: The Unholy Trinity – Satan, False Messiah, and False Prophit

Satan assumes much the same power and prerogatives as God the Father.

The world ruler, who is Satan's masterpiece as a counterfeit of Christ, is the actual supreme dictator of the entire world and in a sense is Satan incarnate. He is undoubtedly a brilliant man intellectually and a dynamic personality, but he is completely dominated by Satan. In keeping with the satanic approach of imitation and counterfeit of God's program, the world ruler is Satan's king of kings and lord of lords. Many students of Scripture assign the term "antichrist" to this person for this reason, although in the Bible none of the references to antichrist clearly indicate the personage in view (cp. I John 2:18, 22; 4:3; II John 7).

THE BEAST AND THE FALSE PROPHIT

The third member of the unholy trinity is the "beast coming up out of the earth" (Revelation 13:11) who assists the world ruler, performing satanic miracles and causing all men to worship the image of the beast (Revelation 13:12-15). He apparently also is instrumental in linking the economic and religious life of the world in that only those who worship the beast can buy or sell (Revelation 13:16). This personage is undoubtedly the same as "***the false prophet***" (Revelation 19:20) and in every respect he is the right-hand man and expediter for the world ruler. In his activities he corresponds to some extent to the ministry of the Holy Spirit on behalf of Christ and thus forms the third member of the trinity of evil. The world situation is therefore firmly in the grasp of Satan, Satan's man who is the ***world dictator***, and the ***false prophet*** who heads up the satanic world religion of the great tribulation. In spite of the satanic control of the world by divine plan (Revelation 16:16), as the great tribulation moves on to its close, major sections of the world rebel against their ruler, and this sets the stage for the final great world war.

The Gathering of The Armies of The World.

THE ARMIES OF THE WORLD

The armies of the world which converge upon the Middle East according to Revelation 16:13 are induced to engage in the final conflict by satanic influences. This is introduced in the statement of John the Apostle:

"*I saw three unclean spirits like frogs come out of the mouth of the dragon, and out of the mouth of the beast, and out of the mouth of the false prophet. For they are the spirits of devils, working miracles, which go forth unto the kings of the earth and of the whole world, to gather them to the battle of that great day of God Almighty*" (*Revelation 16:13, 14*).

There has been endless speculation as to the identity of the three unclean spirits like frogs. The passage itself indicates plainly that they are spirits of devils or demons and unquestionably are fallen angels under the command of Satan who are sent forth to draw the kings of the world into this final conflict. Humanly speaking, they are gathering to wrest the world rulership from the Roman ruler.

In the satanic purpose, however, the armies of the world are gathered to fight the armies of heaven which will accompany Christ at His second coming. As in so many undertakings of Satan, such as is supremely illustrated in the ***crucifixion of Christ***, the very program of Satan is its own destruction, and although Satan is inevitably impelled to gigantic opposition to Christ, he only sets the stage for the triumph of God. It is to facilitate the gathering of these armies that the Euphrates River is dried up that the armies from the east may converge without difficulty upon the Middle East.

Three major armies are mentioned in the Bible, namely, the army from the ***north,*** the army from the **east**, and the army from the ***south***. These three armies are combining their efforts to wrest power from the Roman ruler who may be considered as the king of the west, although this title is never given to him in the Scriptures. The focal point for their gathering is declared in Revelation 16:16 to be:

"a place called in the Hebrew tongue Armageddon."

Although various explanations have been given of this title, it seems to refer to the valley of Esdraelon also known as the valley of Jezreel located to the east of Megiddo in northern Israel. The word Armageddon actually means Mount of Megiddo from har meaning mount and Megiddo.

The broad valley that is here described is approximately fourteen miles wide and twenty miles long and historically has been the scene of many great battles of the past. In modern times the area became a great swamp, but with the revival of the area under the state of Israel the water has now been drained, and it is a fruitful and beautiful plain well suited for a great army. It is obvious, however, that this is only the central staging area for the war as actually the size of the armies involved preclude the possibility of confining them to this valley. As Scripture indicates, the war rages for some two hundred miles north and south thereby engulfing the entire Holy Land.

THE EARLY BATTLES OF THE FINAL WORLD WAR

Scripture does not provide much detail on the characteristics of the final world conflict. The main significance is that they are assembled in the Holy Land at the time of the second advent and oppose Christ in His return to the earth. However, some indication of the nature of the battles preceding the second coming of Christ is given in Daniel 11:40-45. If the order of introduction of events is taken chronologically, it appears that the first stage of the battle is an attack by the king of the south. According to Daniel 11:40, "And at the time of the end shall the king of the south push at him."

In rapid succession an attack also comes from the north which apparently is successful. The Scriptures state that "the king of the north shall come against him like a whirlwind, with chariots, and with horsemen, and with many ships; and he shall enter into the countries, and shall overflow and pass over." This king of the north may be Russia. The force of his invasion is such that he proceeds through the Holy Land and conquers Egypt at least temporarily. (Some expositors, however, interpret the passage beginning in Daniel 11:42 as referring to the Roman ruler who is naturally to the north of Africa, rather than to Russia as the king of the north, i.e., north of Palestine, as this seems to be the main theme of the passage.) The warfare brought about by the invasion of the king of the north and the king of the south, however, is now followed by another phase, namely, the arrival of the host from the east.

According to Daniel 11:44, "Tidings out of the east and out of the north shall trouble him: therefore, he shall go forth with great fury to destroy, and utterly to make away many." The arrival of the forces from the Orient described as an army of two hundred million in Revelation 9:16 brings on the last phase of the world struggle, and at the time of the second coming of Christ the war is raging in a number of areas.

At least four geographic locations are mentioned in the Bible as figuring in the final struggle. The center, of course, is ***Armageddon*** where the main forces are located. Another focal point for the battle is the city of Jerusalem itself. According to Zechariah 12:2-10 a siege will be declared against the city of Jerusalem. Jerusalem apparently will be defended to some extent by the power of God by miraculous intervention, for the armies of the world have great difficulty in subduing the city. It is stated in Zechariah 12:3, "And in that day will I make Jerusalem a burdensome stone for all people; all that burden themselves with it shall be cut in pieces, though all the people of the earth be gathered together against it." The passage goes on to say how the horses are smitten with blindness and the riders with madness.

At the time of the second coming of Christ, however, Jerusalem has finally been entered and is in the process of being subdued at the very moment that the glory of Christ in the heavens in His second advent appears. This is stated in Zechariah 14:2, 3: "For I will gather all nations against Jerusalem to battle; and the city shall be taken, and the houses rifled, and the women ravished; and half of the city shall go forth into captivity, and the residue of the people shall not be cut off from the city. Then shall ***the Lord*** go forth, and ***fight against those nations***, as when he fought in the day of battle." From this description it is clear that the nations are engaged in active warfare in relation to Jerusalem at the time of the second advent.

Another geographic location is that of the valley of Jehoshaphat mentioned in Joel 3:2, 12. Although there is some dispute as to its location, it appears to be a valley immediately east of Jerusalem. Here, according to Joel, God declares: "I will also gather all nations, and will bring them down into the valley of Jehoshaphat, and will plead with them there for my people and for my heritage Israel, whom they have scattered among the nations, and parted my land." This valley is the scene of the divine judgment mentioned in Joel 3:12. Whether this gathering has to do with the battle for Jerusalem or is a subsequent event to the second advent, is not entirely clear.

Still another geographic location mentioned is that of Edom in Isaiah 34:1-6 and 63:1-6. Again, it is not entirely clear, however, whether this is part of the battle or a subsequent judgment of God. In Daniel 11:41 Edom, Moab, and Ammon are specifically mentioned as escaping the full brunt of the battle.

BLOOD SHED

The awful bloodshed stemming from this conflict is indicated in Revelation 9:18 where one-third of the armies are declared to be destroyed by the army from the east and the statement in Revelation 14:20 that "blood came out of the winepress, even unto the horse bridles, by the space of a thousand and six hundred furlongs." This is a distance of approximately two hundred miles and seems to indicate the extent of the bloody battle as these armies converge upon the Holy Land.

The Final Showdown

Although the exact deployment of the forces and the precise character of the successive battles which precede the second coming of Christ are not indicated in Scripture, it is sufficient for us to know that the Holy Land will be crowded with the armies of the world in preparation for the dramatic second advent of Christ. This is the final showdown of Gentile power dominated by Satan in blasphemous opposition to the Lordship of Jesus Christ.

The Second Coming of Christ and the Annihilation of the Armies

As the armies of the world are engaged in struggle for power throughout the Holy Land and in the very act of sacking the city of Jerusalem, the glory of the Lord appears in heaven and the majestic procession pictured in Revelation 19:11-16 takes place. At the head of the procession is Christ, described as riding on a white horse coming to judge and make war. His eyes are as a flame of fire and on His head are many crowns. His vesture is dipped in blood. Accompanying Him are the armies of heaven also riding on white horses and clothed in fine linen. In verse 15 it is stated of Christ, "Out of his mouth goeth a sharp sword, that with it he should smite the nations: and he shall rule them with a rod of iron: and he treadeth the winepress of the fierceness and wrath of Almighty God."

The Triumphant King of kings

In contrast to His lowly birth in Bethlehem where He was laid in a manger, this advent is the triumphant King of kings and Lord of lords coming to claim the world for which He died and over which He is now going to exercise His sovereign authority in absolute power. The verses which follow invite the fowls of the earth to feed upon the carnage of the flesh of kings and mighty men and of their horses (Revelation 19:18).

Fight Against Christ

According to Revelation 19:19 the armies of the world, which have previously been fighting each other, forget their differences and unite to ***fight against Christ*** in His second advent to the earth. John writes:

"*I saw the beast, and the kings of the earth, and their armies, gathered together to make war against him that sat on the horse, and against his army.*"

Their struggle against such an adversary, however, is useless. It is apparent that they are put to death not by ordinary military struggle, but by the word of authority proceeding out of His mouth described as "*a sharp sword*" (Revelation 19:15).

All the armies and their horses apparently are put to death at one stroke, but the beast (the world ruler-Antichrist) and the false prophet (the religious ruler of the world) are taken alive, and according to Revelation 19:20,

"*These both were cast alive into a lake of fire burning with brimstone.*"

However, the doom of the rest is sealed in Revelation 19:21:

"*And the remnant were slain with the sword of him that sat upon the horse, which sword proceeded out of his mouth: and all the fowls were filled with their flesh.*"

Thus, ends in one dramatic blow the power of the Gentiles which had controlled Jerusalem from the time of Nebuchadnezzar, 606 B.C. Thus, ***ends also the satanic control*** of the Gentiles who had been a demonstration of satanic power, guilty of blasphemy and of the blood of countless martyrs, especially in oppressing the nation, Israel.

SATAN CAST INTO THE ABYSS

Satan, their unseen leader, is also dealt with, and according to Revelation 20:1-3 he is cast into the abyss where he is rendered inactive for the entire period of the thousand-year reign of Christ on earth. Then he is destined to join the beast and the false prophet in the lake of fire (Revelation 20:10). The Gentile population of the world as a whole is judged at a separate judgment which follows and is a part of the establishment of Christ's kingdom on earth.

The inglorious end of Gentile power is precisely that which was anticipated in Daniel 2 where the great image disintegrates into chaff when struck by the stone cut out without hands. The same dramatic end is contemplated in the destruction of the beast (Daniel 7:11) followed by the inauguration of the everlasting kingdom in Daniel 7:13, 14. Jerusalem was no longer to be trodden under the feet of the Gentiles and once again Israel was to be exalted.

You are Absolutely Safe – If You Choose!

Prominent Atheists have propagated several hypotheses to the "Origin of the Universe." The leading unscientific-unproven theory is the Big Bang. These atheists are steadily attempting to erase the creator of the Universe from His existence, hoping to inject their own unscientific ideologies. Additionally, well known evolutionists conjectured a sacred life evolved from a non-life natural-materialistic substance, also hoping to erase the Creator of all life in the Universe.

Atheists and evolutionists alike are diligently attempting to CANCEL God from our evolved western culture. While their plans have well succeeded to delude the minds of multitudes of people and even nations, they have particularly aimed to pollute the innocent minds of school children.

However, their debased plans and futile diligence would not change the Wholesomeness of the presence of the Creator to continually Hold and Sustain the entire Universe in the Palm of His Holy Hands.

Both the Big Bangers together with the Evolutionists, supported by Atheistic Media, proponent Research Institutions as well as several Government Departments engaged in spreading their hypothetical demise of the Universe in total destruction through few means, leading to a total annihilation, death, decay and catastrophically ending. The universe of the big banger, e.g., will end up in big chill, big crunch or big rip. While, the evolutionary Universe believe it will expand forever and will eventually run out of hydrogen; e.g., the stars eventually burn out, and the Universe cools down to a vast frozen graveyard of dead stars. The evolutionists will continue their death ceremonies and decay sermons condemning the vulnerable and the helpless.

Albeit the present Universe is polluted by sin, death and decay, the Creator has promised restoring it to its pure condition prior to the advent of sin and decay. Since the dawn of creation sin against God's desire wreaked havoc on all creation. The Creator, who never failed a promise has assured His believers to revitalize this universe to a New Heaven and a New Earth.

According to Bible statistics, God has made approximately 7,500 promises to us as we add up his words in both Old and New Testaments. Of these scriptures, God has never broken a single promise. We can count on God's word always and forever.

With respect to the end of the universe, extensive studies have been conducted to provide guidance to those who put their trust in the Creator as well as those who may not have comprehended yet the value to acknowledge the creator of the Universe.

The author of this book hopes and prays for such people that somehow, one day they may acquire the knowledge and the wisdom to recognize that the Creator is living and awaiting each individual to have a true and meaningful personal fellowship with Him.

Among the 7,500 promises of the Creator, one promise was given to both Daniel, who was the prime minister of Babylon during the reign of King Nebuchadnezzar few centuries prior to the birth of the Lord Jesus Christ and given to the apostle John 95 years after the birth of the Lord Jesus Christ. The assured prophecy stated that the Lord Jesus Christ who have executed all the act of the entire creation of the Universe will descend from Heaven to take the believers with Him to join His glory in Heaven.

The believers who died will be resurrected first to join the believers who are still alive. Both believers will be raptured in a twinkle of an eye to transfigure to an identical form as the form of the Lord Jesus Christ prior to entering the holiness threshold of Heaven.

This rapture signals the beginning of traumatic events, called the tribulation period for 7 years, leading to the second coming of the Lord Jesus Christ descending with all His believers to His created Earth.

THE END OF THE WORLD

The Middle East – The Center of Controversies

End-time prediction focuses on studied events in the Middle East, but can one distinguish the signs leading to the return of Jesus Christ? The Middle East has been in chaos for decades, and many who follow world news question what will happen next. Yet very few understand that Bible prediction explicates *in advance* the momentous events that will lead to amplified conflict and greater tragedies in the region—eventually and most likely World ultimate catastrophe and then the return of Christ!

The Bible outlines end-time events leading up to Jesus Christ's return. What happens in the Middle East, in particular as His return draws near, will affect the lives of every person on earth. Comprehending those events will help you take action for your family's wellbeing. One *may* escape the foretold disasters and judgments (Luke 21:36) upon a warring and God-rebelling world.

The Middle East look as if to bring the world a constant stream of bad news. Bible prediction demonstrates that this bad news will continue until Jesus Christ's return. But once the world endures "Armageddon" in the Middle East, a time of *good news* will follow. Jerusalem will then live up to its name: "City of Peace." A Millennium of peace and prosperity is coming, under the rulership of the Messiah. As King of kings, Jesus Christ will rule the earth from Jerusalem with justice and uprightness, and all nations on earth will learn a new way of life—the way to peace.

Nonetheless, first, the Middle East must go through difficult times ahead. Through history, human nature has beheld to war as the final arbitrator of disputes. The Apostle Paul made a philosophical statement: "The way of peace they have not known. There is no fear of God before their eyes" (Romans 3:17–18). More than a million were killed in the Iran-Iraq War from 1980 to 1988. Thousands upon thousands more died in the Persian Gulf War and Operation Desert Storm in 1990–91. Barely more than a decade later, the United States knotted itself financially and militarily in another Iraq conflict, leading to questions about military and political motivations for U.S. action, and bringing on thousands more deaths and a severe loss of international prestige for the nation. Then the "Arab Spring" sparked riots, rebellions, terrorism, and wars, from Egypt, to Syria, to Iraq. By 2015, the United Nations projected that more than 220,000 Syrians had died in the Syrian conflict. And ongoing atrocities by al-Qaeda and the Islamic State (ISIS) have entangled large swaths of the Middle East as ISIS tries to create an Islamic Caliphate.

Will there be more wars in the Middle East? Dejectedly, the answer is "yes." The Bible discloses that there will be even more regional wars there, leading to what is commonly called "Armageddon." Jerusalem will be a focal point of conflict.

Arabs and Israelis have been at war since the modern state of Israel was formed. Peace has often been decreed in the region, but has never been grasped. More than 25 years after Egyptian President Anwar Sadat offered hope to the world, flying to Israel to bid a dramatic proposition of peace, the process he trusted to encourage has stalled, if not stopped. At present, there is little human vision for lasting peace among Middle Eastern nations.

The Bible divulges that this millennia-old conflict *will* come to a peaceful resolution in the not-too-distant future. But before this determination is grasped, much blood will be shed and mankind will go through a time of war and grief on a scale previously unknown in human history.

Careful Consideration to Bible prophecy about the Middle East can convey us optimism and ease even as the current crisis intensifies. The Bible is filled with information that can help us understand the present Middle East struggle in the light of God's design for end-time world events. It clarifies many details mysterious to the world's politicians and legislators. As we will understand, it exposes what the ancient contention between Abraham's grandsons Jacob and Esau means for the Middle East conflict today. Scripture also demystifies many widespread but false teachings regarding Russia's role in end-time prophecy. It gives an impression of the "key players" in end-time prophecy—including the beast, the false prophet, and the King of the North—and the final phase of World-War, as described later.

Bible prophecy has revealed in advance many key facts and particulars that can help us formulate for the treacherous and stimulating times ahead, as we endeavor to obey Jesus Christ's command to "watch and pray" as end-time proceedings grow closer (Mark 13:33). And it discloses the vital role of Jerusalem—a city with a remarkable past and an even more remarkable future—at the center of many end-time events and prophecies.

THREE RELIGIONS IN THE HOLY CITY OF JERUSALEM

Hot Bed of Terror

Whether the ongoing struggle between the Palestinians and Israel leading to October 7, 2023 or the growing threat of the Islamic State and the aftermath of the "Arab Spring," the West's perceptions of Middle East violence and its relationship to world terrorism were changed forever on September 11, 2001. On that day, terrorists hijacked four planes, which they used as missiles to abolish the twin towers of the World Trade Center in New York and destroy the Pentagon in Washington, DC.

On September 20, 2001, U.S. President George W. Bush, speaking before a joint session of Congress, avowed war on international terrorism. A U.S.-led coalition joined with Afghan militias in destroying and expelling al-Qaeda forces. Afghanistan's Taliban government was ousted, and the world's attention quickly shifted to Iraq. Saddam Hussein, who had for years been accused of manufacturing weapons of mass destruction, was conquered in a war that surprised many with its speed. But "winning the peace" has demonstrated to be a far greater test, as the U.S. remains stuck in a conflict that Defense Department analysts say could last for many years. Through it all, regional terrorism continues, and U.S. agencies are preparing for future attacks by al-Qaeda and other terror cells.

The West was then further stunned—and many people's awarenesses further hardened—by the ruthless violence of the Islamic State and its wars in Syria and Iraq, its decapitations, and its burning hostages alive. Westerners have also been stunned by cumulative Muslim extremism, often connected to ISIS or al-Qaeda. Attacks have occurred in various places, but perhaps the most appalling to date was the January 2015 attack on the Charlie Hebdo offices in Paris, France, which French president François Hollande called "the worst spasm of terrorism in France since the 1954-62 Algerian War" ("French Police Storm Hostage Sites, Killing Gunmen," *The New York Times*, January 9, 2015).

As *The Wall Street Journal* reported, although U.S. and allied airstrikes put Islamic State troops on the defensive in some areas in 2015, Syria and large parts of Iraq remained safe havens and recruiting grounds for Islamic terrorists, and the Obama administration considered "whether the U.S. should embrace more aggressive ideas for containing Islamic State forces" ("Months of Airstrikes Fail to Slow Islamic State in Syria," January 14, 2015). Most Muslims do not consider themselves terrorists, nor do they support the terror carried out by ISIS or al-Qaeda. There are more than 1.6 billion Muslims around the world, between one quarter and one third of whom live in the Middle East. Yet, while the vast majority of Muslims consider themselves peace-loving people, their religion has been used as a rallying point by a small number of extremists who wish to destabilize nations in order to achieve their goals.

Instability appears to be a fact of life in the twenty-first century. Even as it faces reduced international credibility in light of its Iraq involvement, the U.S. now seeks to lead an alliance to counter potential terror threats sponsored by what President George W. Bush once called an "axis of evil" that includes two Middle East nations—Iran and Iraq—but which has now spread to include additional powerful terrorist organizations such as ISIS.

So not only do the United States and the West face an increasingly volatile Middle East, but also these Muslim nations and supranational terrorist organizations are already reaching out beyond the region. Pakistan and India, like Israel, are known to have nuclear weapons. Other countries are widely suspected of nuclear capability. In today's unstable environment, many of the world's experts consider nuclear war a distinct possibility—even a probability. If total war breaks out in the Middle East, might future missiles deliver nuclear, biological, and chemical destruction to Israel and surrounding nations? Will the Holy Land become the focal point for another world war?

WHAT LIES AHEAD?

Many observers think that it is foolish, or even impossible, to predict the future. In the Bible, however, God has given us prophecies that can help us understand what will happen in and around Jerusalem in the end times. What does the Bible say will eventually happen? "Behold, the day of the LORD is coming, and your spoil will be divided in your midst. For I will gather *all the nations* to battle against Jerusalem; the city shall be taken, the houses rifled, and the women ravished. Half of the city shall go into captivity, but the remnant of the people shall not be cut off from the city" (Zechariah 14:1–2).

"All the nations" will fight in this conflict, during that period of time known as the Day of the Lord—the year just before Jesus Christ's prophesied return. But what major prophetic events will lead up to this battle in Jerusalem? "For then there will be great tribulation, such as has not been since the beginning of the world until this time, no, nor ever shall be. And unless those days were shortened, no flesh would be saved; but for the elect's sake those days will be shortened" (Matthew 24:21–22).

The book of Revelation, or the book of the Apocalypse, reveals the dramatic end-time events leading up to the glorious return of Jesus Christ. Revelation 6 describes the events leading up to the Day of the Lord. All of humanity will see dramatic and frightening activity in the heavens. Revelation 6:12–14 describes these heavenly signs as the "sixth seal." Jesus said that "Immediately after the tribulation of those days the sun will be darkened, and the moon will not give its light; the stars will fall from heaven, and the powers of the heavens will be shaken" (Matthew 24:29). These heavenly signs precede the Day of the Lord (Joel 2:30–31), the year of God's judgment on the nations. Then, after these events, the Messiah—Jesus—will rule all nations on earth (Revelation 11:15).

What role will Jerusalem play in that coming kingdom? Will it be the future capital of the world? Perhaps in your lifetime? Scripture gives us a wonderful and encouraging forecast: "The word that Isaiah the son of

Amoz saw concerning Judah and Jerusalem. Now it shall come to pass in the latter days that the mountain of the LORD's house shall be established on the top of the mountains, and shall be exalted above the hills; and all nations shall flow to it" (Isaiah 2:1–2).

Mountains are a biblical symbol for kingdoms or governments (Jeremiah 51:24–25). Isaiah stated plainly that the Lord's Kingdom would be established in Jerusalem and that "all nations shall flow to it." This will not be government in the hands of selfish human beings. Because it has rejected the rulership and the sovereignty of God over the affairs of man, humanity does not know the way to lasting peace.

God's government on earth will be administered from Jerusalem, the future capital of the world, as Zechariah confirms with a beautiful description of life there in the future: "Thus says the LORD: 'I will return to Zion, and dwell in the midst of Jerusalem. Jerusalem shall be called the City of Truth, the Mountain of the LORD of hosts, the Holy Mountain.' Thus says the LORD of hosts: 'Old men and old women shall again sit in the streets of Jerusalem, each one with his staff in his hand because of great age. The streets of the city shall be full of boys and girls playing in its streets'" (Zechariah 8:3–5).

Today, however, the streets of Jerusalem are filled with bitter enemies who share a rich heritage and history—but also share a history of conflict.

BITTER FAMILY RIVALRY

To understand the present Middle Eastern conflict, we need to understand the history of those in conflict—the Jews and the Arabs. Historians often note the "family" relationship between these two peoples; some call them "cousins" because both trace their roots to the great patriarch Abraham.

Judaism, Christianity, and Islam all honor Abraham as a patriarch. Calling him "Religion's Superstar," *Time* magazine asked, "Abraham changed history by espousing just one God, and is sacred to Muslims, Jews and Christians. Can this crossover biblical hero help heal the world?" (September 30, 2002, p. 5).

The answer is, *Not in this age!* But in tomorrow's world, as "heir of the world" (Romans 4:13) under Jesus Christ and as one of the saints in God's Kingdom, Abraham *will* help teach the way to peace when all nations come to Jerusalem (see Isaiah 2:2–4).

In the nineteenth century BC, the Creator God began to work with Abraham (then called "Abram"). God tested his faith many times, even promising him a son in his old age—a promise that seemed to be delayed. Abram's wife, Sarai, unable to bear children, encouraged Abram to have a child through her Egyptian handmaid, Hagar. Eighty-six years old when the child was born, Abram gave the name Ishmael to the son whom Hagar bore (Genesis 16:15–16). To this day, Muslims acknowledge Ishmael as the ancestor of the Arabs, as described in their scripture, the Quran.

When he was 99 years old, God made a covenant with Abram that he would be "a father of many nations." God then changed Abram's name to Abraham, meaning "father of a multitude." God promised that Abraham and Sarai, whose name was changed to Sarah, would have a son named Isaac. God said to Abraham, "I will establish my covenant with him [Isaac] for an everlasting covenant, and with his descendants after him" (Genesis 17:19). One of Isaac's grandsons, Judah, would be the progenitor of the Jews, who now occupy the modern land called Israel.

After Isaac was born, Sarah insisted that Abraham send Hagar and Ishmael away. Sarah said to Abraham, "Cast out this bondwoman and her son; for the son of this bondwoman shall not be heir with my son, namely with Isaac" (Genesis 21:10). Abraham was very upset at this request, but God told him, "Do not let it be displeasing in your sight because of the lad or because of your bondwoman. Whatever Sarah has said to you, listen to her voice; for in Isaac your seed shall be called. Yet I will also make a nation of the son of the bondwoman, because he is your seed" (Genesis 21:12–13).

The envy and rivalry between these two peoples continue to this day. Another dimension that helps explain today's ongoing conflict is the rivalry between Jacob and Esau, the twin sons of Isaac. Esau was the firstborn and the heir of the family birthright. But in a time of hunger, Esau sold his birthright to Jacob for a bowl of soup. Later, Jacob disguised himself as Esau and deceived his aging father into giving him the birthright blessing (see Genesis 25; 27). What was Esau's reaction? "So Esau hated Jacob because of the blessing with which his father blessed him, and Esau said in his heart, 'The days of mourning for my father are at hand; then I will kill my brother Jacob'" (Genesis 27:41).

This bitter rivalry would continue throughout history. Esau is also called Edom (see Genesis 36). The Edomites intermarried with the Ishmaelites, beginning with Esau himself (see Genesis 28:9). The descendants of these peoples continued to be in conflict with Israel. The leading tribe of Edom, the Amalekites, continually harassed Israel (see Exodus 17). God said that He would have "war with Amalek from generation to generation" (Exodus 17:16).

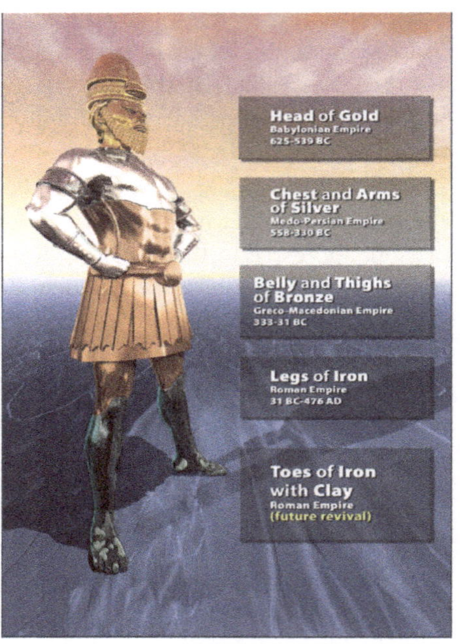

***Fig.32.61:** Daniel's Prophesy of The Rise and Fall of World Empires Curtsy "Tomorrow's World Publications"- Illustration Curtsy Bruce Long.*

One of the psalms also confirms both historically and prophetically that Edom, Ishmael, and Amalek are joined in a confederacy against Israel. "They have said, 'Come, and let us cut them off from being a nation, that the name of Israel may be remembered no more.' For they have consulted together with one consent; they form a confederacy against You: the tents of Edom and the Ishmaelites; Moab and the Hagrites; Gebal, Ammon, and Amalek; Philistia with the inhabitants of Tyre; Assyria also has joined with them; they have helped the children of Lot. Selah" (Psalm 83:4–8).

What nations have continuously committed themselves to destroying the name of Israel?

In July of 1968, Palestinian leaders signed the "Palestinian National Charter," which called for the destruction of the State of Israel. In 1993, shortly before signing the "Declaration of Principles" with Israel, PLO chairman Yasser Arafat signed an agreement renouncing this demand and agreeing to change the charter, but it remains an item of hot dispute to this day. As Psalm 83 indicates, Assyria (identified as modern-day Germany) will be joined by an end-time "confederacy" of Arab nations. Look on a Bible map for the ancient location of Moab, Ammon, and Edom, nations mentioned in Psalm 83 that are also prominently mentioned in Daniel's end-time prophecy. These are the Muslim peoples of the Middle East!

Understanding this bitter family rivalry, beginning with Jacob and Esau, gives us a better understanding of their descendants' present-day wars and conflicts, particularly in the Middle East.

DANIEL'S FAMOUS PROPHECY

Historically, the Roman Empire in Jesus' day enforced a *Pax Romana*—"Roman Peace"—in the Holy Land. The prophet Daniel reveals that there will also be an end-time revival of the Roman Empire, which will play a crucial role in end-time conflicts in the Middle East. Daniel 2 records that King Nebuchadnezzar was going to execute all the wise men of Babylon if they did not tell him his dream, as well as its interpretation, Figure 32.61. The Creator God used the prophet Daniel to reveal the rise and fall of great empires, even down to our modern times. Here is what Daniel told the king: "But there is a God in heaven who reveals secrets, and He has made known to King Nebuchadnezzar what will be in the latter days" (Daniel 2:28). Daniel told Nebuchadnezzar that God had made him a king of kings, and described the image envisioned by the king: A great statue with a head of gold. Daniel told the king, "You are this head of gold" (v. 38). The remaining elements of the great image symbolized the great empires to follow after Babylon.

Who were these empires? All reputable Bible scholars agree on their identity and their prophetic fulfillment. Daniel continued to describe what would eventually happen to this great image. The head of gold represented the Babylonian Empire from 625 to 539 BC. Next came the Medo-Persian Empire from 558 to 330 BC, represented by the chest and arms of silver. The belly and thighs of bronze represented the Greco-Macedonian Empire of Alexander the Great from 333 to 31 BC. The two legs of iron represent the Roman Empire from 31 BC to 476 AD. Finally, the ten toes on two feet of iron mixed with ceramic clay represent a future revival of the Roman Empire.

How do we know that the feet of the image will continue into our modern times? Daniel describes a great stone that smashes the image on its feet. He also describes the stone's meaning: "And in the days of these kings [symbolized by the ten toes] the God of heaven will set up a kingdom which shall never be destroyed; and the kingdom shall not be left to other people; it shall break in pieces and consume all these kingdoms, and it shall stand forever" (Daniel 2:44). The Kingdom of God that will "consume all these kingdoms" is yet future!

The end-time power that will eventually dominate the Middle East will be a revived Roman Empire, also symbolized in the prophecies of Revelation and Daniel as a *beast* (cf. Daniel 7:17–18; Revelation 17:9–14; 13:1–18). This great power will be political, military, and economic. The current European Union is headed toward completing all three dimensions of that power. It could have the potential of being a force for good. But history teaches us that great empires and political alliances can also revert to dictatorial power for the purpose of global expansion and domination without regard for the welfare of conquered nation-states. The Creator God will use such an empire in the future, just as He has in the past. God used Assyria to punish the rebellious kingdom of Israel. In the opening verses of Isaiah 10, God indicts Israel for all her sins, then states, "Woe to Assyria, the rod of My anger and the staff in whose hand is My indignation. I will send him against an ungodly nation, and against the people of My wrath I will give him charge, to seize the spoil, to take the prey, and to tread them down like the mire of the streets. Yet he does not mean so, nor does his heart think so; but it is in his heart to destroy, and cut off not a few nations" (Isaiah 10:5–7).

God will once again use the descendants of Assyria to punish the descendants of the ancient House of Israel. God will use modern-day Germany to conquer the American and British-descended peoples. Germany is presently the leading nation of the European Union, and will play a central role in end-time prophecy. To learn more, please request our free reprint article "Resurgent Germany: A Fourth Reich?" or read it online at *TomorrowsWorld.org*. Be sure to watch developments in Europe.

In Daniel 11, we find two antagonists: the king of the North and the king of the South. In ancient times, Syria was considered to be the king of the North. But the Roman Empire conquered Syria, which became a province of Rome in 64 BC. Bible prophecy shows that by the time of the end, the king of the North will be identified with the revived Roman Empire. Then, "At the time of the end the king of the South shall attack him; and the king of the North shall come against him like a whirlwind, with chariots, horsemen, and with many ships; and he shall enter the countries, overwhelm them, and pass through. He shall also enter the Glorious Land [or the Holy Land], and many countries shall be overthrown; but these shall escape from his hand: Edom, Moab, and the prominent people of Ammon. He shall stretch out his hand against the countries, and the land of Egypt shall not escape" (Daniel 11:40–42).

Notice that the king of the North enters into the "Glorious Land"—the location of the modern state of Israel. The fact that Ammon (which Bible scholars identify as modern-day Jordan) escapes out of his hand, along with Edom and Moab, leads many to believe that Jordan will be allied with the European power. Psalm 83 identifies those peoples who will be confederate with Assyria against Israel. That alliance includes Moab, Ammon, and Edom.

Daniel 11:43 gives us an indication of those nations allied with the king of the South. These are the peoples of North Africa. Verse 40 states that the king of the South attacks the king of the North. What will provoke this attack? Will it be the passion of Islamic fundamentalists? Will it be a sudden interruption of oil flow to Europe? You need to watch world events in Europe and the Middle East.

Who will invade the Middle East? Look for a revived Roman Empire—a political entity now in its formative stages with the European Union as its nucleus. There will be a forced peace upon modern Israel in the Middle East. But, as we will see later, the "beast power" that occupies the Middle East will be overthrown by the returning Commander of heaven's armies, Jesus Christ. The King of kings and Lord of lords, revealed in Revelation 19, will see that the beast and the false prophet are cast into a lake of fire. Jerusalem will be occupied by Earth's new King, Jesus Christ. The name of the city will even be called "THE LORD IS THERE" (Ezekiel 48:35). Peace *will* finally come to the Middle East and to the City of Peace, Jerusalem.

WILL RUSSIA CONTROL THE MIDDLE EAST?

Some teachers of biblical prophecy wrongly believe that Russia will take over the Middle East before Christ's return. Where does that mistaken idea come from? It comes mainly from a misunderstanding of Ezekiel 38, which begins, "Now the word of the LORD came to me, saying, 'Son of man, set your face against Gog, of the land of Magog, the prince of Rosh, Meshech, and Tubal, and prophesy against him'" (vv. 1–2). Most biblical scholars identify these peoples as those from northern Russia. Wilhelm Gesenius, a nineteenth-century Hebrew scholar, noted that the city of Moscow is derived from the Hebrew name Meshech. Ezekiel continues, "Thus says the LORD God: 'Behold, I am against you, O Gog, the prince of Rosh, Meshech, and Tubal. I will turn you around, put hooks into your jaws, and lead you out, with all your army, horses, and horsemen, all splendidly clothed, a great company with bucklers and shields, all of them handling swords. Persia, Ethiopia, and Libya are with them, all of them with shield and helmet; Gomer and all its troops; the house of Togarmah from the *far north* and all its troops; many people are with you'" (vv. 3–6).

Notice that the house of Togarmah is from the far north—that is, north of the Holy Land. When you look at a map, Russia is in the far north. Who will join Russia? Persia is mentioned. Most scholars agree that Persia is modern-day Iran. The Hebrew word for Ethiopia is *Cush*, the eastern branch of which has sometimes been identified with India. The Hebrew word for Libya is *Put* (Put was the third son of Ham, according to Genesis 10:6), whose eastern branch, like Cush, is also identified with India. Magog is identified with the Mongols and Gomer with Indochina. Togarmah is most likely identified with Siberia.

We can generally agree on the identity of these peoples mentioned in Ezekiel 38. And Bible prophecy shows that these armies certainly will *eventually* enter the Middle East. But the question is, *When*? After Jesus' second coming, He will work to bring world peace and reeducate the nations (cf. Isaiah 2; Micah 4). But that will take time. All nations will *not* automatically accept the rule of Jesus Christ when He returns. Even Egypt may apparently refuse the Lord's command to keep the Feast of Tabernacles in Jerusalem (cf. Zechariah 14:18–19). The Russo-Chinese alliance, mentioned in Ezekiel 38, will attack *after* the Millennium begins—*after* the second coming of Christ and after peace is established in the Holy Land. Notice what God says of the invading armies: "You will say, 'I will go up against a land of unwalled villages; I will go to a peaceful people, who dwell safely, all of them dwelling without walls, and having neither bars nor gates'; to take plunder and to take booty, to stretch out your hand against the waste places that are again inhabited, and against a people gathered from the nations, who have acquired livestock and goods, who dwell in the midst of the land" (Ezekiel 38:11–12).

Does Israel today "dwell safely" in unwalled villages, that is, without military protection? Of course not! Ezekiel is speaking of a time yet future, when Christ establishes peace in the Holy Land. Notice: "Therefore, son of man, prophesy and say to Gog, 'Thus says the Lord God: "On that day when My people Israel dwell safely, will you not know it?"'" (v. 14).

God will destroy this eastern alliance and its armies when they attack restored Israel (cf. Ezekiel 39). There will be so many dead bodies that it will take months to bury them. "For seven months the house of Israel will be burying them, in order to cleanse the land" (Ezekiel 39:12). Almighty God has all the power in the universe to fulfill His will. It is folly for nations to fight against Him and His purpose. But the nations will learn who is in charge. As God states, "Thus I will magnify Myself and sanctify Myself, and I will be known in the eyes of many nations. Then they shall know that I am the Lord" (Ezekiel 38:23).

WHO WILL CONTROL JERUSALEM?

Since 1967, Israel has fought wars and negotiated peace treaties with several Arab states. But Israel has never been able to negotiate a permanent peace agreement with the Palestinians. In July 2000, Israeli Prime Minister Ehud Barak did what no Israeli negotiator had done before—he proposed that Israel would grant administrative self-governance to Palestinians in East Jerusalem. But Barak wanted Israel to retain security control over East Jerusalem, and the Palestinians rejected his proposal. Negotiations broke down, leading to ongoing conflict. What caused the impasse in negotiations? The one issue that appears unsolvable is, *Who will control Jerusalem?*

In 1980, Israel declared Jerusalem as its "united and eternal capital." The Palestinians also desire Jerusalem as their capital. Before his death in 2002, Faisal Husseini, a representative of the Palestinian Authority in Jerusalem, stated the Palestinian position that Israel "must withdraw from all of east Jerusalem to the pre-'67 borders…. all settlements and Israeli neighborhoods in east Jerusalem must be dismantled…. Israel must compensate the Palestinians for the damages it has inflicted including the changes in the character of the city and the lives of its citizens" (*The Jerusalem Post*, November 19, 1999).

Israel has maintained that Jerusalem must remain its undivided capital. As Israeli spokesman Gadi Baltiansky stated in 1999, "There will be no negotiations over Jerusalem, and if the Palestinians think otherwise—too bad. This is a red line that Prime Minister Barak will never cross" (*The Jerusalem Post*). Even Barak's proposal for "sharing sovereignty" did not concede that Jerusalem was not a united city under the control of Israeli security. In the years since, facing the *intifada*, Barak's successors have expressed no eagerness to revisit his proposal, which Israelis considered generous but which Palestinians considered unacceptable.

The U.S. has backed Israel in this dispute, and has thus found itself the target of international anger. On September 30, 2002, then-President Bush signed congressional legislation that required his administration to recognize Jerusalem as Israel's capital. What was the reaction? Cairo's leading newspaper reported, "The new US legislation requiring all official American documents to identify the occupied city of Jerusalem as the capital of Israel has triggered angry reactions throughout the Islamic world. Thousands of demonstrators marched through Gaza, threatening to intensify suicide bombings inside Israel, while popular and religious leaders called on Arab and Islamic parliaments to adopt a strong stance against this latest outrage, calling, among other things, for a boycott of American goods" (*Al-Ahram*, October 16, 2002).

The gap between the two sides' views is wide and deep! What solutions has the international community proposed? In 1980, when Israel called Jerusalem its "united and eternal capital," the Vatican strongly objected. In 1984, Pope John Paul II wrote an apostolic letter, *Redemptoris Anno*, calling for Jerusalem to have a "special internationally guaranteed status." In 1975, U.S. Secretary of State Henry Kissinger had made a similar proposal—that Jerusalem become an international city with the control of holy places and the religious administration being given to the pope.

Pope John Paul II spoke to the new Jordanian ambassador to the Vatican on May 28, 1998. *Reuters* reported his comments: "The long history of the city of Jerusalem, filled with tribulations, will reach a new threshold in the year 2000, with the dawn of the third millennium of Christianity.... It is my fervent hope that this may lead to a formal recognition, with international guarantees, of the unique and sacred identity of the Holy City."

And heading into 2015, Pope Francis "made a heartfelt plea to Israelis and Palestinians to put an end to the violence which has plagued Jerusalem and other parts of the Holy Land in recent weeks. His appeal follows an attack on a Jerusalem synagogue Tuesday in which 2 Palestinian men armed with a meat cleaver and a gun killed three U.S.-Israeli rabbis and a British-Israeli man. An Israeli policeman later died of his wounds" (*Vatican Radio*, November 19, 2014).

Will non-Israelis control Jerusalem? The Apostle John said that they will! "Then I was given a reed like a measuring rod. And the angel stood, saying, 'Rise and measure the temple of God, the altar, and those who worship there. But leave out the court which is outside the temple, and do not measure it, for it has been given to *the Gentiles*. And they will tread the holy city underfoot for forty-two months'" (Revelation 11:1–2).

Gentiles will control the "holy city"—Jerusalem—for 42 months. The book of Revelation describes a 42-month period leading up to the return of Christ. Two prophets of God will witness with great power during these three-and-a-half years, or 1,260 days (Revelation 11:3–15). These two witnesses will contend with the forces of a revived Roman Empire, which will dominate the Middle East at that time.

TEMPLE SACRIFICES RESTORED?

Many religious teachers understand *part* of Bible prophecy. They expect that the Jewish orthodox authority in Israel will once again establish temple sacrifices. Groups such as the Temple Institute in Jerusalem are making preparations today to train priests and assemble the items required for the reestablishment of sacrifices. Will the Jewish priesthood once again offer animal sacrifices? Some Jewish groups are going to great lengths trying to make this happen.

A few religious groups and individuals have gone overboard in their desire to see prophecy fulfilled. The Israeli authorities placed their security on special alert during the months of December 1999 and January 2000 because several religious groups, referred to as "apocalyptic," allegedly plotted to "help prophecy along" by their own efforts and threatened to blow up the Dome of the Rock in order to make possible the construction of a third temple.

But true Christians recognize that *God is in charge*. He does not need our help to bring these prophecies to pass! Besides, true Christians do not go around causing violence and blowing up buildings. Even some Jewish extremists have threatened violence around the Temple Mount to get their way. Any such violence on the Temple Mount could cause an Arab uprising, or even a *jihad* ("holy war") that would have international repercussions.

The Dome of the Rock, a mosque holy to Muslims, stands on the Temple Mount, where there is no Jewish temple. Yet many Jews, and some Christians, are expecting some kind of divine intervention to make possible the building of a third temple on the Temple Mount.

What will happen? Prophecy indicates that the sacrifices will be reinstituted. The Jews have not offered temple sacrifices since Roman armies destroyed Jerusalem and the second temple in 70 AD. This leads to another important question: *Is a temple necessary for the revival of sacrifices?*

The Babylonian Empire of the sixth century BC conquered the kingdom of Judah. The Babylonians transported thousands of captive Jews to Babylon. The armies of King Nebuchadnezzar also destroyed the first temple at Jerusalem in 586 BC. Eventually, under the Persian Empire, King Cyrus sponsored a group of exiles to return to Jerusalem. When the returning Jews eventually laid the foundation of the second temple, there was both rejoicing and sadness. Joy because they now had a center of worship, and sadness because this second temple was very simple by comparison to the former glorious temple built by King Solomon. But notice the attitude of the Jews. They actually began sacrificing before the foundation of the temple was laid. "From the first day of the seventh month they began to offer burnt offerings to the LORD, although the *foundation of the temple* of the LORD *had not been laid*" (Ezra 3:6). The Jews had observed the Feast of Tabernacles in their unfinished temple with daily sacrifices.

Notice another important point. Where do the exiles return? Ezra referred to the location of the holy place as the house of God, when the temple had not yet been built: "Now in the second month of the second year of their coming to *the house of God* at Jerusalem" (v. 8). This took place around 536 BC. But there was no building as yet. The third chapter of Ezra goes on to describe the laying of the foundation of the temple. Sacrifices must be presented in a holy place. But in this case, the sacrifices were made daily *without a physical building called a temple*.

The holiest place under the control of the Jewish religious authorities today is the "Wailing Wall"—the western retaining wall of the Temple Mount. Jews are not currently allowed to publicly worship on the Temple Mount; only Muslims now have that privilege. So, it remains to be seen where the Jews will begin sacrificing. It will probably take a national crisis to precipitate that event.

Jesus spoke of a time when enemy armies would invade the Holy Land and surround Jerusalem. "But when you see Jerusalem surrounded by armies, then know that its desolation is near" (Luke 21:20). Remember, this is in the context of Jesus' Olivet Prophecy, also described in Matthew 24 and Mark 13. A powerful alliance of nations will invade the Middle East and dominate the region, as foretold by the prophet Daniel (cf. Daniel 11:41).

There will be international guarantees protecting Jerusalem, enforced by a superpower. But these guarantees will also enforce the pseudo-Christianity of a revived Roman Empire. There will be a revived *Pax Romana* ("Roman Peace"), imposed by what the Bible calls the end-time "beast" (see Revelation 13).

When the "beast" superpower occupies Jerusalem, this will signal the time of the great "tribulation" and "the days of vengeance, that all things which are written may be fulfilled" (Luke 21:22).

Although the word "Jerusalem" means "city of peace, Figure 32.62" Jerusalem has experienced bloody fighting and wars over the millennia. But soon, this great city will truly be the city of peace for all peoples. That is good news you can look forward to!

It is important to note that the prophesied "great tribulation" is primarily a time of trouble for the U.S. and British-descended peoples, who are descended from the ancient patriarch Joseph. When he was near death, Jacob blessed two of his grandsons—Joseph's sons Ephraim and Manasseh—and said, "The Angel who has redeemed me from all evil, bless the lads; let my name [Jacob—i.e., Israel] be named upon them, and the name of my fathers Abraham and Isaac; and let them grow into a multitude in the midst of the earth" (Genesis 48:16).

The end-time prophecies relating to Jacob, or Israel, generally refer to the descendants of Ephraim and Manasseh who were given Jacob's name. Scripture gives an end-time warning about a great "tribulation." Upon whom does this tribulation fall? "Alas! For that day is great, so that none is like it; and it is the time of Jacob's trouble, but he shall be saved out of it" (Jeremiah 30:7).

Fig.32.62: *The Turmoil of Israel in The Middle East*

WATCH FOR THE ABOMINATION OF DESOLATION

The Anti Christ and The False Profit

The prophesied "great tribulation" will be a time unique in all of history—a time such "that none is like it." What does the Bible tell us to do? Jesus tells us, "Watch therefore, and pray always that you may be counted worthy to *escape all these things* that will come to pass, and to stand before the Son of Man" (Luke 21:36).

What should you watch for in the Middle East? When Jesus was on the Mount of Olives, He gave an outline of end-time prophecy, with parallel accounts recorded in Matthew 24, Luke 21, and Mark 13. Jesus' disciples had asked Him about the sign of His coming and the end of the world, or the end of the age. He spoke about religious deception, wars, famines, pestilences, and earthquakes. These trends reveal, in sequence, the meaning of the four horsemen of Revelation 6, who, toward the end of this age, intensify their dangerous ride. Jesus declares, "All these are the beginning of sorrows" (Matthew 24:8).

Religious persecution and martyrdom will follow. "Then they will deliver you up to tribulation and kill you, and you will be hated by all nations for My name's sake. And then many will be offended, will betray one another, and will hate one another. Then many false prophets will rise up and deceive many" (Matthew 24:9–11).

But the Truth will still be preached! "And this gospel of the kingdom will be preached in all the world as a witness to all the nations, and then the end will come" (Matthew 24:14). As Jesus Christ opens doors for the Gospel to expand in coverage and power, you will know that the end is drawing near. Do not be asleep to world conditions and the prophetic trends intensifying before your very eyes!

Jesus gave us a key sign to watch for, which will signify the beginning of the great tribulation and trouble in the Holy Land. He warned, "'Therefore when you see the "abomination of desolation," spoken of by Daniel the prophet, standing in the holy place' (whoever reads, let him understand) 'then let those who are in Judea flee to the mountains'" (Matthew 24:15–16).

What is this abomination of desolation?

Centuries before Christ, an angel gave the prophet Daniel a vision concerning end-time events. When Daniel asked the meaning of the message, the angel told him, "Go your way, Daniel, for the words are closed up and sealed till the time of the end" (Daniel 12:9). More than 2,500 years later, we can now understand these prophecies. Notice another important detail: "And from the time that the daily sacrifice is taken away, and the abomination of desolation is set up, there shall be one thousand two hundred and ninety days" (Daniel 12:11). At the time of the end, shortly before the return of Christ, God reveals that the daily sacrifice will be stopped or "taken away." This obviously implies that the sacrifices must be started before they can be stopped! But what exactly is this "abomination of desolation"?

Historically, the Greek ruler Antiochus Epiphanes issued a decree in 167 BC that prohibited further Jewish sacrifices in the temple at Jerusalem. "And forces shall be mustered by him, and they shall defile the sanctuary fortress; then they shall take away the daily sacrifices, and place there the abomination of desolation" (Daniel 11:31).

In addition to stopping the daily sacrifices in the temple, Antiochus erected a statue of Jupiter Olympus in the Holy of Holies and sacrificed a pig upon the altar, desecrating it with the pig's blood. This abomination, also described in Daniel 8, was also the prophetic forerunner (or *antitype*) of an end-time event—the end-time sacrifices being cut off 1,290 days before Christ's return. Jesus warns us to be alert to an *end-time* abomination of desolation. Just as Antiochus Epiphanes profaned the temple in 167 BC and prohibited the sacrifices, so in the future another profane authority will prohibit sacrifices. In fact, the Apostle Paul warns of a great false prophet that will stand in the holy place: "Let no one deceive you by any means; for that Day will not come unless the falling away comes first, and the man of sin is revealed, the son of perdition, who opposes and exalts himself above all that is called God or that is worshiped, so that he sits as God in the temple of God, showing himself that he is God" (2 Thessalonians 2:3–4).

A great false prophet will work amazing miracles and deceive millions of people. He will cultivate worship toward himself, and claim that he is divine. Scripture describes a symbolic creature that appears like a lamb (the symbol for Christ) but speaks as a dragon (the symbol of Satan, the devil). This is the great religious power that will guide an end-time revival of the Roman Empire. "Then I saw another beast coming up out of the earth, and he had two horns like a lamb and spoke like a dragon. And he exercises all the authority of the first beast in his presence, and causes the earth and those who dwell in it to worship the first beast, whose deadly wound was healed. He performs great signs, so that he even makes fire come down from heaven on the earth in the sight of men. And he deceives those who dwell on the earth by those signs which he was granted to do in the sight of the beast, telling those who dwell on the earth to make an image to the beast who was wounded by the sword and lived" (Revelation 13:11–14).

This Antichrist will even call down fire from heaven! These deceiving miracles will excite millions—but will also lead many away from the true Jesus Christ of the Bible. You must not let yourself become deceived.

The prophet Isaiah explained a test by which we can recognize God's true servants: "To the law and to the testimony! If they do not speak according to this word, it is because there is no light in them" (Isaiah 8:20). In other words, true servants of Christ will speak according to the word of God, the Bible! You need to know and study your Bible. Prove and test all things, as it tells us in 1 Thessalonians 5:21!

The end-time religiopolitical power dominating Jerusalem will cause the sacrifices to cease. Remember that the Jews have not offered animal sacrifices since 70 AD, when the Romans destroyed the temple in Jerusalem. In order for the end-time sacrifices to be eventually stopped, they need to begin! When animal sacrifices begin once again in Jerusalem, you will know that the prophecies Jesus spoke about are heading for a climax!

THREE PROPHETIC MILESTONES

When the abomination of desolation prophesied by Jesus is set up, three prophetic milestones will follow over a period of three-and-a-half years: the "great tribulation," the heavenly signs, and the "day of the Lord." This prophetic framework helps us understand end-time developments in the Middle East and how they affect the world.

Notice that Jesus Himself gives the sequence of events. He warns us about the abomination of desolation in Matthew 24:15, then explains the *first* prophetic milestone in verses 21–22: "For then there will be great tribulation, such as has not been since the beginning of the world until this time, no, nor ever shall be. And unless those days were shortened, no flesh would be saved; but for the elect's sake those days will be shortened." This is a time unique in all history. It is the time of Jacob's trouble (Jeremiah 30:7), the time of God's punishment on the Western nations—including the United States, Britain, Canada, Australia, and New Zealand. The "great tribulation" is the time Daniel announces: "And there shall be a time of trouble, such as never was since there was a nation, even to that time" (Daniel 12:1).

This is a time of war, persecution, and martyrdom. It is the time when the four horsemen of Revelation 6 intensify their ride. They are given power over "a fourth of the earth, to kill with sword, with hunger, with death, and by the beasts of the earth" (Revelation 6:8). The fifth seal of Revelation describes the martyrdom of Christians (Revelation 6:9–11).

We then see a *second* prophetic milestone. The sixth seal reveals frightening, dramatic activity in the sky, or what a *New King James Version* heading calls "Cosmic Disturbances." The Apostle John wrote, "I looked when He opened the sixth seal, and behold, there was a great earthquake; and the sun became black as sackcloth of hair, and the moon became like blood. And the stars of heaven fell to the earth, as a fig tree drops its late figs when it is shaken by a mighty wind. Then the sky receded as a scroll when it is rolled up, and every mountain and island was moved out of its place" (Revelation 6:12–14).

Jesus also confirms this sequence of tribulation followed by great signs in the heavens: "Immediately *after* the tribulation of those days the sun will be darkened, and the moon will not give its light; the stars will fall from heaven, and the powers of the heavens will be shaken" (Matthew 24:29).

How will human beings react to these heavenly signs? "And the kings of the earth, the great men, the rich men, the commanders, the mighty men, every slave and every free man, hid themselves in the caves and in the rocks of the mountains, and said to the mountains and rocks, 'Fall on us and hide us from the face of Him who sits on the throne and from the wrath of the Lamb! For the great day of His wrath has come, and who is able to stand?'" (Revelation 6:15–17).

The heavenly signs announce the time of God's judgment on the nations, "the great day of His wrath." This is the seventh seal; the period of time also called the "day of the Lord"—a *third* prophetic milestone. In vision, the Apostle John refers to it in Revelation 1:10 as "the Lord's day." John is referring *not* to a day of the week, but to the time period leading up to the return of Christ—as mentioned in more than 30 prophecies! The prophet Isaiah describes it this way: "For it is the *day* of the LORD's vengeance, the *year* of recompense for the cause of Zion" (Isaiah 34:8).

The prophet Joel also confirms that the heavenly signs *precede* the "day of the Lord." "And I will show wonders in the heavens and in the earth: blood and fire and pillars of smoke. The sun shall be turned into darkness, and the moon into blood, *before* the coming of the great and awesome day of the LORD" (Joel 2:30–31).

The seventh seal, or the "day of the Lord," consists of seven trumpet plagues: "When He opened the seventh seal, there was silence in heaven for about half an hour. And I saw the seven angels who stand before God, and to them were given seven trumpets" (Revelation 8:1–2). The first six trumpets sound during the year preceding Christ's second coming. The seventh trumpet announces the good news of Christ's return to rule all nations on planet Earth: "Then the seventh angel sounded: And there were loud voices in heaven, saying, 'The kingdoms of this world have become *the kingdoms* of our Lord and of His Christ, and He shall reign forever and ever!'" (Revelation 11:15). While this is good news to genuine Christians, the nations will be angry (v. 18) and will actually gather their military forces to fight against the returning King (Revelation 19:19). As we will see, that final battle will be fought in the Middle East, near Jerusalem.

The good news is that the returning Jesus Christ will conquer all the rebellious armies and nations of planet Earth. The returning King of kings will set up His Kingdom in Jerusalem, conquer His enemies, and enforce lasting world peace (cf. Zechariah 14:1–3). Jesus Christ will come back to Jerusalem. He will rule over all the kingdoms, nations, and governments of this earth. "And the LORD shall be King over all the earth. In that day it shall be; 'The LORD is one,' and His name one" (v. 9). The Messiah *will* be King over all the earth, and will govern from Jerusalem. He will establish *true religion* over all the earth, replacing the *false* religion that will have dominated the revived Roman Empire.

WHO RIDES THE BEAST?

We have seen that the Middle East will be the location of a modern-day fulfillment of Daniel's prophecy concerning the "abomination of desolation." When that end-time abomination comes, what religious system will dominate it? The book of Revelation refers to the great false end-time religion as Mystery Babylon. In Revelation 17, the Apostle John sees in vision the judgment of a great harlot, which symbolizes a great false church or religious system, Figure 32.63. "Come, I will show you the judgment of the great harlot who sits on many waters, with whom the kings of the earth committed fornication, and the inhabitants of the earth were made drunk with the wine of her fornication" (Revelation 17:1–2). This woman rides a "beast" that has seven heads and ten horns (v. 3). That "beast," as we have seen earlier, is the political and military power that will invade the Middle East.

John continues, describing this great harlot: "And on her forehead a name was written: MYSTERY, BABYLON THE GREAT, THE MOTHER OF HARLOTS AND OF THE ABOMINATIONS OF THE EARTH" (Revelation 17:5), Figure 32.63. That system will dominate the great tribulation period. The leader of that system, the false prophet, will even call down fire from heaven, as we read in Revelation 13:13. Do not let yourself be deceived by such "miraculous" displays. Revelation 13:11 tells us that this religious figure symbolically looks like a lamb (the symbol of Christ) but speaks as a dragon! He will stand in Jerusalem's holy place and even claim that he is God!

In Revelation 17, we saw that the harlot rides the "beast." But who is this "beast"? The answer will tell us who will invade the Middle East. "The beast that was, and is not, is himself also the eighth, and is of the seven, and is going to perdition. The ten horns which you saw are ten kings who have received no kingdom as yet, but they receive authority for one hour as kings with the beast. These are of one mind, and they will give their power and authority to the beast. These will make war with the Lamb, and the Lamb will overcome them, for He is Lord of lords and King of kings; and those who are with Him are called, chosen, and faithful" (Revelation 17:11–14).

Fig.32.63: The False Religion in Harmony with The Beast of Revelation Curtsy by Derek London – Curtsy CodeX Medium

The Evangelical Churches have for years pointed out that the end-time "beast power" of Revelation will be a final revival of the ancient Roman Empire. That empire continued from about 31 BC to 476 AD. There were six revivals, including the Imperial Restoration under Justinian in 554 AD and another revival led by Charlemagne in 800 AD.

Other commentators have also identified the beast of Revelation as the Roman Empire. We just read in Revelation 17:11 of the final end-time revival of "the beast that was, and is not"—the final beast that goes into "perdition" or destruction. The *New American Catholic Edition* of the Holy Bible makes this comment concerning the beast of verse 11: "The beast spoken of here seems to be the Roman Empire, as in chapter 13."

The book of Revelation, Jesus' Olivet prophecy, and the book of Daniel together reveal a coming revival of the Roman Empire influenced and ridden by a woman. The woman is symbolic of a church. The "beast" power eventually turns on the woman: "And the ten horns which you saw on the beast, these will hate the harlot, make her desolate and naked, eat her flesh and burn her with fire, Figure 32.63. For God has put it into their hearts to fulfill His purpose, to be of one mind, and to give their kingdom to the beast, until the words of God are fulfilled. And the woman whom you saw is that great city which reigns over the kings of the earth" (Revelation 17:16–18). Again, the *New American Catholic Edition* comments on the ultimate destruction of the woman by the beast: "ten other kingdoms are allies of the beast and battle against the Church. But their dominion is short, typified as an hour." While the *New American Catholic Edition* identifies the beast with the Roman Empire and the woman with the Church, it does not admit that the great city mentioned in verse 18 is Rome itself. However, many Bible commentaries plainly agree that the great city of Revelation 17:18 is, in fact, Rome!

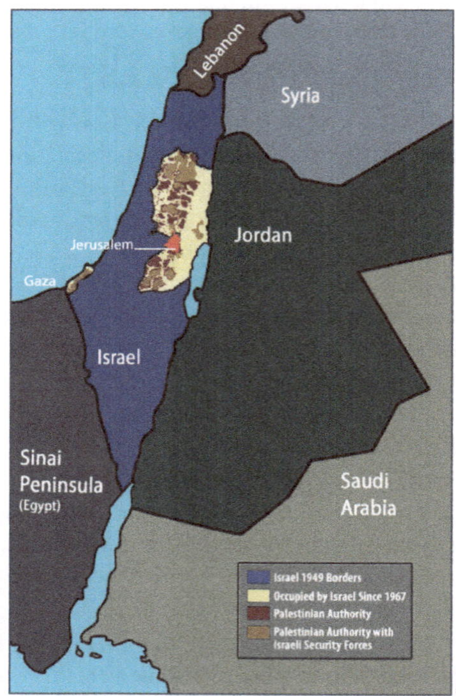

Fig.32.64: Map of Israel in The Middle East

THE KING OF THE NORTH INVADES THE MIDDLE EAST

The armies of Benito Mussolini—a twentieth-century "king of the North"—invaded North Africa during World War II. In contrast, the end-time king of the North will be *provoked* by the king of the South. We have already seen war in the Middle East over the issue of oil, Figure 32.64. Will there be another Arab oil embargo like that of 1974?

When fuel costs began to rise in the summer and fall of 2000, European protests of skyrocketing petroleum prices and shortages were even more vigorous than in the United States. In 2000, imports accounted for 53 percent of U.S. oil consumed. Roughly 24 percent—almost half of all oil imported—came from the Middle East. Europe is even more dependent on Middle Eastern oil, importing approximately 90 percent of its oil—well over half from the Middle East.

When gasoline prices spiked in mid-2008, millions were reminded again of their dependence on Middle Eastern oil. Will Europe let Middle Eastern oil suppliers hold their economies hostage? Whatever the provocation, the king of the North will invade the Middle East.

Luke's account of the Olivet prophecy follows the same general sequence as the account in Matthew that we reviewed earlier. But he adds one important detail—Jesus' warning sign concerning the end-time fate of Jerusalem: "But when you see Jerusalem surrounded by armies, then know that its desolation is near" (Luke 21:20). Whose armies are these? Who will eventually invade the Middle East and control Jerusalem—its most significant city?

As we saw earlier, the king of the North—the "beast power" or revived end-time Roman Empire—will enter the Glorious Land (Daniel 11:41). Jesus said that when you see armies encompassing or surrounding Jerusalem, you will know that the city is about to be captured. In 70 AD, Titus and the Roman armies destroyed the city and the temple. The Roman siege of the city caused famine and disease within it. The Jewish historian Josephus wrote that "no fewer than six hundred thousand were thrown out at the gates, though still the number of the rest could not be discovered" (*Wars of the Jews*, bk. 5, chap. 13, sec. 7). Which army brought disaster upon Jerusalem? The army of Rome did. Will another Roman army repeat history in the twenty-first century? Bible prophecy says, *Yes!*

The book of Luke also emphasizes that Jerusalem will be controlled by another superpower. Notice Jesus' powerful statement: "Then let those who are in Judea flee to the mountains, let those who are in the midst of her depart, and let not those who are in the country enter her. For these are the days of vengeance, that all things which are written may be fulfilled" (Luke 21:21–22).

This clearly is the end-time fulfillment of this prophecy. It continues, "But woe to those who are pregnant and to those who are nursing babies in those days! For there will be great distress in the land and wrath upon this people. And they will fall by the edge of the sword, and be led away captive into all nations. And Jerusalem will be trampled by Gentiles until the times of the Gentiles are fulfilled" (vv. 23–24). Revelation 11 shows that Christ will intervene 42 months after Jerusalem is captured.

Some Bible commentators mistakenly believe that Luke's prophecy applies *only* to the first century AD, when the Roman commander Titus besieged Jerusalem and destroyed the temple. Jesus' warning in Luke 21 certainly *did apply* to this first-century desolation. But as Bible students recognize, prophecy is often dual. There is type and antitype. There is a former fulfillment and a later fulfillment of the prophecy. Jesus told His disciples that, following the desolation of Jerusalem, many would see *His return to earth.* "And there will be signs in the sun, in the moon, and in the stars; and on the earth distress of nations, with perplexity, the sea and the waves roaring; men's hearts failing them from fear and the expectation of those things which are coming on the earth, for the powers of the heavens will be shaken. Then they will see the Son of Man coming in a cloud with power and great glory" (Luke 21:25–27).

This prophecy was not fulfilled in 70 AD by Titus' destruction of the temple. Jesus was clearly outlining the events preceding His second coming. The context of Jerusalem's final desolation, as presented here, is the end time.

THE FINAL PHASES OF WORLD WAR III

The king of the North's armies will occupy northern Africa and the holy land. But developments in the northeast so disturb this king that he takes military action. "But news from the east and the north shall trouble him; therefore he shall go out with great fury to destroy and annihilate many. And he shall plant the tents of his palace between the seas and the glorious holy mountain; yet he shall come to his end, and no one will help him" (Daniel 11:44).

This action appears to be described in Revelation 9 as the first woe or the fifth trumpet plague (cf. Revelation 8:13; 9:1). The Apostle John describes twenty-first-century warfare in first-century imagery: "The shape of the locusts was like horses prepared for battle. On their heads were crowns of something like gold, and their faces were like the faces of men. They had hair like women's hair, and their teeth were like lions' teeth. And they had breastplates like breastplates of iron, and the sound of their wings was like the sound of chariots with many horses running into battle. They had tails like scorpions, and there were stings in their tails. Their power was to hurt men five months. And they had as king over them the angel of the bottomless pit, whose name in Hebrew is Abaddon, but in Greek he has the name Apollyon" (Revelation 9:7–11).

World War III!

Notice that this action takes up five months during the year-long "day of the Lord." What follows next? A massive 200-million-man army from the east crosses the Euphrates River. The Euphrates River runs from Turkey through Syria and Iraq to the Persian Gulf. Whether or not Iraq even continues to exist as a nation in its present form, the geographical area of Iraq *will* play a major role in the final war to come: World War III! The second woe, or the sixth trumpet, now sounds.

One woe is past. Behold, still two more woes are coming after these things. Then the sixth angel sounded: And I heard a voice from the four horns of the golden altar which is before God, saying to the sixth angel who had the trumpet, "Release the four angels who are bound at the *great river Euphrates*." So, the four angels, who had been prepared for the hour and day and month and year, were released to kill a third of mankind. Now the number of the army of the horsemen *was* two hundred million; I heard the number of them. And thus, I saw the horses in the vision: those who sat on them had breastplates of fiery red, hyacinth blue, and sulfur yellow; and the heads of the horses *were* like the heads of lions; and out of their mouths came fire, smoke, and brimstone. By these three plagues *a third of mankind was killed*—by the fire and the smoke and the brimstone which came out of their mouths. For their power is in their mouth and in their tails; for their tails *are* like serpents, having heads; and with them they do harm (Revelation 9:12–19).

This phase of World War III will kill one third of all humanity, over two billion human beings! The "fire, smoke, and brimstone" evokes images of nuclear devastation. Unless Jesus Christ intervenes, "no flesh would be saved," as He said in Matthew 24:22!

The final and seventh trumpet announces the return of Jesus Christ to rule the earth (Revelation 11:15). The nations are angry, as we saw earlier (Revelation 11:18), and will combine their armies to fight a common foe, the Commander of heaven's armies (see Revelation 19:19)! The seventh trumpet or the third woe consists of "the seven last plagues" that complete "the wrath of God" on rebellious nations (see Revelation 15:1, 8; 16:1).

At this point, eastern armies will move westward over the Euphrates River on their way to Megiddo—Armageddon. "Then the sixth angel poured out his bowl on the great river Euphrates, and its water was dried up, so that the way of the kings from the east might be prepared" (Revelation 16:12). Satan and his demons influence warring nations to gather their forces against the "invader from outer space," the King of kings, Jesus Christ. The Apostle John writes, "And I saw three unclean spirits like frogs coming out of the mouth of the dragon, out of the mouth of the beast, and out of the mouth of the false prophet. For they are spirits of demons, performing signs, which go out to *the kings of the earth* and of *the whole world*, to gather them to the battle of that great day of God Almighty" (Revelation 16:13–14).

As the final phase of World War III begins, Satan will have influenced the kings and armies "of the whole world" to prepare for the great battle against Christ. Warring, rebellious nations will fight against the Savior of the world! Where will they gather? "And they gathered them together to the place called in Hebrew, Armageddon" (Revelation 16:16). The name Armageddon comes from the Hebrew *har Magedon*, which means "the hill of Megiddo." Megiddo is located in modern Israel about 55 miles north of Jerusalem. The hill or mount of Megiddo overlooks the largest valley in Israel, the Plain of Esdraelon or the Valley of Jezreel.

The developing European religiopolitical power will eventually exert controlling influence over the Holy Land. As we have seen, the leader of that union, the king of the North, will ultimately dominate Jerusalem and the Middle East. The book of Revelation also predicts a conflict between Asian forces and the European "beast" power, the king of the North (Daniel 11:44). Shockingly, just before Christ's return, these armies will combine as they are "gathered together to make war against Him" (Revelation 19:19). God Almighty says, "For I will gather all the nations to battle against Jerusalem" (Zechariah 14:2). Then "the LORD will go forth and fight against those nations, as He fights in the day of battle. And in that day His feet will stand on the Mount of Olives, which faces Jerusalem on the east" (Zechariah 14:3–4). Jesus Christ will intervene! He will stop World War III and ultimately bring lasting peace to the world!

Notice that the climactic end-time battle that is popularly referred to as Armageddon is actually called "the battle of that great day of God Almighty" (Revelation 16:14). But *where* will that battle be fought? The prophet Joel tells us that this climactic battle will take place near Jerusalem: "For behold, in those days and at that time, when I bring back the captives of Judah and Jerusalem, I will also gather all nations, and bring them down to the Valley of Jehoshaphat; and I will enter into judgment with them there" (Joel 3:1–2).

The Valley of Jehoshaphat is also called the Kidron Valley (between the Mount of Olives and the Temple Mount). Here, God will judge the nations. The valley widens to the south, and through it the world's armies will move southward from Megiddo to fight Jesus Christ at Jerusalem. At this great climactic battle of all ages, the Creator God will prevail, defeating puny humans in their attempt to conquer God.

Foolish military leaders will quickly learn how powerless they are against the omnipotent, divine power of God! Notice the horrible punishment wrought on those rebellious, evil armies: "And this shall be the plague with which the LORD will strike all the people who fought against Jerusalem: Their flesh shall dissolve while they stand on their feet, their eyes shall dissolve in their sockets, and their tongues shall dissolve in their mouths" (Zechariah 14:12).

Jesus will totally conquer the greatest military combine ever assembled. He will return as King of kings and Lord of lords to bring this war-torn planet a thousand years of peace. "And the LORD shall be King over all the earth" (v. 9).

Christ will conquer all the armies opposing Him! He will usher in the Millennium of peace on earth and teach all nations the way to peace. Then the Arabs and Israelis will be reconciled. "In that day there will be a highway from Egypt to Assyria, and the Assyrian will come into Egypt and the Egyptian into Assyria, and the Egyptians will serve with the Assyrians. In that day Israel will be one of three with Egypt and Assyria; a blessing in the midst of the land, whom the LORD of hosts shall bless, saying, 'Blessed is Egypt My people, and Assyria the work of My hands, and Israel My inheritance'" (Isaiah 19:23–25).

Jerusalem The World Capital

After the world's armies have been defeated, the Creator God will establish Jerusalem as the capital of the world. All nations will submit to the Kingdom of God on earth and finally learn the way to peace. "Now it shall come to pass in the latter days that the mountain of the LORD's house shall be established on the top of the mountains, and shall be exalted above the hills; and peoples shall flow to it. Many nations shall come and say, 'Come, and let us go up to the mountain of the LORD, to the house of the God of Jacob; He will teach us His ways, and we shall walk in His paths.' For out of Zion the law shall go forth, and the word of the LORD from Jerusalem" (Micah 4:1–2).

The King of kings, assisted by the saints, will reeducate the nations to live the way of love, peace, and prosperity (Daniel 7:18, 27). Under the rulership of Jesus—the Messiah—the way of sin, suffering, and war will pass away. "He shall judge between many peoples, and rebuke strong nations afar off; they shall beat their swords into plowshares, and their spears into pruning hooks; nation shall not lift up sword against nation, neither shall they learn war anymore" (Micah 4:3).

WATCH THE MIDDLE EAST

Jesus Christ warns us to be spiritually alert and watchful (Luke 21:36). But *for what should we be watching* in the turmoil and conflict of the Middle East?

Watch for increasing demand for international intervention and control, not only in Gaza and the West Bank, but in Jerusalem itself. Palestinian National Authority president Mahmoud Abbas supported calls for a United Nations peacekeeping force in Gaza and the West Bank—a proposal the rival Hamas rejected. As violence escalates in the region around Jerusalem, watch for more world leaders to join in demanding international control of this ancient city held dear by Jews, Christians, and Muslims alike.

Watch for continuing conflicts between Israelis and Palestinians. The Palestinians may be divided in their allegiance between the more militant Islamic Hamas organization and the more secular Fatah group, but they are united in their opposition to Israel's exertion of power in a region they consider their own.

Watch for increasing unity among Arab nations in their stand against Israel. Psalm 83, which we read earlier, lists those peoples who comprise Arabic and Muslim nations in the Middle East. They are supported by European groups in their strong opposition to Israel. "They have said, 'Come, and let us cut them off from being a nation, that the name of Israel may be remembered no more.' For they have consulted together with one consent; they form a confederacy against You" (Psalm 83:4–5). Watch for an end-time Arab/Muslim union or confederacy forming against Israel.

Watch for the European Union's growing economic, political, and military unification. As we saw in Daniel's prophecy, the king of the North will eventually occupy the Holy Land. Europe is moving to coordinate and unify its various military and security forces, seeking an alternative to its long-standing dependence on its NATO alliance, which includes the U.S. as a dominant partner. In July 2005, EU foreign ministers agreed to establish a "European Security and Defense College" to organize and conduct multinational training that would bring Europeans from different countries under one consistent policy and procedure framework. In November 2004, after five years of planning, the EU officially established a Rapid Reaction Force that would allow the multinational entity to carry out military objectives. A small deployment occurred in Bosnia. And 2014 and 2015 Russian aggression in Ukraine prompted Germany to head a new "high-readiness spearhead force" ("NATO creates rapid-reaction force to protect Eastern Europe," *Los Angeles Times*, January 14, 2015). Will these forces eventually play a significant part in the Middle East conflict? Watch the development of Europe's own military forces.

In January 2015, *Stratfor* reported, "The question about Europe now is not whether it can retain its current form, but how radically that form will change. And the most daunting question is whether Europe, unable to maintain its union, will see a return of nationalism and its possible consequences"—including, eventually, renewed war on the continent ("The New Drivers of Europe's Geopolitics," *Forbes*, January 27, 2015).

Watch for the preparation and eventual implementation of animal sacrifices by the Jews in Jerusalem. Many believe a physical temple will also be built, but history and the book of Ezra confirm that sacrifices can be conducted with just an altar at the holy place.

Watch for the increasing consolidation of religious power in Europe. Visiting Austria in 1983, Pope John Paul II appealed for European unity, stating, "Europeans should overcome the menacing international confrontations of states and alliances, and create a new united Europe from the Atlantic to the Urals" (Adrian Hilton, *The Principality and Power of Europe*, p. 36). Throughout his papacy, Pope Benedict XVI called for religious unity in Europe and reached out to the Eastern Orthodox churches to a degree that surprised many observers. Pope Francis has continued this trend with his entreaties toward European leaders, urging them to rediscover their Roman Catholic roots.

ULTIMATE PEACE

Jerusalem has a glorious prophesied future ahead! It will be the capital of planet Earth under the rulership of the Prince of Peace and King of kings, Jesus Christ. That is good news for all the nations—for you, your children, and your grandchildren. But we need to remember that true peace can only come about in the lives of men and women when their human nature gives way to God's Spirit. In the Millennium and beyond, the vast majority of people will accept God's gift of His Spirit. Then they shall learn how to find the way to peace.

The whole world will learn God's ways. "Many people shall come and say, 'Come, and let us go up to the mountain of the Lord, to the house of the God of Jacob; He will teach us His ways, and we shall walk in His paths.' For out of Zion shall go forth the law, and the word of the Lord from Jerusalem. He shall judge between the nations, and rebuke many people; they shall beat their swords into plowshares, and their spears into pruning hooks; nation shall not lift up sword against nation, neither shall they learn war anymore" (Isaiah 2:3–4).

Jerusalem will not only be the governmental capital of the world—it will also be the educational capital of the world. The great Educator and Teacher, the Lord Jesus Christ, will teach true knowledge based on the word of God. All nations will learn the eternal principles, laws, and values that guarantee peace and prosperity to all.

This is the good news of your future and the future of the world. At the heart of the Middle East—and of the entire world—Jerusalem will be the center of world government, education, and religion. The Creator God will guarantee future world peace, "for the mouth of the Lord of hosts has spoken" (Micah 4:1–4). We all look forward to the peace and reconciliation that only Jesus, the Messiah, can bring to the Middle East—and to the whole world. In the meantime, Jesus tells us to watch and pray (Mark 13:33). He exhorts us in Mark 13:37, "And what I say to you, I say to all: Watch!"

APPENDIX

i. According to the technical discussion the author had with Dr Ratnakant Sanjay, M.D., of Bangalore, India.

ii. Kepler quoted in Tiner, Ref. 1, p. 178. Kepler's Third Law of Planetary Motion states: $P^2 \propto a^3$ where P is the orbital period of planet and a is the semimajor axis of the orbit, while $\propto$ means "is proportional to".

iii. The evolutionary ideas of spontaneous generation and of species being able to change from one to another were not originated by Charles Darwin. These ideas had been around since the time of the ancient Greeks. Darwin merely put various existing ideas together into his theory of evolution and popularized them.

iv. Adapted from Henry C. Thiessen, *Lectures in Systematic Theology*, Eerdmans, Grand Rapids, 1979, pp. 384–85.

v. Tim LaHaye, Pre-Trib Perspectives (Vol. V, Num. 5; Aug. 2000), p. 1.

vi. Thomas Ice, Pre-Trib Perspectives (Vol. V, Num. 5; Aug. 2000), p. 4.

vii. Evelyn Leopold, Reuters Internet News Service, "UN, NY gear for largest-ever meet of world leaders," August, 2000.

viii. Josephus, Antiquities of the Jews

REFERENCES

1. Mozur, P., Google's A.I. program rattles Chinese Go master as it wins match, *New York Times*; nytimes.com, 25 May 2017.
2. Aleksander, Igor: Artificial neuro-consciousness: An update, IWANN, 1995.
3. Showalter, B., Artificial intelligence, transhumanism and the church: How should Christians respond? christianpost.com, 27 Jan 2018.
4. Ruse, M., How evolution became a religion: creationists correct? *National Post*, pp. B1, B3, B7, 13 May 2000; creation.com/ruse.
5. Karr, J.R. *et al.*, A whole-cell computational model predicts phenotype from genotype, *Cell* 150(2):389–401, 12 Jul 2012. See also Cellular complexity "nearly unbelievable", Focus, *Creation* 35(1):11, 2013; creation.com/focus-351#128-computers.
6. Statham, D., The remarkable language of DNA, *Creation* 36(2):52–55, 2015; creation.com/dna-remarkable-language.
7. Hoyle, F., The big bang in astronomy, *New Scientist* 92(1280):527, 1981, quoted in Batten, D., Cheating with chance, *Creation* 17(2):14–15; creation.com/cheating-with-chance.
8. Harris, M., God Is a Bot, and Anthony Levandowski Is His Messenger, wired.com, 27 September 2017.
9. Harris, M., Inside the First Church of Artificial Intelligence, wired.com, 15 November 2017.
10. McFarland, M., Elon Musk: 'With artificial intelligence we are summoning the demon.', washingtonpost.com, 24 October 2014.
11. Korosec, K., Anthony Levandowski, former Google engineer at center of Waymo-Uber case, charged with stealing trade secrets, techcrunch.com, 27 August 2019.
12. Shepherd, G.M., *Neurobiology*, Oxford University Press, London, p. 577, 1983.
13. Tortora, G.J. and Anagnostakos, N.P., *Principles of Anatomy and Physiology*, Harper & Row, New York, p. 290, 1981.
14. Gitt, W., *The Wonder of Man*, CLV Publishing, Germany, p. 82, 1999.
15. Restak, R.M., *The Brain*, Bantam Books, New York, pp. 34–35, 1984.
16. Math Equity Toolkit; equitablemath.org.
17. https://equitablemath.org/wp-content/uploads/sites/2/2020/11/1_STRIDE1.pdf.
18. Bertrand Russel, *The Principles of Mathematics*; Cambridge University Press, 1903; people.umass.edu.

19. "The right answer", posted on *The Atheist Conservative;* theatheistconservative.com
20. Sergiu Klainerman, *There Is No Such Thing as "White" Math,* bariweiss.substack.com, 1 Mar 2021.
21. The Medieval Problem of Universals, plato.stanford.edu, 31 Oct 2017.
22. Vern S. Poythress, *A Biblical View of Mathematics*, 4 Jun 2012; frame-poythress.org.
23. Vern Poythress, *Creation and Mathematics; or What Does God Have to do with the Numbers?* frame-poythress.org, 21 May 2012.
24. Other examples of beauty in mathematical equations are provided by Clara Moskowitz, The 11 Most Beautiful Mathematical Equations, livescience.com, 1 Jun 2017.
25. Ramos, S.C., *The Discovery of Calculus: Leibniz vs. Newton*; stmuhistorymedia.org, 3 Nov 2017.
26. Johannes Kepler, quoted in: Tiner, J. H., *Johannes Kepler-Giant of Faith and Science*, Mott Media, Milford, Michigan (USA), p. 193, 1977.
27. *Encyclopaedia Britannica*, 15th ed., vol. 22, p. 506, 1985.
28. *Encyclopaedia Britannica*, vol. 22, p. 507, 1985.
29. Crowther, J.G., *Founders of British Science: John Wilkins, Robert Boyle, John Ray, Christopher Wren, Robert Hooke, Isaac Newton*, Cresset Press, London, p. 94, 1960.
30. *McGraw-Hill Encyclopedia of World Biography*, 9:118, McGraw-Hill, New York, USA, 1973.
31. Asimov, I., *Biographical Encyclopedia of Science and Technology: The Lives and Achievements of More Than 1000 Great Scientists from Ancient Greece to the Space Age*, 3rd edition, Doubleday & Co. Inc., Garden City, New York, USA, p. 137, 1982. Return to text.
32. Ray, J., quoted in Ref. 1, p. 126.
33. Ray, J., quoted in: H.M. Morris, *Men of Science, Men of God*, Master Books, El Cajon, CA., USA, p. 18, 1982.
34. Derham, W., quoted in Ref. 1, p. 129.

Printed at Repro India Ltd.